SOLENT SCIENCE – A REVIEW

UNESCO

ELSEVIER
2000

Amsterdam - Lausanne - New York - Oxford - Shannon - Singapore - Tokyo

Proceedings in Marine Science, 1

SOLENT SCIENCE – A REVIEW

Edited by

Michael Collins
*School of Ocean and Earth Science
University of Southampton
Southampton, UK*

Kate Ansell
*Solent Forum
Winchester, UK*

ELSEVIER
2000

Amsterdam - Lausanne - New York - Oxford - Shannon - Singapore - Tokyo

ELSEVIER SCIENCE B.V.
Sara Burgerhartstraat 25
P.O. Box 211, 1000 AE Amsterdam, The Netherlands

© 2000 Elsevier Science B.V. All rights reserved.

This work is protected under copyright by Elsevier Science, and the following terms and conditions apply to its use:

Photocopying
Single photocopies of single chapters may be made for personal use as allowed by national copyright laws. Permission of the Publisher and payment of a fee is required for all other photocopying, including multiple or systematic copying, copying for advertising or promotional purposes, resale, and all forms of document delivery. Special rates are available for educational institutions that wish to make photocopies for non-profit educational classroom use.

Permissions may be sought directly from Elsevier Science Global Rights Department, PO Box 800, Oxford OX5 1DX, UK; phone: (+44) 1865 843830, fax: (+44) 1865 853333, e-mail: permissions@elsevier.co.uk. You may also contact Global Rights directly through Elsevier's home page (http://www.elsevier.nl), by selecting 'Obtaining Permissions'.

In the USA, users may clear permissions and make payments through the Copyright Clearance Center, Inc., 222 Rosewood Drive, Danvers, MA 01923, USA; phone: (978) 7508400, fax: (978) 7504744, and in the UK through the Copyright Licensing Agency Rapid Clearance Service (CLARCS), 90 Tottenham Court Road, London W1P 0LP, UK; phone: (+44) 171 631 5555; fax: (+44) 171 631 5500. Other countries may have a local reprographic rights agency for payments.

Derivative Works
Tables of contents may be reproduced for internal circulation, but permission of Elsevier Science is required for external resale or distribution of such material.
Permission of the Publisher is required for all other derivative works, including compilations and translations.

Electronic Storage or Usage
Permission of the Publisher is required to store or use electronically any material contained in this work, including any chapter or part of a chapter.

Except as outlined above, no part of this work may be reproduced, stored in a retrieval system or transmitted in any form or by any means, electronic, mechanical, photocopying, recording or otherwise, without prior written permission of the Publisher.
Address permissions requests to: Elsevier Science Rights & Permissions Department, at the mail, fax and e-mail addresses noted above.

Notice
No responsibility is assumed by the Publisher for any injury and/or damage to persons or property as a matter of products liability, negligence or otherwise, or from any use or operation of any methods, products, instructions or ideas contained in the material herein. Because of rapid advances in the medical sciences, in particular, independent verification of diagnoses and drug dosages should be made.

First edition 2000

Library of Congress Cataloging-in-Publication Data

Solent science : a review / Michael Collins and Kate Ansell, (editors).
    p. cm. -- (Proceedings in marine science)
  Papers from a conference.
  Includes bibliographical references.
  ISBN 0-444-50465-6
    1. Oceanography--England--Solent Channel--Congresses. 2. Solent Channel
  (England)--Congresses. I. Collins, M. B. II. Ansell, Kate. III. Series.

  GC343 .S65 2000
  551.46'136--dc21
                                                            00-055106

ISBN: 0 444 50465 6

⊗ The paper used in this publication meets the requirements of ANSI/NISO Z39.48-1992 (Permanence of Paper).
Printed in The Netherlands.

Preface

Sound scientific knowledge is fundamental to the long-term planning and management of the Solent and its coastline. As such, it is appropriate to bring together here, the results and observations of various researchers, to establish the present level of understanding of the Solent System.

A previous 'up-to-date' assessment of the scientific knowledge of the area was published in 1980, by the Natural Environment Research Council (upon the initiative of, and edited locally by, Professor Dennis Burton (Department of Oceanography, now SOES)), as part of a Series relating to major UK estuarine systems. The publication was the culmination of meetings held between an *ad hoc* group of scientists, with active interests in the area. This publication has been recognised widely as a useful tool within the scientific/planning sectors. A broader-based earlier review, on a diversity of subjects, was published in 1964 (*Survey of Southampton and its Region*, by the Southampton University Press, for the British Association for the Advancement of Science).

The overall aim of the Solent Science Conference was to clarify the extent of our existing knowledge of science in the Solent and identify information gaps. The Conference enabled a wide range of organisations, including both the suppliers and users of scientific information, to review how existing information is utilised in the making of 'real life' decisions; likewise, how this approach and interaction can be improved. This Conference is the first in what is intended to be a series of conferences; these will facilitate the dissemination of information on the Solent. Such meetings will also encourage dialogue between the various scientific disciplines and between the scientists and the wider community.

These Proceedings incorporate scientific papers dealing with four different topics, covered over the two days of the Conference: coastal processes; water quality and chemistry; biodiversity and conservation; and integrative Solent case studies. The papers are supplemented by a range of short contributions (presented originally as posters), which examine more specific pieces of research, and the findings of the Conference Workshops. Additionally, two maps have been included in the Preface, showing the general location and main sites referenced.

Recently, in parallel with the Conference, Colin Tubbs (sadly, now deceased) produced a delightfully written book on *The Ecology Conservation and History of the Solent*. Based upon his own personal experience (over some 33 years) of planning, development and management issues affecting the Solent, this particular publication includes (for example) sections on: the wealth of the estuary; the congenial shore; the exploitation of natural resources; meadows in the sea; and the Brent Goose Story. As such, where appropriate, it should be read in conjunction with the contents of the present volume.

Attempting to integrate multidisciplinary science through interaction with various decision-making authorities, is not new (see, for example, Collins, *et al.* 1979). Following this earlier publication, J.A. Steers (see Tomalin, *this volume*) was complimentary of this approach and considered it to be 'ahead of its time'. The present volume provides further recognition of the 'systems approach' to science and management (see Townend, *this volume*), which may well have broader applicability at local, regional, national and international levels (Poulos *et al.*, *in press*). In particular, a strong scientific base is required to support those working in the Solent (see, for example, Solent Forum, 1997), especially with the increasing number of European Directives and the need for their implementation.

REFERENCES

Collins, M.B. 1980. *Industrialised embayments and their environmental problems: a case study of Swansea Bay*. Proceedings of an Interdisciplinary Symposium held University College, Swansea, 26-28 September 1979. Oxford: Pergamon Press, 616 pp.

Burton, J.D.(Ed.) 1980. *The Solent Estuarine System. An assessment of present knowledge*. N.E.R.C. Publications Series C No.22., 100 pp.

Poulos, S.E., Chronis. G. Th., Collins, M.B. and Lykousis, V. (in press). Thermaikos Gulf coastal-System, NW Aegean Sea: An overview of water/sediment fluxes in relation to air-land-ocean interactions and human activitites. *Journal of Marine Systems*.

Solent Forum, 1997. *Strategic Guidance for the Solent*. Solent Forum (Hampshire County Council), Winchester, 200 pp.

Tomalin, D. (this volume). Wisdom of hindsight: Palaeo-environmental and archaeological evidence of long-term processual changes and coastline sustainability. In: *Solent Science - A Review*. Collins, M.B. & Ansell, K. (Eds.), Proceedings in Marine Science Series, Elsevier, Amsterdam.

Townend, I. (this volume). Developing a research agenda for the Solent. In: *Solent Science - A Review*. Collins, M.B. & Ansell, K. (Eds.), Proceedings in Marine Science Series, Elsevier, Amsterdam.

Tubbs, C. (1999). *The Ecology Conservation and History of the Solent*. Packard Publishing Limited, Chichester, 184 pp.

Editors

Michael Collins
Southampton Oceanography Centre

Kate Ansell
Solent Forum

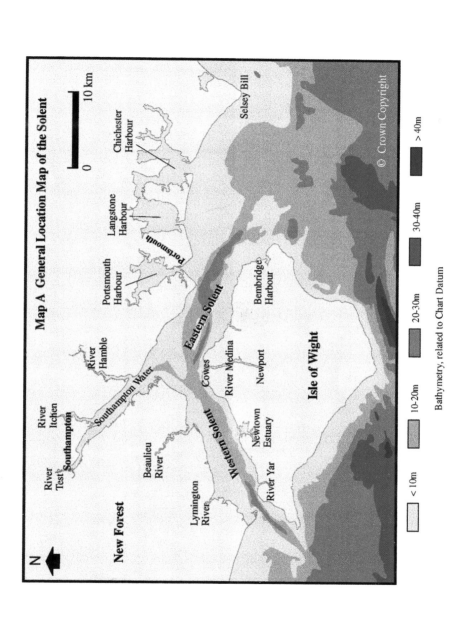

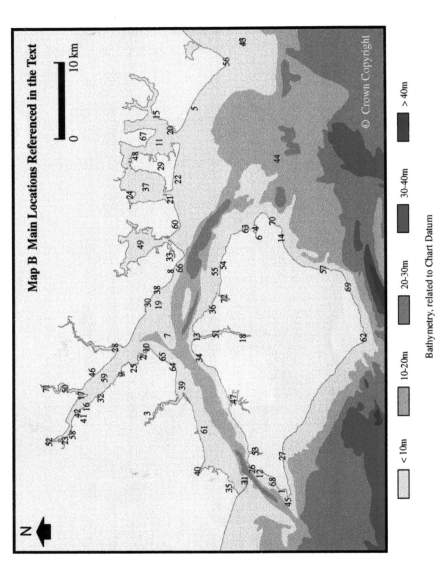

Map B Main Locations Referenced in the Text

Key to Map B - Main Locations Referenced in the Text.

1. Alum Bay
2. Ashlett Creek
3. Beaulieu River
4. Bembridge Harbour
5. Bracklesham Bay
6. Brading Harbour
7. Brambles Bank
8. Browndown
9. Cadland Creek
10. Calshot Spit
11. Chichester Harbour
12. Colwell Bay
13. Cowes
14. Culver Cliff
15. Dell Quay
16. Dibden Bay
17. Dockhead
18. Dodnor
19. East Brambles
20. East Head
21. Eastney
22. East Winner Bank
23. Eling Marsh
24. Farlington Marshes
25. Fawley
26. Fort Albert
27. Freshwater Bay
28. Hamble Estuary
29. Hayling Island
30. Hill Head
31. Hurst Spit
32. Hythe
33. Gosport
34. Gurnard
35. Keyhaven River
36. King's Quay
37. Langstone Harbour
38. Lee-on-the-Solent
39. Lepe Middle Bank
40. Lymington River
41. Marchwood
42. Marchwood Yacht Club
43. Mixon Hole
44. Nab Tower (The)
45. Needles (The)
46. Netley
47. Newtown Estuary
48. Northney
49. Portsmouth Harbour
50. River Itchen
51. River Medina
52. River Test
53. River Yar
54. Ryde Harbour
55. Ryde Sands
56. Selsey Bill
57. Shanklin
58. Slowhill Copse
59. Southampton Water
60. Southsea
61. Sowley Ground
62. St Catherine's Point
63. St Helen's Fort
64. Stansore Point
65. Stanswood Bay
66. Stokes Bay
67. Thorney Island
68. Totland Bay
69. Ventnor
70. Whitecliff Bay
71. Woodmill
72. Wootton Creek

Acknowledgements

Firstly, as Chairman of the Conference Organising Committee, I would like to thank all those individuals who put in an enormous amount of effort and commitment over a number of months, to bring to fruition the Conference and its subsequent publication. Particular thanks are extended to Members of the Conference Steering Committee (see overleaf) who helped to devise the programme, and present, chair and co-ordinate the Conference findings. In particular, during meetings held at the Southampton Oceanography Centre (SOC), discussions on the mechanism of integration of scientific findings and management issues were both stimulating and rewarding. On the basis of these discussions, the format of the Conference was decided i.e. scientific presentations, with associated science/management workshops.

Secondly, I would also like to thank the various organisations that sponsored the Conference, together with those who brought along displays and posters. In addition, special thanks are due to Alan Inder and Jacqui Hall (Hampshire County Council), Zoe Hughes (School of Ocean and Earth Science) for their sterling efforts in ensuring the smooth running of the Conference. Likewise, to Sara Hallet (Johnson Controls, Southampton Oceanography Centre) for assisting with the overall logistical planning of the meeting.

Subsequently, various Members of the Steering Committee are acknowledged for their assistance in the collating and editing of the contributions; in particular, Jane Taussik (University of Portsmouth).

Thirdly, I would like to acknowledge the organisations that fund and steer the work of the Solent Forum, for their dedication and support in achieving better management of the Solent estuarine system.

Finally, on behalf of Kate Ansell and myself, it is a pleasure to thank Georgina Owrid and Kate Davis (School of Ocean and Earth Science) for their invaluable and exceptional assistance with the preparation of the text and drawings for the Proceedings, respectively.

Michael Collins
Chair, Solent Science Conference Steering Group
2000

Solent Science Conference Steering Committee

Michael Collins (Chair)	SOES, University of Southampton
Kate Ansell (Secretary)	Solent Forum
Malcolm Bray	University of Portsmouth
Sarah Fowler	Nature Conservation Bureau
David Hydes	Challenger Division, SOC
David Johnson	Southampton Institute
Alan Inder	Hampshire County Council
Duncan Purdie	SOES, University of Southampton
Heidi Roberts	Isle of Wight Council
Lawrence Talks	Environment Agency
Jane Taussik	University of Portsmouth
Ian Townend	ABP Research & Consultancy

Solent Forum Funding Partners

Associated British Ports
English Nature
Environment Agency
New Forest District Council
Hampshire and Isle of Wight Wildlife Trust
Hampshire County Council
Isle of Wight Council
Portsmouth City Council
Queen's Harbour Master, Portsmouth
Royal Society for the Protection of Birds
Solent Protection Society
Southampton City Council

SOLENT SCIENCE - A REVIEW

Contents

Preface v

Preface maps vii

Acknowledgements xi

Section 1 - Introduction

An introduction to Solent Science
Maldwin Drummond 3

Opening address
Peter Brand 5

Section 2 - Coastal Processes

Geomorphological evolution of the Solent seaway and the severance
of Wight: A review
David Tomalin 9

Geology, geomorphology and sediments of the Solent System
Adonis Velegrakis 21

Water circulation in Southampton Water and the Solent
Jonathan Sharples 45

Hydrodynamic, sediment process and morphological modelling
Darren Price and Ian Townend 55

Wisdom of hindsight: palaeo-environmental and archaeological
evidence of long-term processual changes and coastline sustainability
David Tomalin 71

Shoreline management plans - A science or an art?
Jonathan McCue 85

Short Contributions

Late Pleistocene/Holocene evolution of the upstream section of the Solent River
Velegrakis, A.F., Dix, J.K. and Collins, M.B. 97

Sea level rise in the Solent region
Bray, M.J., Hooke, J.M. and Carter, D.J. 101

Littoral sediment transport pathways, cells and budgets within the Solent
Bray, M.J., Hooke, J.M., Carter, D.J. and Clifton, J. 103

Residual circulation and associated sediment dynamics in the eastern approaches to the Solent
Paphitis, D., Velegrakis, A.F. and Collins M.B. 107

Seabed mobility studies in the Solent region
Velegrakis, A.F., Brampton, A.H., Evans, C.D.R., and Collins, M.B. 111

Lee-on-the-Solent coast protection scheme
Banyard, L. and Fowler, R. 115

Findings of the Coastal Processes Workshops 121

Section 3 - Water Quality and Chemistry

Nutrients in the Solent
David Hydes 135

Trace metals in water, sediments and biota of the Solent system: A synopsis of existing information
Peter J. Statham 149

Microbiological quality of the Solent
David Lowthion 163

Behaviour of organic carbon in Southampton Water
Mark Varney 175

Sewage contamination of bathing waters: health effects
Gareth Rees 187

Short Contributions

M,tidally-induced water mass transport and water exchange
in Southampton Water and the Solent
Shi, L. and Purdie, D.A. 199

Fluxes of dissolved inorganic phosphorous to the Solent from the
River Itchen during 1995, 1996 and 1998
Wright, P. N., Xiong, J. and Hydes, D.J. 205

Evaluation of the environmental risk of the use of preservative
treated wood in Langstone Harbour
Cragg, S.M., Brown, C.J., Praël, A. and Eaton, R.A. 209

The prevention of biofilm formation and marine settlement
on protective coatings prepared from low-surface energy materials
*Graham, P., Stone, M., Thorpe, A., Joint, I., Nevell, T.
and Tsibouklis, J.* 213

Findings of the Water Quality and Chemistry Workshops 217

Section 4 - Biodiversity and Conservation

Viewpoint: conservation, policy and management of maritime
biodiversity
Dan Laffoley 225

Coastal habitats of the Solent
Sarah Fowler 231

Marine habitats and communities
Ken J. Collins and Jenny J. Mallinson 247

Ornithology of the Solent
Dave Burges 261

Fisheries of Southampton Water and the Solent
Antony Jensen 271

Short Contributions

Underwater light in tidal waters: possible impact on macro-algae communities
Charrier, S., Weeks, A., Lewey, S. and Robinson, I. 283

Phytoplankton - annual sequences in the Hamble Estuary
O'Mahony, J. and Weeks, A. 287

Truncatella subcylindrica (Mollusca: Prosobranchia) in the Solent area: its distribution, status and conservation
Light, J.M. and Killeen, I.J. 295

Evolution and current status of the saltmarsh grass, *Spartina anglica*, in the Solent
Raybould, A.F., Gray, A.J. and Hornby, D.D. 299

Saltmarsh monitoring studies adjacent to the Fawley refinery
May, S. 303

Use of the dog-whelk, *Nucella lapillus*, as a bio-indicator of tributyltin (TBT) contamination in the Solent and around the Isle of Wight
Herbert, R.J.H., Bray, S. and Hawkins, S.J. 307

The biology and distribution of the kelp, *Undaria pinnatifida* (Harvey) Suringar, in the Solent
Farrell, P. and Fletcher, R. 311

The Sussex *Seasearch* project
Irving, R. 315

Analysis of the numbers and distribution of wildfowl and waders as an aid to estuarine management
De Potier, A. 319

Findings of the Biodiversity and Conservation Workshops 329

Section 5 - Integrative Solent Case Studies

Aggregate extraction
Elizabeth Dower 339

Main channel deepening - Port of Southampton - 1996/7
Colin Greenwell 347

Evaluating the intertidal wetlands of the Solent
David Johnson 355

Section 6 - Concluding Remarks

Developing a research agenda for the future
Ian Townend 367

Conclusions and close
Maldwin Drummond 371

Location Index 375

Subject Index 377

SECTION 1

Introduction

An Introduction to Solent Science

Maldwin Drummond, OBE, JP, DL, Hon.DSc.

Chairman, Solent Forum

c/o Hampshire County Council Planning Department, The Castle, Winchester, Hampshire, SO23 8UE, U.K.

The Solent has had a fundamental influence on many important sectors of the United Kingdom's history. In the past, we have taken this enclosed area of water as a gift from nature with the Island being a particularly beautiful wind and wave break, placed exactly, to our good fortune to create our valuable double tide.

As a nation, we have made good use of these waters for ports and as a seaway. Portsmouth has long been home to the Royal Navy and now to the cross channel ferry trade. Southampton has, historically, serviced the commercial deep sea, the liners of yesterday, cruise ships of today, tankers at Fawley and the vast and growing trade in container shipment.

The Solent is also the most valued sailing and recreational water in Northern Europe. Yet alongside all this activity, the natural environment, too, is acclaimed in European wide terms, in part at least as a Special Area of Protection and a candidate Special Area of Conservation.

Until the 1960's, although our understanding of the conflicts within this marine system may have been known, efforts to co-ordinate or study the demands of competing interests were unstructured. Planning and Management on the land and on the foreshore was better achieved. Gathering information was undertaken by the agencies, the local and harbour authorities and the universities, but it was solely for their own management or study purposes.

In 1992, this began to change with the formation of the Solent Forum. This new body quickly recognised that good scientific information is fundamental to the long term planning and management of the Solent. The production and acceptance of a Management Strategy gave a common incentive. We see gaps in our knowledge and through this conference and other methods, we hope to fill them. To this end, these two days will be devoted to *Reviewing the Future of Science in the Solent.*

The establishment of the Southampton Oceanography Centre, part of the University of Southampton and with strong links to the Departments, has created a world class establishment. Portsmouth University and the Southampton Institute have added to this local, national and international ability.

1998 has the added bonus of having been declared the International Year of the Ocean.

I remember some 30 years ago coming across Clement Reid's sketch map of *The Basin of the Ancient River Solent* which first appeared, believe it or not, in a publication called *The Geology of Ringwood,* published in 1902. The creation of today's Solent, by what must now be termed "unmanaged retreat", at the end of the Pleistocene into the recent period, has overtones for the debates on "sea level rise" today. I live just to the east of the last supposed land bridge between Egypt Point on the Island and Stone Point on the mainland. This was eventually overwhelmed in the comparatively rapid, but not continuous movement of depression and sea level rise. I, therefore, declare an interest in today's increasing millimetres, caused by a continuing rhythm much accentuated by global warming.

We face a considerable challenge and one that can only be responded to by gaining and sharing knowledge and working together. Our future depends upon it.

Opening Address

Dr Peter Brand, MP

House of Commons, London, SW1A 0AA, U.K.

I have been involved with coastal matters for many years, having been the first chairman of SCOPAC (Standing Conference on Problems Associated with the Coastline) and the MP for the Isle of Wight, the constituency with the longest coastline in the country. This experience has led me to believe that coastal defence planning has been based more on tradition than science.

I welcome moves towards 'evidence-based government' that recognises the growing need for sharing research based on sound scientific information. There are many topical coastal issues, including erosion, dredging, water quality, conservation and fisheries. Good quality research is needed to provide a sound basis for policymaking, and to help in the making of difficult decisions with confidence. I congratulate the Solent Forum and the Southampton Oceanography Centre for taking the joint initiative to organise this Conference. I hope that clear messages will come from the Conference, and I look forward to the Proceedings.

SECTION 2

Coastal Processes

Geomorphological Evolution of the Solent Seaway and the Severance of Wight: A Review

David Tomalin

County Archaeologist, Isle of Wight Council, 61 Clatterford Road, Newport, Isle of Wight, PO30 1NZ, U.K.

EARLY QUESTIONS CONCERNING THE SEVERANCE OF WIGHT

It was more than 400 years ago when the first historians and geographers began to enquire into the nature and origins of the Solent as an open east-west seaway and the date at which it had precipitated the severance of the land of Wight. The first recorded questions are those of William Camden, whose first edition of *Britannia* (published in 1586) included the mischievous speculation that the Isle of Wight, with its Roman name of *Vectis*, might perhaps be equated with a prehistoric island, otherwise known as *Ictis*.

A British island called *Ictis* had been cited in the 1st century BC, by the classical writer *Diodorus Siculus*; however, we should note that in describing Britain or "*Prettanike*," this classical historian commonly used the expressions "we are told" or "they say". The style of *Diodorus* indicates that he was relating the accounts of others and, unlike the earlier Greek explorer *Pytheas*, who had visited the Cornish coast in the 3rd century BC, it seems that he could offer no personal experience. His gatherings tell of an island close to the shore of southern Britain where the natives could cross at low tide whilst drawing wagons loaded with tin ore or ingots. These consignments were loaded into visiting ships bound for the Atlantic seaboard of Gaul (Rivet & Smith, 1979). *Diodorus* added that it was people dwelling near the promontory of *Belerion* (Land's End) who prepared this tin and transported it to the tied island of *Ictis*.

Archaeologists are mostly agreed that the most credible claimant for the island of *Diodorus* is St Michaels Mount, near Penzance. This granite islet is accessible at low tide; it is very close to the tin stream deposits at Marazion Marsh and it has even yielded scattered fragments of tin ore (Skinner, 1797; Hencken, 1932; Maxwell, 1972; Herring, 1992; and Penhallurick, 1997). A less convincing contender has been the rock promontory of Mount Batten, in Plymouth Sound. This area once served as an Iron Age port and it is sited not far from the spot where a prehistoric tin ingot was found on the seabed (Cunliffe, 1988).

Since Camden's day, the Ictis-Vectis theory has claimed many champions, but it has remained a specious argument, readily flawed by the restriction of tin lodes to the Cornish peninsula. The best that might be postulated is that, whilst *Diodorus* was gathering descriptions of all islands on the south coast of Britain, he applied the name used in Wight to a trading post in Cornwall. Indeed, he reminds us that he is speaking of "the neighbouring islands lying between Europe and Prettanike". Such a fusion of accounts permits *Vectis* to be *Ictis,* whilst excluding the impracticalities of a Vectensian tin trade.

A very different attempt to pursue historical explanations for the severance of Wight was made in the 18th century by Dr Thomas Short, a physician of Sheffield. This author produced an extraordinary and unsubstantiated classical reference to a great earthquake which had allegedly severed the Isle of Wight from the mainland, in the year AD 68 (Short, 1749). The source of this claim has never been substantiated. It has been observed, however, by Charles Thomas (1985) that where near-shore islands are concerned, folklore and credulity have always favoured a catastrophic event, rather than a slow processual one.

THOMAS WEBSTER AND THE WIGHT-PURBECK RIDGE

The first scientific enquiries into the configuration of the Solent began in the early 19th century. Thomas Webster opened the discussion in 1816, when examining the Chalk stacks, including Old Harry Rock in the Isle of Purbeck. Here, he observed that "the Chalk at Handfast Point which being in a line with that of the Isle of Wight appeared like the continuation of the same strata" (Englefield, 1816). He considered that these stacks "like those of the Needles had resisted, longer than the rest, the destroying effects of the waves" and he concluded that "these two places were once united".

Unsettled by a contemporary regard for biblical time-scales, Webster demurred over the means by which the Chalk ridge had been lost or submerged. In 1840, however, such reservations were confidently expelled. Charles Lyell now announced that the severance of the Chalk at Handfast Point, together with the present shape of the Isle of Wight, was due to encroachment and continued action of the sea. Lyell cited Hurst Spit as proof and product of long incremental coastal processes. He drew persuasive comparisons with other British examples, including Chesil Beach. In each case, he argued that the beach shingle was derived from the slow and on-going erosion of flint-bearing cliffs. This argument was reinforced in 1852, when Redman observed that the deep-cutting of the Hurst Channel was further evidence of protracted coastal processes.

The mid-19th century was a time when the processual history of the Earth and the antiquity of man were subjects of hot contention, in both public bar and pulpit. A progressive figure in this debate was the Rector of Brighstone, the Reverend William Fox. Fox had been an ardent investigator the remains of the fossil reptile in the cliffs of his parish. Consequently, his tea parties had become highly scientific events. More eminent guests had included Sir Charles Lyell, the palaeontologists Gideon Mantell and John Owen, and the poet and philosopher Alfred Lord Tennyson.

It was William Fox who was first to pose the direct question *"when and how was the Isle of Wight separated from the mainland?"* This was the title of his short and perceptive paper published in the Volume 5 of the new journal, the *Geologist* (1862). Fox first considered the old historical Ictis-Vectis explanation currently championed by his non-conformist contemporary, the Reverend Edmund Kell of Newport. He then applied his geological knowledge, to calculate where the Island's last overland connection or umbilical may have been. Kell, meanwhile, was postulating a fanciful overland staging route which could enable Iron Age people to carry Cornish tin to the Isle of Wight, so that it might be laden on to ships which would return past the Cornish coast, en route for the Atlantic seaboard of Gaul (Kell, 1866).

Fox used the observations of Webster to advocate a lost Wight-Purbeck ridge. He then applied the case presented by Lyell and Redman, to demonstrate that the Solent seaway was sufficiently old to allow for the slow accumulation of shingle deposits at Hurst Spit and Shingles Bank. Fox now challenged Lyell's intimation that the severance of Wight had been an incremental event. For him, the Wight-Purbeck had formerly been a continuous range of Chalk hills; its breaching by the sea had been a catastrophic event, leading almost directly to the severance of Wight.

WILLIAM FOX AND THE SOLENT RIVER THEORY

For Fox, his Wight-Purbeck ridge was a protective wall of Chalk, sheltering the Dorset and Hampshire lowlands from the sea. If this had been the case, however, then an earlier seaward exit seemed necessary to allow the substantial Dorset rivers of Frome, Stour, Piddle and Avon to reach the ocean. Fox developed the ingenious idea of a lost *Solent River* that might formerly have passed behind the land of Wight to drain the river catchments of east Dorset. The exit of this river might lie somewhere off the Bembridge Ledges whilst its inland tributaries, including the Test and the Hamble, could also include a *great westward arm* passing beneath the present Western Solent to serve the east Dorset catchment (Figure 1).

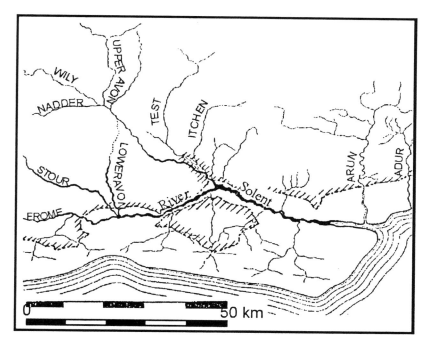

Figure 1 The supposed configuration of the Solent river and its Dorset headwaters. As perceived by William Fox (1862) and illustrated by Clement Reid (1905).

Some twenty years later, the implication of such changes for the Ice Age geography of north western Europe were ably summarised by the pioneer of Pleistocene landscape studies, James Geikie.

> *..The Solent is an old land valley. At what particular point the ancient river discharged into the sea, or whether or not it really joined the Seine can only be conjectured.*

Geikie's ensuing comment confirmed that he largely supposed the formation of the Solent seaway to be a Holocene event.

> *But that Palaeolithic man saw the Isle of Wight as part of the mainland, there cannot be any reasonable doubt* (James Geikie, 1881).

Despite Fox's predictions, the presence of buried river channels or *palaeovalleys* gained little interest during the 19th century. When engineers and investors gathered in the year 1900, to propose a plan for an Isle of Wight railway tunnel, the dangers of encountering a buried channel seem to have been entirely disregarded. Civil engineering experiences gained during the cutting of the Severn Tunnel, in 1879, should have rung a cautionary bell. Yet, when a more modest Medina tunnel was planned, between East and West Cowes in 1892, a clearance of a mere eleven feet below the seabed was considered to be quite adequate.

In 1889, the construction of the great Graving Dock at Southampton revealed new and significant discoveries (Shore & Elwes, 1889). Navvies labouring in their coffered caisson encountered a deep buried shoreline, overlain by freshwater terrestrial peat. Later, it was realised that flint implements in this peat signified the activities of Middle Stone Age communities, who had ventured out on a land surface which was now some 6.7 m below present highest astronomical tide and 1.8 m below Ordnance Datum (O.D.) (Godwin, 1940; Oakley, 1943).

CLEMENT REID'S UMBILICAL OR ISTHMUS

In 1905, Fox's model for the severance of Wight was taken up by the geologist Clement Reid. Reid first pursued the old and perennial antiquarian hare let loose by Camden, considering how the island of *Ictis/Vectis* had remained connected to the mainland as late as the 1st century BC. Camden had observed that *"we may well think this Vecta* [Vectis] *to be that Icta which as Diodorus Siculus writteth, seemed at every tide to be an island but when it was ebbe, the ancient Britaines were wont that way to carry tinne thither by carts."* (Camden [English ed.], 1637). With this seed implanted, there was much fruitless ground to be repetitively harrowed.

Clement Reid wished to consider how the land of Wight might remain connected to the mainland after the sea had breached the Wight-Purbeck ridge. For this, he was prepared to accept the date offered by *Diodorus*. Reid rightly dismissed Hurst Spit as a possible umbilical, for he required firm ground on which to set the groaning carts, the heaped tin ingots and the sweating Britons.

Reid identified his potential umbilical in the region of Hamstead and Yarmouth, where hard outcrops of the Bembridge Limestone Formation led offshore in the direction of Lymington (see Figure 2). On the Hampshire shore, no trace of the limestone could be found, but Reid suggested ingeniously that such an outcrop had since been eradicated by coastal erosion, prior to the formation of the Lymington salt marshes.

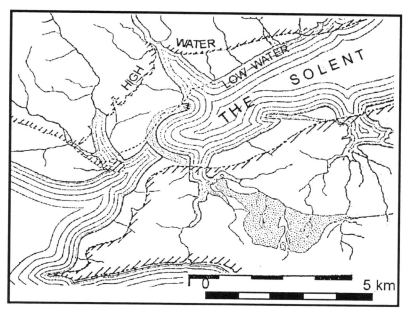

Figure 2 Clement Reid's conjectured limestone isthmus between Yarmouth and Pennington Marsh.

Unfortunately, more recent geological mapping in the Western Solent (Hamblin & Harrison, 1989) has demonstrated that the Yarmouth-Hamstead limestone outcrop reaches no more than 1.8 km from the Isle of Wight shore. However, if we accept Reid's postulation that a greater area of marshland once extended from the Lymington shore, his case for an isthmus remains tenable.

MARINE GEOPHYSICAL PROSPECTION

During the 1970's, marine geophysical surveys were commenced by the University of Southampton over the site of Fox's lost Solent River (Dyer, 1972 & 1975). A sparker sonar unit, towed behind a boat, soon revealed the course of Fox's deep and elusive channel as it passed beneath the Eastern Solent. A tributary channel was also detected on the floor of Southampton Water. In the Western Solent, the results obtained were less rewarding. Here, the supposed 'great western arm' of the Solent River remained elusive; however, the seismic signal had been dissipated by the presence of sand and this left the matter unresolved (Dyer, 1975).

Contemporary with these early geophysical explorations were interpretations of the raised and submerged gravels and sediments of the Solent region (Curry *et al.*, 1968; Hodson & West, 1972). These were later summarised in a masterly synthesis, in which bore-hole records and sonar traces were used to compile two basic cross-sections through the floor of the modern Solent (West, 1980). At Fawley, the floor of Southampton Water was transected to reveal a bedrock floor at -24 m O.D. This had been covered by a spread of some 5 m of Pleistocene gravel. In post-glacial times, the Fawley side of this valley had been infilled with marine silts and salt marsh sediments; these, it seemed, had steadily accumulated in dynamic equilibrium, as the sea level progressively rose.

In the Eastern Solent, a transect was drawn between Gosport and Ryde. Here, the valley floor was traced to around -30 m O.D., above which a spread of Pleistocene gravel was found to be covered by a thick mass of post-glacial shingle and sand. These transects confirmed the presence of a deep flat-bottomed river valley which had, eventually, been invaded by the sea. These details have since been confirmed and enhanced by more recent sub-bottom seismic (CHIRP) surveys, revealing more detail including much current bedding in the upper fills (Justin Dix, University of Southampton, pers. com.).

After these early geophysical investigations in the Solent, there came reports from farther offshore. It was soon realised that some 25 km to the south of Isle of Wight, the course of the lost Solent River could still be traced to the point where it entered a larger complex of palaeovalleys, which were associated with the former course of the Seine (Larsonneur, *et al* 1982; Hamblin & Harrison, 1989). As Geikie had predicted in 1881, between the French and English coasts lay a drowned riverine landscape capable of accommodating populations of Pleistocene animals and hunters.

THE LOSS OF AN ARM

During the 1990's, a new generation of sub-bottom profilers has enhanced the stratigraphical knowledge of Fox's Solent River in the floor of the Eastern Solent. These studies have also largely discredited his case for a 'great western arm' extending throughout and beyond the length of the Western Solent. In 1986, a palaeochannel was reported beneath Hurst Spit and Pennington Marshes (see Figure 3, locations 11 & 12), but Nicholls was quick to observe that its depth of -7 m O.D. seemed insufficient to provide a downstream outlet for the Fox's Dorset catchment. This, he observed, had already attained a depth of -15 m O.D., in a deep palaeochannel at Hamworthy (Nichols & Clarke, 1986; Nichols, 1987).

In 1994, a new study of the terrestrial and submarine gravel terraces of the Solent region was presented (Allen & Gibbard, 1994). This investigation demonstrated how an earlier Pleistocene valley or lowland could form in the area that was, eventually, to accommodate the Western Solent. This study proposed a fluvial history, which generally endorsed the original model proposed by Fox. Unfortunately, this work did not embrace the new seismo-topographical research being conducted by the University of Southampton. A crucial omission was the palaeovalleys, which the oceanographers were discovering in the Wight-Purbeck ridge (Velegrakis, 1994).

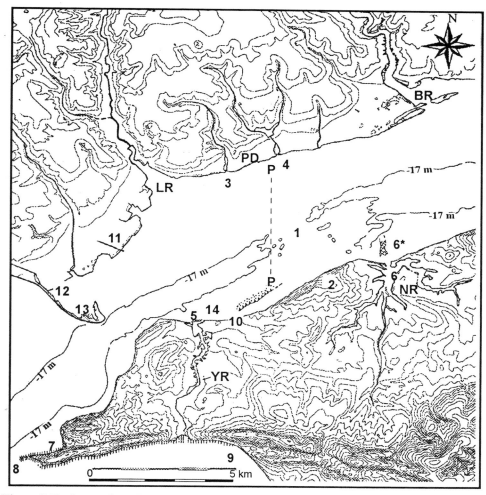

Figure 3 Bathymetric and sub-bottom indications of a submerged West Solent land-bridge or umbilical (1) at -17 O.D. Other numbered sites are cited in this and related text. P-P Sub-bottom transect after J. Dix; YR Western Yar River; NR Newtown River; BR Beaulieu River; LR Lymington River; PD Pitts Deep.

The completed seismic study of the submerged topography of Bournemouth and Christchurch Bays was soon to confirm the suspicions raised by Nichols (Velegrakis, 1994). This revealed seven major palaeovalley complexes cutting through the submerged Wight-Purbeck ridge and escaping across the floor of the English Channel (Velegrakis *et al.*, 1999; Velegrakis, *this volume*). It was now apparent that well before the close of the Pleistocene, the East Dorset catchment had no need for an exit route to the north of the land of Wight. Moreover, it now seemed that the Hurst-Pennington palaeochannel was probably part of a westward-flowing

PRESENT KNOWLEDGE, OUTSTANDING LACUNAE

With a southward pattern of drainage now identified in Christchurch Bay, there arises the possibility that prior to the formation of the Western Solent, a similar southerly exit was provided by a single river formed by the Lymington River and the Western Yar. This possibility has yet to be fully explored, yet sub-bottom profiles examined by the present writer across Freshwater Bay seem to exclude the presence of a discernable palaeochannel in the sandstone bedrock of this embayment.

Meanwhile, the relationship between the catchments of the lower Yar and Lymington rivers, together with that of the Newtown river, has become critical to our enquiry. Between these two ancient catchments, at Bouldnor, the Island's cliff line rises as high as 50 m OD but the opposing topography along the Hampshire coastline remains low and offers ready opportunity for past river capture. This might be easily affected between the lower and lost pre-inundation courses of the Lymington and Sowley rivers. Such capture may have been the triggering mechanism for the severance of Wight.

Recent sediment cores extracted at Yarmouth and Newtown seem well-placed to explore this issue. Some marine influences are evident at Yarmouth prior to 7000 *cal*. BC but there follow two episodes of peat accretion which attest the persistence of fresh or brackish conditions in the river Yar (Scaife, pers. comm.). It seems most unlikely that these peat-forming conditions could occur during a process of marine inundation or during the genesis of an open West Solent seaway. After an hiatus in the 6th millennium BC, a second episode of peat accretion commences at Yarmouth and this persists until the mid 4th millennium BC when it terminates at a height of -5.50 m OD. It is after the cessation of this peat, around 3650-3360 *cal* BC (GU-5419), that marine conditions seem to make an unremitting appearance.

At Newtown Spit, east of the supposed Reid isthmus, diatom and palynological evidence has yet to be fully collated but here, as in the Yar river, there is evidence of marine conditions after 3500-3040 *cal* BC (OXA-4778). It is after the close of the 4th millennium BC that the construction of Neolithic intertidal wooden trackways is evident on the Isle of Wight shore and these seem to attest a human response to worsening tidal conditions. At Newtown, 'platform B' was constructed around 2920-2500 *cal* BC (GU-5341) when flooding seems to have been occurring at -1.60m OD. At Yarmouth a somewhat similar structure has been dated at 2920-2620 *cal* BC (GU-5260). We might speculate that it was around this time that the sea was first able to advance on the high land at Bouldnor. Having weighed this information, the speculative might wish to postulate a severance event during or after the mid 4th millennium BC but the wise will await the results of further study and the emergence of a more definitive picture. Some of these possibilities are further discussed in the research agenda, which is presented separately in this publication (see Biodiversity and Conservation Workshop Findings: Archaeology).

This short review has shown that past studies of the geomorphological history of the Solent seaway were largely driven by antiquarian curiosity and by early interests in processual geology. Marine geophysical survey, as demonstrated by Dyer (1972 & 1975), offered a geology. Marine geophysical survey, as demonstrated by Dyer (1972 & 1975), offered a major advance in the investigation of this topic, yet this was to be followed by a comparative dearth of follow-up activity during the 1980's.

At the close of the twentieth century, fresh needs and new opportunities have emerged. These include a new generation of geophysical instruments, which have been used so successfully in Christchurch Bay (Velegrakis et al., 1999). New archaeological and palaeo-environmental investigations have begun to trace the later history of Holocene sea-level rise, as it is attested in the intertidal zone of the Solent shore and creeks (Tomalin, 1993; Allen et al., 1993; Loader et al., 1997). Concurrent with these investigations is a demand by coastal managers and planners for an understanding of long-term coastal processes and the severity of regional sea-level change. Behind these requirements lurks a disturbing and outstanding lacuna. Before we proceed to coastal development strategies, environmental management plans and predictions of coastal change, we have yet to resolve just how old is the Solent seaway and just how young are its channel-forming processes? Some contributions to a research agenda, addressing these problems, are outlined separately in this volume.

ACKNOWLEDGEMENTS

The review of this topic arises out of work which has been promoted by the Isle of Wight Council, English Heritage and the Hampshire and Wight Trust for Maritime Archaeology. I am particularly indebted to Adonis Velegrakis, Rob Scaife, Anthony Long, Justin Dix and Garry Momber, who have generously provided original research data and instructive discussions on this topic. Support in production has been provided by my colleagues Frank Basford, Rebecca Loader, David Motkin and Janet Abernethy.

REFERENCES

Allen, L.G. & Gibbard, P.L. 1993. Pleistocene evolution of the Solent River of southern England, *Quaternary Science Review* **12**: 503-528.

Camden, W. 1586, 1587 Latin editions, 1610 & 1637 English editions. *Britannia, a chorographicall description of the most flourishing kingdomes, England, Scotland & Ireland.*. London, 1109 pp.

Cunliffe, B. 1988. *Mount Batten, Plymouth. A prehistoric and Roman port.* Oxford University Committee for Archaeology, monograph **26**, Oxford, 108 pp.

Curry, D. Hodson, F. & West, I.M. 1968. The Eocene succession in the Fawley transmission tunnel *Proceedings of the Geological Association,* **79**: 179-206.

Dyer, K.R. 1972. Recent sedimentation in the Solent Area. *Memoire du Bureaux des Recherches des Mines (BRGM)* No. **79**: 271-280.

Dyer, K.R. 1975. The buried channels of the Solent River, Southern England *Proceedings of the Geological Society,* **86**: 239-245.

Englefield, H.C. 1816. *A description of the principle picturesque beauties, antiquities and geological phenomena of the Isle of Wight.* London, 292 pp.

Fulford, M. Champion, T. & Long, A. 1997. *England's coastal heritage; a survey for English Heritage and the RCHME*. EH/RCHME, London, 268 pp.

Geikie, J. 1881. *Prehistoric Europe: a geological sketch*. Edward Stanford. London.

Godwin, H. 1940. Pollen analysis and forest history of England and Wales. *New Phytologist*, **39**: 270-400.

Hencken, H.O'N. 1932. *An archaeology of Cornwall and Scilly*. Methuen. London, 340 pp.

Hamblin, R.J.O & Harrison, D.J. 1989. *Marine aggregate survey, Phase 2: South Coast*. British Geological Survey. Kidworth, Nottingham, 30 pp.

Herring P.C. 1992. *St Michaels Mount*. National Trust archaeological report. Cornwall Archaeological Unit. Truro, 194 pp.

Hodson, F. & West, I.M. 1972. Holocene deposits at Fawley, Hampshire and the development of Southampton Water, *Proceedings of the Geological Association*, **83**: 421-444.

Kell, E. 1866. An account of a Roman building at Gurnard Bay in the Isle of Wight, and its relation to the ancient British tin trade in the Island, *Journal of the British Archaeological Association*, **22**: 351-368.

Larsonneur, C. Bouysee, P. & Auffret, J-P. 1982. The superficial sediments of the English Channel and its Western Approaches, *Sedimentology*, **29**: 851-864.

Lyell, C. 1840. *The principles of geology or the modern changes in the earth and its inhabitants*. London.

Maxwell, I.S. 1972. The location of Ictis. *Journal of the Royal Institute of Cornwall*, **6** (4): 293-319.

Nicholls, R.J. 1987. Evolution of the upper reaches of the Solent River and the formation of the Poole and Christchurch Bays. In: *Field guide to Wessex and the Isle of Wight*, Barber, K.E. (Ed.) Cambridge University Press: 99-114.

Nicholls, R.J. & Clarke, M.J. 1986. Flandrian peat deposits at Hurst Castle Spit, *Proceedings of the Hants Field Club & Archaeological Society*, **42**: 15-21.

Oakley, K.P. 1943. A note on the post-glacial submergence of the Solent margin, *Proceeding of the Prehistoric Society*, **9**: 56-59.

Penhallurick, R.D. 1997 The evidence for prehistoric mining in Cornwall. In: *Prehistoric extractive metallurgy in Cornwall*, Budd, P. & Gale, D. (Eds.) Cornwall Archaeological Unit. Truro, 57 pp.

Redman, J.B. 1852. The alluvial formations and local changes in the south coast of England, *Proceedings of the Institute of. Civil Engineering*, **11**: 162-223.

Reid, C. 1905. The island of Ictis. *Archaeologia*, **59**: 218-288.

Rivet, A.L.F & Smith, C. 1979. *The place names of Roman Britain*. Batsford, London, 526 pp.

Shore, T.W. & Elwes, J.W. 1889. The New Dock excavations at Southampton. *Proceeding Hampshire Field Club & Archaeological Society*, **1**: 43-56.

Short, T. 1749. *A general chronological history of the air, weather, seasons, meteors etc. in sundry places and different times more particularly for the space of 250 years*. Longman, London.

Skinner, J. 1797. This diarist cites tin ore on St Michael's Mount in his ms journal of a West Country tour through Somerset, Devon and Cornwall. Reproduced in: West *Country Tour*, Jones, R (Ed.) 1985 (p60). Bradford on Avon. 96 pp.

Short, T. 1749. *A general chronological history of the air, weather, seasons, meteors etc. in sundry places and different times more particularly for the space of 250 years.* Longman, London.

Skinner, J. 1797. This diarist cites tin ore on St Michael's Mount in his ms journal of a West Country tour through Somerset, Devon and Cornwall. Reproduced in: West *Country Tour*, Jones, R (Ed.) 1985 (p60). Bradford on Avon. 96 pp.

Tomalin, D.J. 1993. Maritime archaeology as a coastal management issue. In: *Proceedings of the Standing conference on problems associated with the coastline; seminar on regional coastal groups-after the House of Commons Report*, SCOPAC, Isle of Wight County Council, 93-112.

Thomas, C. 1985. *Exploration of a drowned landscape; archaeology and history of the Isle of Scilly*. Batsford. London, 320 pp.

Velegrakis, A.F. 1994. Aspects of morphology and sedimentology of a transgressional embayment system: Poole and Christchurch Bay, Southern England. Unpublished Ph.D. thesis. University of Southampton, 376 pp.

Velegrakis, A.F. (this volume). Geology, geomorphology and sediments of the Solent. In *Solent Science - A Review.* Collins, M.B. & Ansell, K. (Eds.), Proceedings in Marine Science Series, Elsevier, Amerstam.

Velegrakis, A.F, Dix, J.K. & Collins, M. B. 1999. Late Quaternary evolution of the upper reaches of the Solent River, Southern England, based upon marine geophysical evidence. *Journal of the Geological Society, London,* **156**: 73-87.

West, I.M. 1980. Geology of the Solent estuarine system. In: *The Solent Estuarine System; an assessment of present knowledge*. N.E.R.C. Publications. Series C. No. 22, 6-19.

Geology, Geomorphology and Sediments of the Solent System

Adonis Velegrakis

School of Ocean and Earth Science, Southampton Oceanography Centre, University of Southampton, European Way, Southampton, SO14 3ZH, U.K.

INTRODUCTION

The Solent (Figure 1) forms the largest estuarine system of the southern coast of the UK. The constituent components of the system include the West and East Solent and their approaches, Southampton Water, Portsmouth, Langstone and Chichester Harbours and other smaller tributary river estuaries (e.g. Beaulieu, Lymington and Yar) found along the southern Hampshire and Isle of Wight coastlines. Parts of the coastline are characterised by coastal accumulation forms, such as barrier spits and islands (e.g. Hurst and Calshot Spits in the West Solent and Hayling Island in the East Solent), inter-tidal flats and saltmarshes (e.g. Lymington Flats). Erosional coastal environments (i.e. coastal cliffs) are also present, particularly along the coastline of the Isle of Wight. The offshore areas also show complex morphology (Figure 1), associated with several offshore banks and deeply-incised channels (e.g. Hurst Narrows).

The Solent has attracted a great deal of human economic development, including extensive urban and industrial development, agriculture, shipping, fisheries, recreation, marine aggregate extraction and offshore oil exploration (Shell, 1987). At the same time, the area is associated with important conservation areas such as the National Nature Reserves (NNR), Sites of Scientific Interest (SSSI), Local Nature Reserves (LNR), as well as important archaeological sites. This diverse human activity both influences and is influenced by the physical characteristics and dynamics of the natural environment. Therefore, frequent monitoring is necessary, in order to assess the human impact on the environment and its evolution and, equally important, to understand and predict the influence of such evolution on the regional economic development. The understanding of the dynamic interrelationships between nature and economic development forms the 'backbone' of 'sustainable development' policies, which emerge as the main UK and European Union environmental strategy.

The objective of this contribution is to review the present state of knowledge and identify gaps in information on some of the physical characteristics of the Solent Estuarine system and, particularly, its geology, geomorphology and sedimentology. In this sense, this contribution forms an update of the meticulous reviews of West (1980) and Dyer (1980).

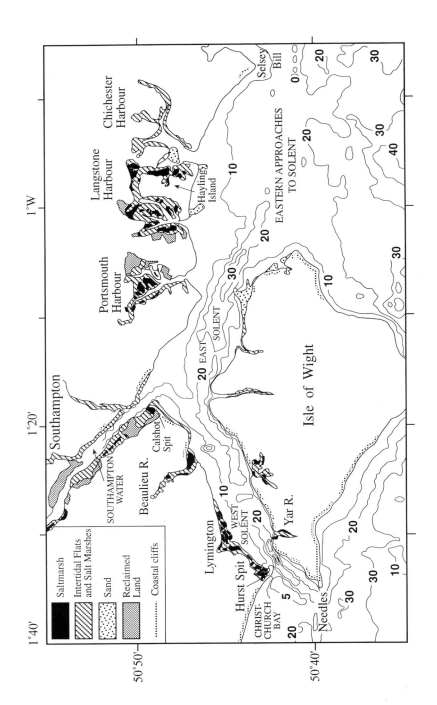

Figure 1 Coastal sedimentary environments, river systems and offshore bathymetry of the Solent Estuarine system. Contours in metres, below Lowest Astronomical Tide (LAT).

GEOLOGY

Lithology

The Solent Estuarine System is located at the southern margin of the Hampshire Basin, an elongated asymmetrical downwarp of mainly Tertiary deposition, the southern limb of which exhibits a near-vertical northern dip, whilst the beds of its northern limb slope gently southward (Melville & Freshney 1982). The Lower Cretaceous Wealden Beds, which consist of freshwater marls and estuarine/shallow marine shales (Stewart, 1981; Stoneley, 1982; Whittaker, 1985; Rawson, 1992; Hamblin *et al.*, 1992; Ruffel & Batten, 1994; Allen, 1998; Radley & Barker, 1998), are the o)ldest strata outcropping in the area (Figure 2). These beds are overlain by the Lower Greensand (comprising shallow marine sands and clays (Casey, 1961)) and the shallow marine clays and sands of the Albian Gault and Upper Greensand, which rest unconformably on the Lower Greensand (Rawson, 1992). The transgressional contact between the Upper Greensand and the overlying Chalk is formed by the Cenomanian Chalk Basement or Glauconitic Marl Beds. The Chalk comprises three distinct units: the Lower, Middle and Upper Chalk (Melville & Freshney, 1982). The Upper Chalk, which is the most important of these units, contains extensive fossils and flint nodule bands (Anderton *et al.*, 1979); its early lithification took place during several progressive stages (Kennedy & Garrison, 1975). The Upper Chalk forms the vertical cliffs and stacks at the Needles, outcropping with an E-W strike; it underlies the Solent System (Dyer *et al.*, 1969; Whittaker, 1985), but onshore outcrops are found only on the Isle of Wight and over the northern part of the area. Offshore, narrow Chalk bands outcrop at the seabed, both to the east and the west of the Isle of Wight (the Purbeck-Wight outcrop) (Hamblin *et al.*, 1992; BGS, 1995; Velegrakis *et al.*, 1999).

The position of the Palaeocene/Eocene and Eocene/Oligocene sediment boundaries has been the subject of much discussion in the past, as many local names have been used to describe lithological units in the Hampshire Basin; this has resulted in confusion regarding the nomenclature (Keen, 1977; Curry *et al.*, 1978; Odin *et al.*, 1978; Stinton & Curry, 1979; West, 1980; Freshney *et al.*, 1985; Edwards & Freshney, 1987; Hamblin *et al.*, 1992; BGS, 1995). The older Tertiary sediments in the area are the Palaeocene Reading Beds (Curry *et al.*, 1978; Odin *et al.*, 1978), which outcrop close to the Needles (Alum Bay) in the south, and around Chichester, in the north (Figure 2, for locations, see Figure 1). The Reading Beds lie unconformably on the Upper Chalk and are characterised by variable lithology (clays, red-mottled clays, clay-breccias, cross-bedded sands, pebbles) and a basal bed composed of glauconitic sands, clays and flint pebbles. Stratigraphic evidence suggests that the sediments have been deposited in an environment characterised by low salinities and unstable geochemistry (Buurman, 1980; Melville & Freshney, 1982). The Eocene London Clay (Gilkes, 1966; Curry *et al.*, 1972; Burnett & Fookes, 1974), which rests with a sharp and slightly erosive boundary on the Reading Beds, consists of bioturbated silty and sandy clays, clayey silts and sandy silts; its thickness decreases towards the west of the study area (Edwards & Freshney, 1987). Five individual coarsening-upward sequences have been recognised in the London Clay (Freshney *et al.*, 1985), which also exhibits (as a whole) a general coarsening-upward trend.

The Eocene Bracklesham Group, as defined here, corresponds to the Bangshot Beds and the Bracklesham Beds of the earlier studies over the area (Curry *et al.*, 1978; West, 1980; Melville & Freshney, 1982). Lithologically, the Group consists of clayey sands, sandy clays, glauconitic sands with molluscan remains and calcareous nodules (Curry, 1976), which were deposited in various sedimentary environments (meandering rivers, flood basins, estuaries, lagoons and coastal waters). These sedimentary environments had been laterally intergradational (from mainly marine in the east to alluvial in the west); 5 depositional cycles have been recognised (Plint, 1983). The constituent sediments were derived originally from the erosion of several distinct formations, such as the Britanny Massif, the English Midlands Massif and the Cornubian Granite (Walder, 1964).

The Eocene Barton Group lies above the Bracklesham Group (Figure 2); it includes a predominantly argillaceous part (Barton Clay), a predominantly arenaceous part (Barton Sand) and the Lower Headon Beds (Melville & Freshney, 1982). The Barton Clay consists of blue-grey clays, rich in illite and montmorillonite (Gilkes, 1978). The Barton Sand and the Lower Headon Beds consist of sands, glauconitic sandy clays, thin bands of limestone and lignitic silts and clays; they have been formed as regressive coastal sequences and show an upward transition, from shallow-marine to fluvio-lacustrine environments (Plint, 1984).

Finally, the Solent Group (Upper Headon, Osborne and Bembridge Formations), which overlie the Barton Group, consist of some 150 m of freshwater sediments (Keen, 1977) interbedded with thin deposits; these represent two brief marine incursions. The first two formations in the Group consist of blue/greenish clays, whilst the Oligocene Bembridge Formation comprises limestones and marls.

Structure

In the Solent area, geophysical evidence has indicated the presence of three zones in the crust (Whittaker & Chadwick, 1984): Zone 3, found at depths from 15-20 to 30 km, consisting of crystalline basement rocks; Zone 2 or Variscan Orogenic Basement, found at depths from 4 to 15-20 km, comprising the strongly folded and faulted Precambrian and Palaeozoic rocks; and Zone 1, extending to depths down to 4 km, consisting of relatively gently deformed Permian, Mesozoic and Tertiary sediments.

Following the Variscan orogeny, thick sedimentary sequences were deposited in a deep basin within the region (Bacon, 1975; Whittaker, 1985). This basin is divided into sub-basins by faults, which strike mostly ENE-WSW and appear to have a complex history. Movement along them in early post-Variscan times usually shows downthrows to the south; in contrast, later movements show small-scale downthrows mostly to the north (Hamblin *et al.*, 1992; see also Underhill & Paterson, 1998). As a result, the Cretaceous and Tertiary strata appear not to be seriously fractured (but, see Nowell, 1995); rather to drape across the deep faults to form the Late Tertiary Portland-Wight Monocline, which formed the high grounds of the Purbeck-Wight Chalk Ridge, and the Portsdown Anticline. Two hypotheses have been suggested for the stress regime responsible for the formation of the monocline: (a) that it has been formed under a predominantly extensional stress regime (Melville & Freshney, 1982; Edwards & Freshney, 1987); and (b) that it has been formed under a compressional

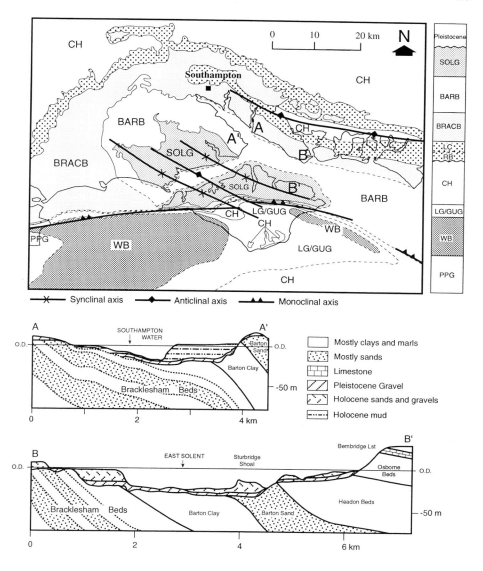

Figure 2 Simplified solid geology of the Solent, based upon IGS (1975a; 1975b), Melville & Freshney (1982), Edwards & Freshney (1987), Hamblin et al. (1992), Velegrakis (1994), BGS (1995). Key: PPG, Portland/Purbeck Group; WB, Wealden Beds; LG/GUG, Lower Greensand/Gault and Upper Greensand; CH, Chalk; RB/LC, Reading Beds and London Clay; BRACB, Bracklesham Beds; BARG, Barton Group; SOLG, Solent Group. The Tertiary formation division follows Hamblin et al. (1992) and BGS (1995). The sections shown are adapted from West (1980). The lithological succession log is not to scale.

stress regime resulting from the reactivation of older normal faults (Arkell, 1947; Hamblin *et al.*, 1992; Velegrakis, 1994). Minor folds, trending mostly NW-SE are also present within the Solent and its approaches (Daley & Edwards, 1971; Curry, 1976; Dixon, 1986; Velegrakis, 1994).

Quaternary Development

Although a considerable amount of research has been undertaken into the Pleistocene deposits in the Solent, no clear stratigraphic picture of the region has emerged yet, as organic sediments (which could provide chronological control) have rarely been preserved within the stratigraphic record (Brown *et al.*, 1975; Wright, 1982; Holyoak & Preece 1983; Munt & Burke 1986; Roberts, 1986; Barber & Brown 1987; Long & Tooley, 1995). In addition, mild tectonic movements may have complicated further the interpretation of the few existing data, as they make correlations based upon comparisons of the altitude of the deposits questionable (Preece *et al.*, 1990).

The Pleistocene evolution of the southern part of the Hampshire Basin may have been controlled by both its structure and the sea-level fluctuations. Wooldridge & Linton (1955) have suggested that the drainage systems of southeastern England have been influenced largely by the Calabrian sea level fluctuations, whilst other investigators have argued that these fluctuations had only a minor effect on the drainage pattern (Jones, 1980).

The present Solent system is the Holocene 'heir' of the middle section of the "Solent River" system, a lowland riverine system which, during the Pleistocene lowstands, integrated all the consequent rivers of the basin (Fox 1862; Reid 1905; Everard 1954; West 1980) and formed one of the major tributaries of the 'English Channel River' (Gibbard, 1988). This Pleistocene Solent River flowed along a large E-W trending watershed, incised on Tertiary sediments and surrounded by high Chalk country. Evidence for the existence of the Solent River system are found not only onshore (e.g. extensive deposits of Pleistocene terrace gravels), but also offshore, where marine geophysical surveys have revealed the presence of systems of buried river valleys under the present seafloor (Figure 3). During the interglacial periods, the area was submerged repeatedly, creating broad estuaries (Brown *et al.*, 1975; Nicholls, 1987). Most of the river's watershed was drowned during the Flandrian Transgression, when Poole and Christchurch Bays and the Solent area were formed (Dyer, 1975; Nicholls, 1987 (but see Kellaway *et al.*, 1975)); only parts of its tributary river systems are still intact, forming the modern drainage network of the area.

It has been suggested widely that the upper reaches of the Solent River (i.e. the part to the west of Hurst Narrows) were disrupted by the breaching of the Purbeck-Wight Chalk Ridge. The dating of this event has been the subject of much discussion. It has been suggested to have taken place: (a) in the Late Pliocene (Reid, 1902); (b) in the mid-Pleistocene (Green, 1946); (c) during the Flandrian Transgression (Everard, 1954; Keen, 1980 and Allen & Gibbard, 1993); and (d) during the Devensian (West, 1980; Wright, 1982; McKay, 1990). Recently acquired marine geophysical evidence has shown that the upper reaches of the Solent River had been breached before the Flandrian Transgression, by three southerly-flowing rivers in the area of the present Poole Bay (Velegrakis *et al.*, 1999 and Velegrakis *et al.*, *this volume*).

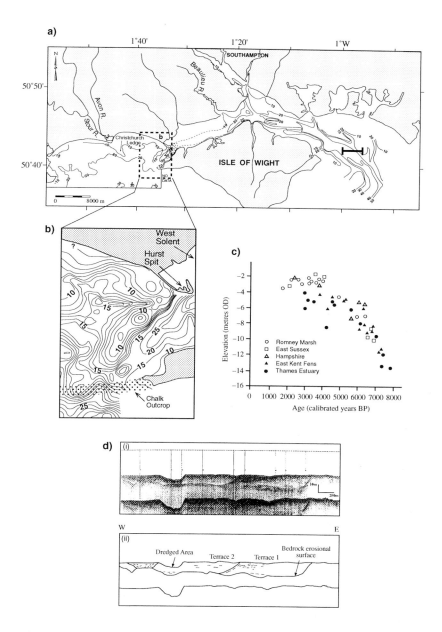

Figure 3 (a) Structure contour map (contours in metres below OD Newlyn) of the base of the unconsolidated sediments (bedrock erosional surface), based upon Dyer (1980) and recent results of various SUSOES (Southampton University, School of Ocean and Earth Science) marine geophysical investigations; (b) Detailed structure contour map; (c) Sea level data from the South Coast (adapted from Long & Tooley, 1995); (d) Shallow seismic section (from the SUSOES data base), from the downstream section of the Solent River buried valley (location is shown in Figure 3a). Note: the different generations of gravel terraces within the buried valley.

The two westernmost buried valleys of these Poole Bay southerly-flowing rivers are infilled by a variety of sediments, which range from lowstand fluvial gravels in the bottom of the river's thalweg to estuarine and coastal transgressive sediments at the top of the sequence. The architecture and type of these infilling sediments suggest that these rivers may have experienced only one cycle of fluvial incision and marine transgression (Velegrakis et al., 1999). In contrast, the buried valleys of the downstream section of the Solent River show different generations of gravel terraces (Figure 3), which may indicate a series of such cycles (see also Bellamy, 1995).

Information on the Holocene sea-level rise over the area is both limited and fragmentary. Pollen and radiocarbon analyses have been used on sediments deposited in estuarine saltmarsh and fenwood environments, in Poole Harbour and the Solent (Devoy, 1972, 1982; Long & Tooley, 1995). The results have provided evidence of a progressive marine inundation of levels above -30 m O.D. after 9000 years BP (Before Present); these indicate that the trend of sea level recovery in the area does not conform to a simple pattern of exponential change (Figure 3). Comparison of the elevation of the basement of the unconsolidated sediments (the upper bedrock erosional surface) in the Solent area, with the regional sea level curve (Figure 3), suggests that large part of the Solent system had been inundated before 7000 years BP and that the Christchurch Bay-Solent link was established during that time. On the basis of these data, it may be suggested that the eastern part of Christchurch Bay and the West Solent were subject to a radical change in the prevailing hydrodynamic conditions, when the sea level rise linked these areas and effectively separated the Isle of Wight from the mainland (see also West (1980)). The presence of strong tidal currents flowing over this part of the system may have eroded away most of the sediments deposited previously in eastern Christchurch Bay and the West Solent. The ability of the marine currents to erode the seabed is demonstrated by the depth of incision at the Hurst Narrows (Figures 1 and 3) and the coarse-grained nature of the surficial sediments in the West Solent (Figure 4). Moreover, Dyer (1969) observed the presence of rolled fragments of clay bedrock, in surficial sediment samples collected in the Hurst Narrows; this indicates that erosion of the seabed is still active in this area.

GEOMORPHOLOGY

The coastline of the area may be divided into a number of geomorphological units (Figure 1): (i) erosional cliffed coasts, found along the open marine coasts of the Isle of Wight and the approaches to the Solent (and, to lesser extent, along parts of the coastline of the West Solent and Southampton Water); (ii) barrier beach coasts found associated mainly with the tidal inlets of Hurst Narrows (Hurst Spit), Southampton Water (i.e. Calshot Spit) and Portsmouth, Langstone and Chichester Harbours, which can be characterised as areas of sediment accumulation and consist mainly of coarse-grained sediments; and (iii) inter-tidal flat/saltmarsh coasts, which are found mainly behind the barriers and along protected parts of the coastline.

Large sections of the coasts of the Solent Estuarine System have experienced extensive residential and industrial development (e.g. land reclamation for harbour and industrial facilities); the effects of this development on the natural evolution of the system have not

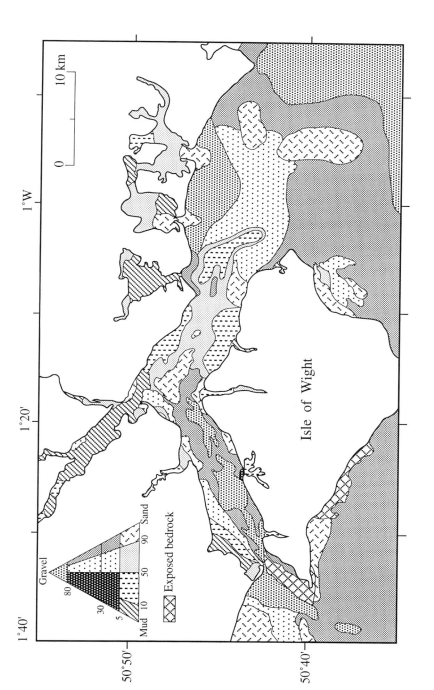

Figure 4 Simplified distribution of the lithological types of surficial seabed sediments in the Solent, based upon the publications of several authors (Dyer, 1980; Brampton, 1993; Algan, 1993; Velegrakis, 1994; Whitcombe, 1995; Paphitis, 1997; Elderfield, 1999).

yet been fully assessed. However, most coasts of the system appear to be presently under erosion (King & McCullagh, 1971; Harlow, 1980; Nicholls & Webber, 1987; Wallace, 1990; Ke & Collins, 1993; Hooke et al., 1996), although large efforts and resources have been spent on coastal protection schemes (e.g. Lacey, 1985; McFarland et al., 1994; Whitcombe, 1995; Bradbury, 1999).

The offshore areas of the Solent system are associated generally with modest water depths, which rarely exceed 20 m (Figure 1); however, localised deeps are present in the vicinity of the West and East Solent entrances (the Hurst Narrows and Spithead, respectively). Other seabed morphological features include narrow tidal channels, ebb tidal deltas and offshore gravel and sand banks (Figure 5). The seabed of the area has also experienced human interference including capital and maintenance dredging along the shipping lanes and within the harbours (Elderfield, 1999) and offshore gravel and sand dredging (e.g. Hydraulic Research Station, 1977; Shell, 1987; Bradbury, 1999).

SEDIMENTS

Type and distribution

The Quaternary drift deposits of the area consist of terrace and valley gravels, interglacial (mud and beach) deposits, head, alluvium, estuarine, coastal and open marine deposits. Terrace gravels occur at several levels, above or under the alluvial plain (Keen 1980; Freshney et al., 1985; Allen & Gibbard, 1993); they mainly consist of flint pebbles (in matrices of brown sands) originating from the erosion of the Upper Chalk flint bands, with some sarsen (silicified sandstone) pebbles also present. These deposits are of fluvial origin, formed during the Pleistocene glacial fluctuations (Green, 1946; Everard, 1952; Nicholls, 1987), and do not show any direct evidence of structural deformation (Keen, 1980; but, see Allen & Gibbard, 1993). Interglacial (mud and beach) deposits have been found to the north of Chichester and Portsmouth Harbours, at Stone (West Solent) and at Bembridge and Newtown (Isle of Wight) (Hodgson, 1964; Brown et al., 1975; West, 1980; Preece et al., 1990). Head deposits occur mainly in the valley bottoms and slopes, formed from soliflucted downwashes of gravels, sands and clays and from material eroded from the solid formations (Freshney et al., 1985). The recently deposited river sediments (alluvium) consist commonly of grey silts and gravel seams, containing shells of freshwater molluscs (West, 1980).

The offshore sediments rest upon a bedrock erosional surface of complex relief (Figures 2 and 3) and may be divided into: (i) buried channel infilling sediments; and (ii) surficial (modern) sediments. The buried channel deposits are generally coarse-grained and their architecture and textural characteristics suggest that they have been deposited within various (fluvial, estuarine and shallow marine) sedimentary environments (see also the Quaternary Development section).

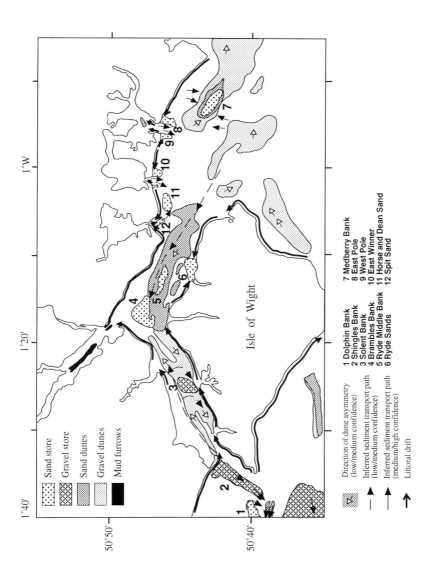

Figure 5 Sediment storage areas, bedform fields and bedload sediment transport pathways in the Solent area. The littoral drift directions are based mainly upon Bray et al. (1995). The degree of confidence on the sediment transport paths varies, depending on the approach adopted.

within various (fluvial, estuarine and shallow marine) sedimentary environments (see also the Quaternary Development section).

The surficial sediments consist of a variety of lithological types, ranging from the muds found in the estuarine environments to the very coarse-grained sediments found in the offshore areas (Figure 4). Within the Solent, the distribution of the lithological types (and mean grain-sizes) appears to follow an interesting pattern; the West Solent is associated with coarse- and very coarse-grained sediments (sandy gravels and gravels), whereas the East Solent is dominated by medium- and fine-grained sediments (see section on Sediment Transport).

The surficial sediments are generally multi-modal (Dyer, 1969; 1980; Langhorne *et al.*, 1986; Brampton, 1993; Whitcombe, 1995; Paphitis, 1997; Brampton *et al.*, 1998). The coarse fraction consists typically of sub-angular flint gravels, covered by brown iron oxide staining. The medium fraction comprises mainly quartz sands (and, to a lesser extent, shell fragments), with some opaque minerals also present (Dyer, 1969; Velegrakis, 1994). Different types of clay minerals (e.g. illite, smectite, kaolinite and chlorite) are present within the fine-fraction, with their relative abundance depending on location (Algan, 1993). Heavy metal pollution is significant in the area, although, once again, the concentration of the heavy metals within the sediments is variable over the area (Algan, 1993; Mercone, 1995).

On the basis of the rather limited (and fragmentary) information available on the thickness of the offshore unconsolidated sediments, it appears that the seabed of the Solent Estuarine System is characterised generally by a rather thin sedimentary cover, which rarely exceeds 2 m in thickness. Thick sedimentary deposits occur, either in association with buried valleys (e.g. Dyer, 1975; Dixon, 1986; Dean, 1995; Velegrakis *et al.*, 1999) or the tidal deltas and offshore sand and gravel banks (e.g. Hydraulics Research Station, 1977; Velegrakis & Collins, 1992).

Sedimentary morphology

Several sedimentary bodies are found within the Solent Estuarine System. Coastal features (spit/barrier beaches), associated with longshore sediment transport, are common with most of the inlets showing well developed spits and tidal deltas (Figure 5).

Offshore, several banks occur; the most important of these banks are the Dolphin and Shingles Banks (within the western approaches to Solent), the Solent, Brambles and Ryde Middle Banks (Solent) and the Horse and Dean and Medmerry Banks (eastern approaches to the Solent). In addition to these sedimentary bodies, several large bedform (gravel and sand dune, see Ashley *et al.*, 1990) fields are observed in the area (Figure 5); some of the bedforms exceed 2 m in height and 25 m wavelength (Dyer, 1980; Langhorne *et al.*, 1986; Velegrakis, 1994). The western part of the area is associated, generally, with higher concentrations of bedform fields, particularly along the southern part of the West Solent and the Shingles Bank (eastern Christchurch Bay). The gravel dunes, in these areas, occur either as more or less homogeneous fields, or in "isolated" trains. In the latter case, bedforms the morphological characteristics of the bedforms are controlled hydrodynamically (e.g. Yalin, 1964) and shallow marine environments are dominated by unsteady tidal and wave-induced currents, the morphology of the bedforms may be spatially and temporally variable (Terwindt & Brouwer, 1986). The cross-sectional profiles of bedforms, in particular, have been observed to be very sensitive to changes in the direction and intensity of the tidal currents (e.g. Harris & Collins, 1984). Therefore, morphological observations on sub-aqueous dunes, obtained at a particular stage of the tide, should not be used as representative of dune asymmetry (and the direction of sediment transport) during all stages of the tidal cycle.

Finally, linear furrows (Figure 5) aligned parallel to the (tidal) flow occur in Southampton Water (Dyer, 1970). If these bedforms are destroyed by dredging activities, they tend to reform in the sediments deposited after the end of the maintenance/capital dredging which takes place in their vicinity (Dyer, 1980; Elderfield, 1999).

Sediment transport

The sediment transport patterns in the Solent Estuarine system are controlled by its complex hydrodynamic regime. Thus, a summary of the hydrodynamics is necessary in order to understand the intricate patterns of sediment transport.

Some limited (and fragmented) investigations into suspended sediment budgets and transport has been carried out in the area (e.g. Srisaenthong, 1982; Velegrakis et al., in press), however, most of the studies undertaken are related to bedload transport.

Hydrodynamics

The tidal regime in the Solent area is semi-diurnal, showing a marked increase in tidal range from west to east (2.2 m at Hurst Point and 4.1 m at Portsmouth). Previous investigations have shown that the complex morphology of the area results in intricate tidal hydraulics (Dyer, 1969; DeMesquita, 1972; Dyer & King, 1975; Blain, 1980; Webber, 1980; Shell, 1987; Brampton, 1993; Velegrakis, 1994; Whitcombe, 1995; Paphitis, 1997; Boxall et al., 1997; Brampton et al., 1998; Sharples, *this volume*). Double high waters occur in the West Solent and Southampton Water, whilst young flood stands occur in Southampton Water and the East Solent. As the duration of the flood is longer (about 7 hours), than that of the ebb (about 5.5 hours) for most of the area, the ebb currents are characterised by greater speeds.

The tidal currents in the West Solent flow almost parallel to the channel axis and reach speeds (at the surface) of up to 1.8 m/s (Shell, 1987); localised flows occur in the vicinity of the Avon, Lymington, Beaulieu, Yar and Newtown estuaries. The currents are very strong in the vicinity of the Hurst Narrows, where near-bed currents of about 1.8 m/s have been recorded on spring tides (Velegrakis, 1994). In the East Solent, the tidal currents are more variable; in the main channel, the currents reach speeds of up to 1.2 m/s, although in certain areas (e.g. off Cowes on the Isle of Wight), localised flow patterns generate stronger currents (near-bed currents up to 2 m/s) (Shell, 1987). One particular characteristic of the tidal flow in the eastern approaches to the Solent is the presence of residual eddies, formed probably as a result of the protrusion of the Isle of Wight into the tidal streams of the English Channel (Boxall & Robinson, 1986; Boxall et al., 1997; Paphitis 1997). In this area, the tidal currents weaken towards the inshore areas (Grontmij, 1973; Harlow, 1979; Whitcombe, 1995), with the exception of the entrances of Portsmouth, Langstone and Chichester Harbours, where surface currents in excess of 3 m/s are observed (Admiralty Chart 3418).

Wave action is a major factor controlling the geomorphology and sedimentology within the study area. As long time-series on wave characteristics are rarely available for the area (Shell, 1987), the wave climate can be established only through the use of indirect evidence i.e. from wind records. This approach is based upon the establishment of empirical relationships between wind stress and resulting waves, taking also into account the shallow water effects. Analysis of long-term wind records available from 4 regional meteorological stations (Portland Bill, Calshot, Lee-on Solent and Thorney Island, see Map B - Preface) show that the wind climate in the Lee-on-Solent is more severe than those in Calshot and Thorney Island (Whitcombe, 1995). Estimations based upon empirical relationships and field observations have shown that significant wave heights are much lower within the Solent. Here, significant wave heights very rarely exceed 1.8 m (Shell, 1987), whereas greater heights are observed at the Eastern and Western Approaches to the Solent (Brampton, 1993; Brampton et al., 1998). An offshore wave climate compiled for Christchurch Bay, using hindcasting techniques (HR Wallingford, 1989a; 1989b) showed that the prevailing wave direction in this area was from WSW. With regard to the wave height, waves exceeding 1 m were predicted for 31% of the time; with those exceeding 3 m over 2.6% of the time. Extreme value analysis for waves with the longest fetch (225-255° N) revealed an offshore significant wave height H_s of 5.8 m (1-year return period) and 7.2 m (20-year return period). Due to wave transformation in the inshore areas, there is a high concentration of wave energy over the middle section of the Christchurch Bay coastline (Henderson & Webber, 1979), an area associated with severe coastal erosion (Lacey, 1985). In the Hurst Spit area, inshore H_s of 3 m (1-year return period) and 3.4 m (20-year period) have been estimated (Halcrow, 1982).

For the Eastern Approaches to the Solent, analysis of wave data from the Owers Light Vessel (Draper & Shellard, 1971) has shown that H_s exceeded 1.2 m over 37% of the time; similarly, that the prevailing waves were those with H_s of 0.6 to 1.2 m and periods of 4 to 6 s (these occurred over some 38% of the time). Extreme waves in the area can reach up to 4.2 m, having periods of up to 6 s. The inshore wave conditions appear to be less severe, with H_s at Hayling Island exceeding 1.2 m only for 8% of the time (Whitcombe, 1995). Wave refraction analysis has shown that the wave energy may concentrate at different sections of the coastline, as might be anticipated, depending upon the direction of the offshore waves. Under southwesterly waves, the wave energy tends to focus upon the eastern part of Hayling Island and on particular sections of the Bracklesham Bay coastline. Under southeasterly waves, in contrast, the wave energy tends to focus upon sections of the Bracklesham Bay coastline, the central section of the Hayling island coastline, the eastern tip of the Isle of Wight and parts of the East Solent coastline (Whitcombe, 1995).

Bracklesham Bay coastline, the central section of the Hayling island coastline, the eastern tip of the Isle of Wight and parts of the East Solent coastline (Whitcombe, 1995).

Bedload sediment transport

Bedload transport in the Solent area has been studied using: (i) bedform asymmetry and trends in the sediment textural characteristics (e.g. Dyer, 1969; 1980); (ii) analysis of sedimentary budgets along the coastline (Bray *et al.*, 1995); (iii) the geochemistry of the sediments (Algan, 1993); numerical modelling (Brampton, 1993; Whitcombe, 1995; Paphitis, 1997; Brampton *et al.*, 1998; Townend & Price, *this volume*); and (v) innovative techniques based upon the combination of (low and high frequency) current measurements with visual and/or hydrophone observations of sediment transport (Hammond *et al.*, 1984; Heathershaw & Langhorne, 1988, Thorne *et al.*, 1989; Williams *et al.*, 1989; Voulgaris & Collins, 1994). In spite all the research activity, it appears, however, that there are many unanswered questions associated with the magnitude and pathways of sediment transport within the Solent area. For example, it is not yet clear why the East and West Solent are so different in terms of surficial sediment distribution (Figure 4). Although the currents in the East Solent are admittedly less intensive than those present in the West Solent, there is some evidence from numerical models (e.g. Paphitis, 1997) to suggest that they are still able to resuspend coarse silt particles (with 40 μm diameter) over a considerable part of the spring tidal cycle (60% of the time). Thus, the presence of the large coarse silt fraction, within the surficial sediments of East Solent, appears to be unusual.

The best documented bedload sediment transport pathways in the area, based upon a variety of research approaches methods, are shown in Figure 5. However, it appears that more detailed and innovative research is needed in order to identify, with confidence, the sediment transport directions and rates within the system.

SUMMARY AND ASSESSMENT OF PROGRESS

The extensive recent research undertaken into the physical characteristics of the Solent Estuarine System cannot be presented within this short contribution. However, it appears that, since the time of the previous review of the relevant information (West, 1980; Dyer, 1980), our knowledge has increased significantly. Certain questions associated with the geology, geomorphology and sediments of the system have been answered; these are summarised below.

Firstly, it appears that the nomenclature of the solid formations of the system has been standardised; their offshore outcrops have been now mapped more accurately. Secondly, our knowledge on the geological structure of the area has been advanced significantly, particularly in relation to the deep structural elements and the stress regime responsible for the formation of the Purbeck - Isle of Wight Monocline.

Thirdly, there is now conclusive evidence to suggest that the upper reaches of the Pleistocene Solent River had been disrupted by river capture, before the Flandrian Transgression. Moreover, the marine geophysical evidence and new information on the sea level rise curve for the area suggest that the Isle of Wight was separated from the mainland around 7,000-7,500 years BP. Finally, there is now substantial information on the geochemistry of the sediments and the littoral sediment budgets and transport. Moreover, the area has been used extensively as a physical laboratory, to test innovative instrumentation and methodology related to the study of sediment transport processes.

Nonetheless, it appears that certain problems remain unsolved, particularly with regard to sediment distribution and transport. The information available on the overall thickness of the unconsolidated sediments within the Solent and Southampton Water is of limited resolution. Such a limitation is likely to have detrimental effects on advancing our knowledge of the system, as it confines our ability to use advanced (hybrid) models to understand (and predict) the morphological development of the system.

In addition, it appears that the hydrodynamic information available is of limited duration and spatial resolution; knowledge on the hydrodynamics of the system is based mainly upon the predictions derived from numerical models, which have been rarely validated and calibrated properly. For suspended material, in particular, one of well-recognised problems in estuarine research is related to the accurate determination of the scalar fluxes in and out of the estuaries (for a review, see Jay et al. 1997). Therefore, accurate estimations and predictions of sediment transport rates and directions are not possible, at present.

In conclusion, there is limited (and, in most cases, fragmentary) information available on the thickness, internal architecture and textural characteristics of the major sediment depositional areas of the system, such as the tidal deltas and offshore banks. This limitation may have significant implications as these areas not only represent the most reliable stratigraphic records of the sedimentary environment, but also because they may constitute significant pollution 'reservoirs'.

REFERENCES

Algan, O. 1993. Sedimentology and Geochemistry of Fine-grained Sediments in the Solent Estuarine System. Ph.D. Thesis, Department of Oceanography, Southampton University.

Allen, L.C.& Gibbard, P.L. 1993. Pleistocene evolution of the Solent River of Southern England. *Quaternary Science Reviews*, **12**: 503-528.

Allen, P. 1998. Purbeck – Wealden (early cretaceous) climates. *Proceedings of the Geologists' Association*, **109**: 197-236.

Anderton, R., Bridges, P.H., Leeder, M.R. & Sellwood, B.W. 1979. *A Dynamic Stratigraphy of the British Isles*. George Allen & Unwin, London.

Arkell, W.J. 1947. *The Geology of the Country around Weymouth, Swanage, Corfe and Lulworth*. Memoir of the Geological Survey of Great Britain. 386pp.

Ashley, G.M., Boothroyd, J.C., Bridges, J.S., Clifton, H.E., Dalrymple, R.W., Elliot, T., Fleming, B.W., Harms, J.C., Harris, P.T., Hunter, R.E., Kriesa, R.D., Lancaster, N., Middleton, G.V., Pocola, C., Rubin, D.M., Smith, J.D., Southard, J.B., Terwindt, J.H.J, Twitchell, D.G. Jr., 1990. Classification of large scale subaqueous bedforms: A new look at an old problem. *Journal of Sedimentary Petrology*, **60**: 160-172.

Barber, K.E. & Brown, A.G. 1987. Late Pleistocene organic deposits beneath the floodplain of the river Avon at Ibsley, Hampshire. In: *Wessex and the Isle of Wight Field Guide*. Barber, K.E. (Ed.). Quaternary Research Association, Cambridge, Special Publication: 65-74.

Bellamy, A.G. 1995. Extension of the British landmass: evidence from shelf sediment bodies in the English Channel. In: *Island Britain: a Quaternary Perspective*. Preece, R.C. (Ed.). Geological Society Special Publication, No 96: 47-62.

BGS 1995. *Wight Sheet (50° N 02° W) 1:250000, Solid Geology*. 2nd Edition. British Geological Survey.

Blain, W.R. 1980. Tidal Hydraulics of the West Solent (Vol 1). Unpublished Ph.D. Thesis, Department of Civil Engineering, University of Southampton. 326 pp.

Boxall, S.R., & Robinson, I.S. 1986. Shallow sea dynamics from CZCS imagery. *Advanced Space Research*, **7**: 237-246.

Boxall, S.R., Bishop, C., Nash, L., Santer, R., Chami, M., Dilligeard, E., Wernand, M. & Matthews, A. 1997. Physical processes and field measurements. In *FLUXMANCHE II, Hydrodynamics Biogeochemical Processes and Fluxes in the Channel*. Final Report to EC, 53-72.

Bradbury, A. 1999. Response of shingle barrier beaches to extreme hydrologic conditions. Ph.D. Thesis, School of Ocean and Earth Science, Southampton University.

Brampton, A.H. 1993. South Coast Mobility Study (East of Isle of Wight): Summary Report. HR Wallingford Ltd Report EX2795. 24 pp.

Brampton, A., Evans, C.D.R. & Velegrakis A.F. 1998. South Coast Mobility Study: West of Isle Wight. CIRIA Report PR 65, CIRIA, London. 218 pp.

Bray, J.M., Crater, D.J. & Hooke, J.M. 1995. Littoral cell definition and budgets for central southern England. *Journal of Coastal Research*, **11**: 381-400.

Brown, R.C., Gilbertson, D.D., Green, C.P. & Keen, D.H. 1975. Stratigraphy and environmental significance of Pleistocene deposits at Stone, Hampshire. *Proceedings of Geological Association*, **86**: 349-365.

Burnett, A.D. & Fookes, P.G., 1974. A regional engineering study of the London Clay in the London and Hampshire Basins. *Quarterly Journal of Engineering Geology*, **7**: 257-295.

Buurman, P. 1980. Palaeosols in the Reading Beds (Palaeocene) of Alum Bay, Isle of Wight, England. *Sedimentology*, **27**: 593-606.

Casey, R., 1961. The stratigraphical palaeontology of the Lower Greensand. *Palaeontology*, **3**: 487-621.

Curry, D., 1976. The age of the Hengistbury Beds (Eocene) and its significance for the structures of the area around Christchurch, Dorset. *Proceedings of the Geological Association of London*, **89**: 401-40.

Curry, D., Daley, B., Edwards, N., Middlemiss, F.A., Stinton, F.C. & Wright, C.W. 1972. *The Isle of Wight. Geologists' Association Guides* No **25**. (3rd edition), 27 pp.

Curry, D., Adams, C.G., Boulter, M.G., Dilley, F.C., James, F.E., Funnel, B.M. & Wells, M.K. 1978. A Correlation of Tertiary Rocks in the British Isles. *Special Report of the Geological Society of London*, No **12**.

Daley, B. & Edwards, N. 1971. Paleogene warping in the Isle of Wight. *Geological Magazine*, **108**: 399-405.

Dean, J.M. 1995. Holocene palaeo-environmental reconstruction for the nearshore Newtown Area, Isle of Wight. B.Sc. Dissertation, Department of Oceanography, Southampton University. 110 pp.

DeMesquita, A.R. 1972. Studies on the Mean-Flow of West Solent. Unpublished M.Phil. Thesis, Department of Oceanography, Southampton University. 122pp.

Devoy, R.J.N. 1972. Environmental Changes in the Solent Area During the Flandrian Era. Unpublished B.Sc. Dissertation, Department of Geography, Durham University. 123 pp.

Devoy, R.J.N. 1982. Analysis of the geological evidence for Holocene sea-level movements in Southern England. *Proceedings of the Geological Association*, **93**: 65-90.

Dixon, M. 1986. A geological Survey of West Solent and Christchurch Bay. Unpublished M.Sc. Dissertation, Department of Oceanography, Southampton University. 88 pp.

Draper, L. & Shellard, H.C. 1971. *Waves at the Owers Light Vessel, Central English Channel*. N.I.O Report, A46.

Dyer, K.R. 1969. Some Aspects of Coastal and Estuarine Sedimentation. Unpublished Ph.D. Thesis, Department of Oceanography, Southampton University. 102 pp.

Dyer, K.R. 1970. Linear erosional furrows in Southampton Water. *Nature*, **225**: 56-58.

Dyer, K.R. 1972. Bed shear stress and the sedimentation of the sandy gravels. *Marine Geology*, **13**: M31-M36.

Dyer, K.R. 1975. The buried channels of the 'Solent River', Southern England. *Proceedings of Geological Association of London*, **86**: 239-246.

Dyer, K.R. 1980. Sedimentation and sediment transport. In: *The Solent Estuarine System*. N.E.R.C. Publications, series C, No **22**: 20-24.

Dyer, K.R., Hamilton, N. & Pingree, R.D. 1969. A seismic refraction line across the Solent. *Geological Magazine*, **106**: 92-95.

Dyer, K.R. & King, H.L. 1975. The residual water flow through the Solent, South England. *Geophysical Journal of Royal Astronomical Society*, **42**: 97-10.

Edwards, R.A. & Freshney, E.C. 1987. *Geology of the Country Around Southampton. Memoir for the 1:50000 geological sheet 315*, British Geological Survey. 111pp.

Elderfield, N. 1999. The Effects of Capital Dredging on the Morphology and Sedimentology of Southampton Water. BSc Dissertation, School of Ocean and Earth Science, University of Southampton. 90 pp.

Everard, C.E., 1952. A Contribution to the Geomorphology of South Hampshire and the Isle of Wight. M.Sc. Thesis, Southampton University College.

Everard, C.E. 1954. The Solent River: a geomorphological study. *Transactions of Institute of British Geographers*, **20**: 41-58.

Fox, W.D., 1862. When and how was the Isle of Wight separated from the mainland? *Geologist*, **5**: 82-102.

Freshney, E.C., Bristow, C.R. & Williams, B.J. 1985. *Geology of the Sheet SZ 09 (Bournemouth-Poole-Wimbourne, Dorset)*. Geological Report for DOE: Land Use Planning. British Geological Survey, Exeter.

Gibbard, P.L. 1988. The history of the great northwest European rivers during the past three million years. *Philosophical Transactions of Royal Society of London*, B **318**: 559-602.

Gilkes, R.J. 1966. The Clay Mineralogy of the Tertiary Sediments of the Hampshire Basin. Ph.D. Thesis, Department of Geology, University of Southampton. 243pp.

Gilkes, R.J. 1978. On the clay mineralogy of Upper Eocene and Oligocene sediments in the Hampshire Basin. *Proceedings of the Geological Association*, **89**: 43-56.

Green, J.F.N. 1946. The terraces of Bournemouth, Hampshire. *Proceedings of the Geological Association*, **57**: 82-101.

Grontmij, N.V. 1973. West Winner Reclamation Feasibility Study. City of Portsmouth Council.

Hamblin, R.J.O., Crosby, A., Balson, P.S., Jones, S.M., Chadwick, R.A., Penn, I.E. & Arthur, M.J. 1992. *The geology of the English Channel. British Geological Survey Report*, HMSO, London. 106 pp.

Hammond, F.D.C., Heathershaw, A.D. & Langhorne, D.N. 1984. A comparison between Shields' threshold criterion and the movement of loosely packed gravel in a tidal current. *Sedimentology*, **31**: 51-62.

Halcrow 1982. Hurst Castle Coastal Protection: Initial Design Report. Unpublished Technical Report, Halcrow & Partners. 32 pp.

Harlow, D.A. 1979. The littoral sediment budget between Selsey Bill and Gilkicker and its relevance to coast protection works on Hayling Island. *Quarterly Journal of Engineering Geology*, **12**: 257-265.

Harlow, D.A. 1980. Sedimentary Processes, Selsey Bill to Portsmouth and a coast Protection Strategy for Hayling Island. Ph.D. Thesis, Civil Engineering Department, Southampton University. 772 pp.

Harris, P.T. & Collins, M.B. 1984. Side-scan sonar investigations into temporal variation in sand wave morphology: Helwick Sand, Bristol Channel. *Geo-Marine Letters*, **4**: 91-97.

Heathershaw, A.D. & Langhorne, D.N. 1988. Observations of nearbed velocity profiles and seabed roughness in tidal currents flowing over sandy gravels. *Estuarine Coastal Shelf Science*, **26**: 459-482.

Henderson, G. & Webber, N.B. 1979. An application of the wave refraction diagrams to shoreline protection, with particular reference to Poole and Christchurch Bays. *Quarterly Journal of Engineering Geology*, **12**: 319-327.

Hodgson, J.M. 1964. The low-level Pleistocene marine sands and gravels of the West Sussex coastal plain. *Proceedings of the Geologists' Association*, **75**: 547-561.

Holyoak, D.T. & Preece, R.C. 1983. Evidence of a Middle Pleistocene sea-level from estuarine deposits at Bembridge, Isle of Wight, England. Proceedings of the *Geologists Association*, **94**: 231-244.

Hooke, J.M., Bray, M.J. & Carter, D.J. 1996. Sediment transport analysis as a component of coastal management - a UK example. *Environmental Geology*, **27**: 347-357.

HR Wallingford 1989a. Christchurch Bay-Offshore Wave Climate and Extremes. HR Wallingford Ltd Technical Report No EX1934, Wallingford. 9 pp.

HR Wallingford 1989b. Wind-wave Data Collection and Analysis for Mildford-on-Sea. HR Wallingford Ltd Technical Report No EX1979. 20 pp.

Hydraulics Research Station 1977. Solent Bank, Pot Bank and Prince Consort Dredging. Unpublished Technical Report EX770, Hydraulics Research Station, Wallingford. 23 pp.

IGS 1975a. *Geological Sheet 330 (Lymington), Geological Survey of Great Britain maps (1:50000 series)*. Ordnance Survey of Great Britain, Southampton.

IGS 1975b. *Geological Sheet 329, Geological Survey of Great Britain maps (1:50000 series)*. Ordnance Survey of Great Britain, Southampton.

Jay, D.A., Uncles, R.J., Largier, J., Geyer, W.R., Vallino J. & Boynton, W.R., 1997. A review of recent developments in estuarine scalar flux estimation. *Estuaries* 20: 262-280

Jones, D.K.C. 1980. The Tertiary evolution of south-east England, with particular reference to the Weald. In: *The Shaping of Southern England*, Jones, D.K.C. (Ed.). Special Publication of the Institute of British Geographers, Academic Press, London, 13-47.

Ke, X. & Collins, M.B., 1993. Saltmarsh Protection and Stabilisation, West Solent. Southampton University Technical Report, SUDO/TEC/93/6C. 81pp.

Keen, M.C. 1977. Ostracod assemblages and the depositional environments of the Headon, Osborne and Bembridge beds (Upper Eocene) of the Hampshire Basin. *Palaeontology*, 20: 405-445.

Keen, D.H. 1980. The environment of deposition of the South Hampshire Plateau Gravels. *Proceedings of the Hampshire Field Club and Archaeological Society*, 36: 15-24.

Kellaway, G.A., Redding, J.H., Shephard-Thorn, E.R. & Destombes, J.P. 1975. The Quaternary history of the English Channel. *Philosophical Transactions of the Royal Society of London*, A279: 189-218.

Kennedy, W.J. & Garrison, R.E. 1975. Morphology and genesis of nodular chalks and hardgrounds in the Upper Cretaceous of Southern England. *Sedimentology*, 22: 311-386.

King, C.A.M. & McCullagh, M.J. 1971. A simulation model of complex recurved spit. *Journal of Geology*, 79: 22-37.

Lacey, S. 1985. Coastal Sediment Processes in Poole and Christchurch Bays and the Effects of the Coastal Protection Works. Ph.D. Thesis, Department of Civil Engineering, Southampton University. 372pp.

Langhorne, D.N., Heathershaw, A.D. & Read, A.A. 1986. Gravel bedforms in the west Solent, Southern England. *Geo-Marine Letters*, 5: 225-230.

Long, A.J. & Tooley, M.J. 1995. Holocene sea-level and crustal movements in Hampshire and Southeast England, United Kingdom. In: *Holocene Cycles: Climate, Sea Levels and Sedimentation*. Frinkl Jr. (Ed.). *Journal of Coastal Research*, Special Issue 17, 299-210.

McFarland, S., Whitcombe, L.J. & Collins, M.B. 1995. Recent shingle beach nourishment schemes in the UK: Some preliminary observations. *Ocean and Coastal Management*, 25: 143-149.

McKay, N.A. 1990. A Shallow Seismic Survey Investigating Buried Channels in Poole Bay. M.Sc. Dissertation, Department of Oceanography, Southampton University. 72 pp.

Melville, R.V. & Freshney, E.C. 1982. *The Hampshire Basin and the Adjoining Areas* (4th Edition). HMSO, London. 146pp.

Mercone, D. 1995. A Chemical and Sedimentological Study on the Sediment at Calshot Deep. B.Sc. Dissertation, Department of Oceanography, Southampton University. 74 pp.

Munt, M.C. & Burke, A. 1986. The Pleistocene Geology and Faunas at Newtown, Isle of Wight. *Proceedings of the Isle of Wight Natural History and Archaeological Society*, **8**: 7-14.

Nicholls, R.J. 1985. The Stability of Shingle Beaches in the Eastern Half of Christchurch Bay. Unpublished Ph.D. Thesis, Department of Civil Engineering, Southampton University. 468 pp.

Nicholls, R.J. 1987. The evolution of the upper reaches of the Solent River and the formation of Poole and Christchurch Bays. In: *Wessex and the Isle of Wight: Field Guide.* Barber, K.E. (Ed.). Quaternary Research Association, Cambridge, Special Publication, 99-114.

Nicholls, R.J. & Webber, N.B. 1987. The past, present and future evolution of Hurst Castle Spit, Hampshire. *Progress in Oceanography*, **18**: 119-137.

Nowell, D.A.G. 1995. A fault in the Purbeck-Isle of Wight monocline. *Proceedings of the Geologists' Association*, **106**: 187-190.

Odin, G.S., Curry, D. & Hunziker, J.C. 1978. Radiometric dates from NW European glauconites and the Pelaeogene time-scale. *Journal of Geological Society of London*, **135**: 481-497.

Paphitis, D. 1997. Residual Circulation and Associated Sediment Dynamics: Eastern Approaches to the Solent. BSc Dissertation, Department of Oceanography, University of Southampton. 106 pp.

Plint, A.G. 1983. Facies, environments and sedimentary cycles in the Middle Eocene, Bracklesham Formation of the Hampshire Basin: evidence for global sea-level changes. *Sedimentology*, **30**: 625-653.

Plint, A.G. 1984. A regressive coastal sequence from the Upper Eocene of Hampshire, Southern England. *Sedimentology*, **31**: 213-225.

Preece, R.C., Scource, J.D., Hougton, S.D., Knudsen, K.L. & Penney, D.N. 1990. The Pleistocene sea-level and neotectonic history of the Eastern Solent, Southern England. *Philosophical Transactions of the Royal Society of London, B* **328**, 425-477.

Radley, J.D. & Barker, M.J. 1998. Stratigraphy, paloeontology and correlation of the Vectis Formation (Wealden Group, Lower Cretaceous) at Compton Bay, Isle of Wight, Southern England. *Proceedings of the Geologists' Association*, **109**: 187-196.

Rawson, P.F., 1992. The Cretaceous. In: *Geology of England and Wales.* Duff, P.M.D. & Smith, A.J (Eds.). The Geological Society of London, 355-388.

Reid, C. 1902. Geology of the Country around Ringwood. *Memoir of the Geological Survey of Great Britain.* 62 pp.

Reid, C. 1905. The island of Ictis. *Archaeologia*, **59**: 281-288.

Ruffel, A.H. & Batten, D.J. 1994. Uppermost Wealden facies and Lower Greensand Group (Lower Cretaceous) in Dorset, southern England: correlation and palaeoenvironment. *Proceedings of the Geological Association*, **105**: 53-69.

Srisaenthong, D. 1982. Suspended Sediment Dynamics and Distribution in the Solent Using LANDSAT MSS Data. Ph.D. Thesis, Department of Oceanography, Southampton University. 153 pp.

Sharples, J. (this volume). Water movement in the Solent region. In: *Solent Science - A Review*. Collins, M.B. & Ansell, K. (Eds.), Proceedings in Marine Science Series, Elsevier, Amsterdam.

Shell 1987. *The Solent Estuary: Environmental Background*. Shell UK Ltd. 54 pp.

Stewart, D.J. 1981. A field guide to the Wealden Group of the Hastings area and Isle of Wight. In: *Field Guides to Modern and Ancient Fluvial Systems in Britain and Spain*. Elliot, T (Ed.). International Fluvial Conference, Keele University, 3.1-3.31.

Stinton, F.C. & Curry, D. 1979. Lithostratigraphical nomenclature of the English Palaeogene succession. *Geological Magazine*, **116**: 66-67.

Stoneley, R. 1982. The structural development of the Wessex Basin. *Journal of the Geological Society of London*, **139**: 545-552.

Terwindt, J.H.J. & Brouwer, M.J.N. 1986. The behaviour of intertidal sand waves during neap-spring tidal cycles and the relevance of palaeoflow reconstruction. *Sedimentology*, **33**: 1-31.

Townend I. & Price D. (this volume). Hydrodynamic, sediment process and morphological modelling. In: *Solent Science - A Review*. Collins, M.B. & Ansell, K. (Eds.), Proceedings in Marine Science Series, Elsevier, Amsterdam.

Thorne, P.D., Williams, J.J. & Heathershaw, A.D. 1989. In situ acoustic measurements of marine gravel threshold and transport. *Sedimentology*, **36**: 61-74.

Underhill, J.R. & Paterson, S. 1998. Genesis of tectonic inversion structures: seismic evidence for the development of key structures along the Purbeck-Isle of Wight Disturbance. *Journal of the Geological Society of London*, **155**: 975-992.

Velegrakis A.F. 1994. Aspects of morphology and sedimentology of a transgressional embayment system: Poole and Christchurch Bays, Southern England. Ph.D. Thesis, Department of Oceanography, Southampton University. 319 pp.

Velegrakis, A.F. & Collins, M.B. 1992. Marine Aggregate Evaluation of Shingles Bank, Christchurch Bay. Southampton University Technical Report, SUDO/TEC/92/14C. 33 pp.

Velegrakis, A.F., Dix, J.K. & Collins, M.B. 1999. Late Quaternary evolution of the upper reaches of the Solent River, Southern England, based upon marine Geophysical evidence. *Journal of the Geological Society, London*, **156**: 73-87.

Velegrakis, A.F., Dix, J.K. & Collins, M.B. (this volume). Late Pleistocene-Holocene evolution of the upstream section of the Solent River, Southern England. In: *Solent Science - A Review*. Collins, M.B. & Ansell, K. (Eds.), Proceedings in Marine Science Series, Elsevier, Amsterdam.

Velegrakis, A.F., Michel, D., Collins, M.B, Lafite, R., Oikonomou, E., Dupont, J.P., Huault, M.F., Lecouturier, M., Salomon, J.C. & Bishop, C. (in press). Sources, sinks and resuspension of suspended particulate matter in the Eastern English Channel. *Continental Shelf Research*.

Voulgaris G. & Collins M.B. 1994. Storm Damage-South Coast Shingle Study. Technical Report SUDO/TEC/94A/6/C, Oceanography Department, University of Southampton. 93 pp.

Walder, P.S. 1964. Eocene sediments of the Isle of Wight. *Proceedings of the Geological Association*, **75**: 291-414.

Wallace, H. 1990. Sea level and Shoreline between Portsmouth and Pagham for the Last 2500 Years. Unpublished Manuscript (Chichester Yacht Basin). 60 pp.

Webber, N.B. 1980. Hydrography and water circulation in the Solent. In: *The Solent Estuarine System: An Assessment of the Present Knowledge*. N.E.R.C. Publications, Series C, No **22**: 25-35.

West, I.M. 1980. Geology of the Solent estuarine system. In: *The Solent Estuarine System: An Assessment of the Present Knowledge*. N.E.R.C. Publications, Series C, No **22**: 6-19.

Whittaker, A. 1985. *Atlas of Onshore Sedimentary Basins in England and Wales: Post-Carboniferous Tectonics and Stratigraphy*. Blackie, Glasgow.

Whittaker, A. & Chadwick, R.A. 1984. The large scale structure of the Earth's crust beneath Southern Britain. *Geological Magazine*, **121**: 621-624.

Whitcombe, L.J. 1995. Sediment Transport Processes, with particular reference to Hayling Island. PhD Thesis, Department of Oceanography, University of Southampton. 299 pp.

Williams, J.J., Thorne, P.D. & Heathershaw, A.D. 1989. Comparisons between acoustic measurements and predictions of the bedload transport of marine gravels. *Sedimentology*, **36**: 973-979.

Wooldridge S.W. & Linton, D.L. 1955. *Structure, Surface and Drainage in South-east England*. George Phillip, London. 176 pp.

Wright, P. 1982. Aspects of the Coastal Dynamics of Poole and Christchurch Bays. Ph.D. Thesis, Department of Civil Engineering, Southampton University. 201 pp.

Yalin, M.S., 1964. Geometrical properties of sand waves. Proceedings A.S.C.E., *Journal of Hydraulics Division*, **90**: 105-119.

Water Circulation in Southampton Water and the Solent

Jonathan Sharples

School of Ocean and Earth Science, Southampton Oceanography Centre, University of Southampton, European Way, Southampton, SO14 3ZH, U.K.

INTRODUCTION

An accurate knowledge of the circulation patterns in the Solent region is important for several reasons. An ability to predict the pathways of materials introduced into the coastal region, (such as oil spills, sewage effluent or other contaminants) is required. The Port Authority needs to be able to quantify the rates and direction of sediment transport, in order to manage dredging operations; likewise they need to provide pilots with information on how flow patterns are likely to affect vessels of different draughts. Rescue services need to have some idea of the likely drift routes of casualties within the water. In order to gain this knowledge, it is necessary to understand a variety of physical mechanisms that can be expected to force water movements in estuaries and coastal regions.

The most visible water movement is that of the twice-daily ebb and flood of tidal currents (see Webber, 1980). However, whilst these currents are obviously very strong, on time-scales longer than a few hours, the tide may not be the only process to consider when attempting to predict longer-term water movement. Broadly, tidal flows bring water into the Solent and Southampton Water during the flood, during the ebb the same water is moved back out again. If it is required to know how currents affect the passage of, for instance, a large container vessel within the confines of the shipping channel around Calshot Spit, then it is critical to measure the behaviour of these flood and ebb currents. If, however, an estimate is required of the length of time a patch of pollutant will remain within Southampton Water, information is needed on the generally weak "residual" flows that are superimposed upon the tide. During a single tidal cycle, the patch of pollutant will be moved up and down the estuary by the "tidal excursion" (typically 5 to 8 km, in Southampton Water), but any *net* movement requires a non-oscillatory component to the flow.

This paper aims to describe some of the water circulation mechanisms operating in Southampton Water and the Solent. Tidal flows are dealt with from the point of view of the requirement for understanding short-term flow patterns i.e. for ship pilotage or search-and-rescue operations. Emphasis is placed upon assessing those processes that result in net water movement, over time-scales longer than a tidal cycle. For all the cases presented, recent observations are used to illustrate both present measurement capabilities, and presented plans for future work. For general location map and main sites referenced, refer to *Map A* and *B* in the *Preface*.

TIDAL CURRENTS

In Southampton Water and the Solent, tidal flows are more understood than any other process. Indeed, the tides can generally be modelled with considerable accuracy over all of the NW European coastal and shelf seas (Kwong *et al.*, 1997; Davies *et al.*, 1997). Tides within the narrow confines of estuaries, such as Southampton Water, are more difficult to model. This limitation arises because the requirement of finer spatial resolution is computationally expensive; likewise there are interesting scientific challenges in dealing with shallow water systems that have significant areas which dry out for part of the tidal cycle. It may be stated, however, that tidal modelling is generally dependable.

Tidal currents can be very strong within Southampton Water, with near-surface currents reaching over 1 m/s during spring tides. Observation and understanding of these currents is important for two main reasons. Firstly, as mentioned above, these are the flows that control the short-term water movements and that are of primary interest to those involved in navigation and search-and-rescue. Secondly, strong tidal currents cause re-suspension of seabed sediments into the water column. Whilst the tide itself may not cause any net movement of these sediments; by retaining them within the water column, the tidal currents are making the suspended sediments available for transport by other, weaker, non-oscillatory flows. An example of the type of information we can now collect on tidal flow patterns is shown in Figure 1. The current data were collected using and Acoustic Doppler Current Profiler (ADCP) towed behind a small boat around a fixed course over a complete tidal cycle. The resulting raw data, on near-surface and near-bed flows, was then analysed for the main tidal constituents. The results obtained show flow changes around the bend of Calshot Spit, over the full tidal cycle.

Calshot Spit is a region of particular interest, as the sharp turn in the shipping channel puts strong constraints on the course that larger vessels can navigate safely. The curvature of the channel results in curvature of the flow, which is clearly visible in the results shown in Figure 1. During the flood and ebb tides, the flow on the inside of the bend is significantly stronger than that farther out in the channel. Also, during strong flows, it is clear that the surface and bottom currents are not in the same direction. This difference is a result of the surface flow being thrown "centrifugally" away from the curve of the shipping channel. Near the seabed the effect of friction against the bed slows the current; thus it is not detected as much away from the bend.

Whilst the data shown in Figure 1 illustrate the sort of water flow information that is of use for navigational purposes, Figure 2 illustrates the kind of problem that time-varying tidal flows present for rescue operations. These data were collected as part of a project funded by the Defence Evaluation and Research Agency, aimed at generating a set of observations simulating the drift path of a person in the water. The observations shown in Figure 2 were made by deploying a drifting buoy several times in the same area, then tracking the position of the buoy over a few hours using differential GPS. The main point to draw from the results is the lack of similarity between any pair of drifter trajectories; this problem is presently being addressed by modellers at ABP Research, with the aim of producing a computer model that could simulate such drift (so be of use in rescue operations).

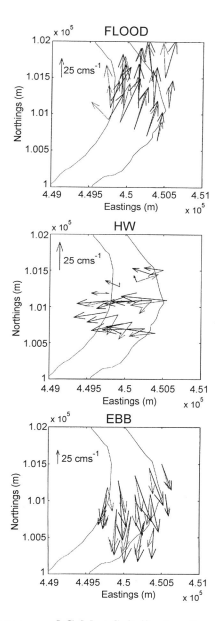

Figure 1 Currents around Calshot Spit (for location, see Figure 2) during maximum flood, high water, and maximum ebb. The bold arrows represent surface currents; the faint arrows are the near-bed currents. The outline shown is the route of the dredged shipping channel. Currents are calculated from a tidal analysis of data, collected using an Acoustic Doppler Current Profiler.

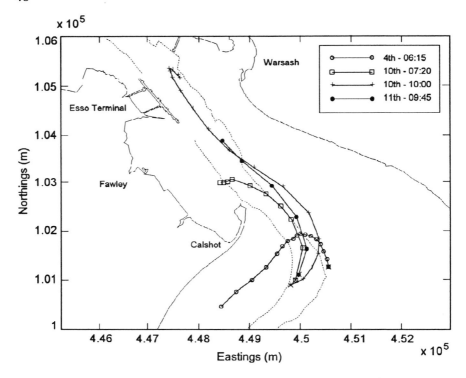

Figure 2 A selection of 4 drifter tracks, for release positions off Calshot Spit, during August 1998. Symbols are drawn every 15 minutes. Positions were logged using a differential GPS system, to an accuracy of approximately ± 5 m.

RESIDUAL CURRENTS

Many coastal management problems require knowledge of the time taken for a particular coastal or estuarine region to flush out any pollutant that has been introduced into it. This requires an understanding of the longer-term residual (i.e. non-tidal) currents that are superimposed upon the strong tidal flows. Because tides are oscillatory, they may not necessarily flush a system efficiently, as the net movement of water over one tidal cycle is approximately zero. An estimate of the residence time of a conservative substance within a coastal system can be made, by considering the time required for the typical freshwater input rate to completely replenish the freshwater within the system. For instance, taking the section of Southampton Water between Dock Head and the Solent as an example, the residence time can be estimated (using this method) to be approximately 5 to 10 days. This provides, of course, a very crude estimate, but its strength lies in the fact that it implicitly takes in to account all of the physical processes governing net water movement within the region. The remainder of this section will describe some of the processes that have been observed recently, as potentially important in determining the flushing time of Southampton Water and the Solent.

Tidal Residuals and Eddies

While tidal currents are oscillatory, they can also generate weak, uni-directional residual flows. Typically, these residuals are generated as a result of interaction of the main tidal currents with coastal or seabed topography. In open shelf sea areas, it is possible to measure these currents from long-term current meter records. For instance, the tidal residual in the Irish Sea is believed to be approximately 1 cm/s northward. A long-term record is required to observe this movement, as the very small residual has superimposed on it a main depth-averaged tidal flow of typically 1 m/s. Thus a considerable amount of information is required before the average flow is observable, above the noise of the measurements. Very little is known about such tidal residuals within the Solent system (within Southampton Water the residual flow is dominated by the freshwater input; see the section on "Estuarine Circulation"). However, recent work during the Fluxmanche experiment has highlighted the existence of numerous coastal eddies along both the English and French sides of the English Channel. These eddies are typically set up by tidal flows, past the headlands. A good local example of such a feature is a semi-permanent eddy in the east Solent (Figure 3).

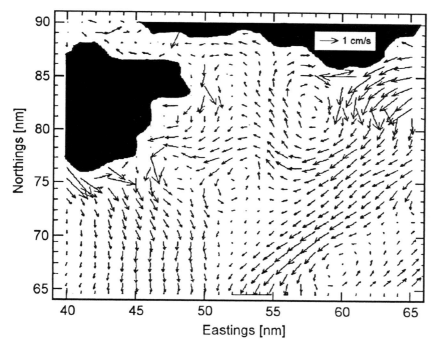

Figure 3 Numerical model predictions of tidally-averaged currents east of the Isle of Wight. The current vectors clearly show a number of small eddies, caused by the interaction of the tidal currents with the seabed and coastal boundaries. Observational evidence has also been collected that supports these model predictions. *[Reproduced, with permission, from Fluxmanche II Final Report, 1997].*

Quantitative knowledge of the effect of these eddies on the flushing of a coastal region is poor. However, it is easy to imagine (looking at model results, such as those in Figure 3), that these eddies will effectively delay the passage of any material through the coastal zone and then into the main flow of the English Channel. Thus, it is reasonable to expect that the eddies will have the overall effect of increasing the residence time of material within Southampton Water and the Solent.

Another form of residual transport in Southampton Water is related to the ebb-flood difference in the maximum tidal stress at the seabed; this may have consequences for sediment transport within the estuary. In general, the total amount of water that enters Southampton Water during the flood phase of the tide, leaves during the ebb phase. However, a particular feature of the distinctive tidal regime in Southampton Water is that the ebb currents are typically twice the amplitude of the flood currents. Stress at the seabed is proportional to the square of the tidal current, so that the stress during the ebb is four times that of the flood. This stronger stress will be able to re-suspend larger particles of sediment, than the flood-induced stress, hence there is the potential for a gradual 'shunting' of larger sediments, towards the Solent at each ebb tide. The same argument suggests that smaller particles, which can be suspended by both flood and ebb currents, should not have any net tidal movement. Of course, this particular observation is totally speculative; research work is planned to test this hypothesis.

Estuarine Circulation

The key feature of all estuaries is that freshwater is input at one end; hence, there must be a transport of this freshwater out of the estuary system into the sea. During the passage of the freshwater down the estuary, it gains salt by mixing with the denser, deeper seawater. Thus, the surface outflow from the estuary tends to contain some salt; the salt removed in this way has to be replaced by a weak inflow of seawater, along the bottom of the estuary. All estuaries, to some degree or other, have this two-way exchange as a mean flow pattern: flow towards the sea near the surface, and towards the land near the bottom of the estuary. Note that 'mean' relates to the fact that observations of currents at different levels of the water column have to be averaged over a long time period, so that the tidal component of the flow averages to zero; the net, non-tidal flow remains. Quantifying this "estuarine" or "gravitational" circulation is critical in the determining the long-term flushing of a particular location, such as Southampton Water. The difficulty in observing this mean flow is illustrated by a theoretical profile, representing the current behaviour between the surface and the bottom of the water column (Figure 4).

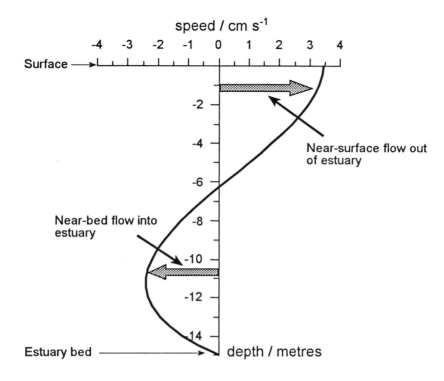

Figure 4 Theoretical vertical profile of the estuarine circulation within the shipping channel of Southampton Water. Water within 6 m of the sea surface has a mean flow out of the estuary; that between 6 m and the estuary bed has a mean flow into the estuary.

The velocity profile, calculated for conditions typical for Southampton Water, suggests that the surface estuarine flow is only 3 or 4 cm/s. Considering that this flow needs to be measured against the tidal background flow of 50 to 100 cm/s, it is clear that observing this net transport requires a long-time series of accurate current measurements. Whilst 3 or 4 cm/s is only a low speed, during one tidal cycle (12.5 hours) this would suggest that a positively buoyant pollutant would move downstream a distance of 1.4 to 1.8 km. Alternatively, a negatively buoyant substance, with large concentrations in the lower half of the water column, would move upstream by 1.4 to 1.8 km during the same tidal cycle. Available data collected over single tidal cycles during summer, shows surface flows of approx. 6 cm/s towards the estuary mouth, and near-bed flows of approx. 4 cm/s in the opposite direction. However, these values are only 'snapshots' from a very limited amount of data. The estuarine circulation certainly going varies on a wide range of time-scales.

An aspect of estuarine circulation that is particularly important in all estuaries is that it acts to stratify the water column. The surface outflow is continually bringing water of lower salinity down the estuary, whilst the near-bed inflow introduces waters of higher salinity. At any one point in the estuary, therefore, the tendency is for the strength of the stratification (quantified, for instance, by the difference in salinity between the near-bed and near-surface water) to increase. Southampton Water is generally classified as "partially mixed", as the tidal currents are able to overcome much of this stratifying tendency by generating turbulence and mixing the water column. The strength of the stratification in Southampton Water is known to be dependent on the river inflow rate, together with the strength of the tides. The stratification can vary substantially on time scales from less than tidal cycle, through the fortnightly spring-neap cycle, up to the seasonality of the freshwater inflows. However, as yet, there are no long-term data sets that illustrate quantitatively this process.

Stratification is important because it inhibits mixing between the upper and lower parts of the water column. This pattern of water movement has well-known consequences for primary production in estuaries, as an increase in the stratification results in the algae being retained longer in the surface waters. The higher levels of light in the surface water can then 'trigger' a bloom of the algae. Thus, for instance, blooms of the diatom *Mesodinium rubrum* in Southampton Water are more likely to occur around neap tides, when the weaker tidal currents are less able to mix against the stratification (Hayes *et al.*, 1989; Garcia *et al.*, 1993).

SUMMARY

This paper has described some, although by no means all, of the known process controlling water movement in Southampton Water and the Solent. It should be remembered that other processes are likely to be operating (for instance, wind-driven flows or the "Stokes drift," associated with waves in shallow water), but only those for which some observational support have been described. It has been emphasised throughout that, in attempting to understand processes, a key consideration is the time-scale over which you wish to predict water movements. From the navigational and search-and-rescue standpoints, it is important to have some understanding of the change in currents over time-scales of a few hours, when the flow is dominated by the tides. From the point of view of an interest in effluent disposal, then knowledge of net transport over several days may be of more concern; in the latter case, the residual flows will be more important. In summary, observations and predictions from numerical models need to take into account all of the processes that are known to be acting, on the time-scale of interest.

This brief description illustrates a good understanding of the general mechanisms driving water movements, on the basis of estuarine processes. The development of high-resolution numerical models will undoubtedly result in a better ability to predict tidal flows; likewise, some observational data are able to provide tests of these predictions. Direct observation of the non-tidal residual flows is presently limited, although work is presently underway; this is aimed at quantifying the variability in these flows, both within single tidal cycles and on longer-time scales. Numerical modelling of these non-tidal flows represents a particular challenge.

REFERENCES

Kwong, S. C. M., Davies, A. M. & Flather, R. A. 1997. A three-dimensional model of the principal tides on the European shelf. *Progress in Oceanography,* **39**(3): 205-262.

Davies, A. M., Kwong, S. C. M. & Flather, R. A. 1997. A three-dimensional model of diurnal and semidiurnal tides on the European shelf. *Journal of Geophysical Research,* **102**(C4): 8625-8656.

Garcia, C. A. E., Purdie, D. A. & Robinson, I. S. 1993. Mapping a bloom of the photosynthetic ciliate *Mesodinium rubrum* in an estuary from airborne thematic mapper data. *Estuarine, Coastal and Shelf Science,* **37**(3): 287-298.

Hayes, G. C., Purdie, D. A. & Williams, J. A. 1989. The distribution of ichthyoplankton in Southampton Water in response to low oxygen levels produced by a *Mesodinium rubrum* bloom. *Journal of Fish Biology,* **34**(5): 811-813.

Webber, N.B. 1980. Hydrography and water circulation in the Solent. In: *The Solent Estuarine System: An Assessment of the Present Knowledge*. N.E.R.C. Publications, Series C, No **22**: 25-35.

Hydrodynamic, Sediment Process and Morphological Modelling

Darren Price and Ian Townend

ABP Research & Consultancy Ltd., Pathfinder House, Maritime Way, Southampton, SO14 3AE, U.K.

INTRODUCTION

Modelling of coastal processes has made significant progress over recent years, with far greater linkages between the various models, which simulate the different processes involved. It remains the case, however, that the value of the output they provide is dependent upon the quality of the data used for their set-up, calibration, and validation. Despite this limitation, there is often extreme pressure to undertake the minimum amount of fieldwork, in order to keep costs down (in the same way that site investigation is always the 'poor relation' in construction projects). Equally, it is all too easy to dismiss models as being incomplete, making simplifying assumptions or not representing all the processes present in the 'real world'. Whilst these are very real limitations, it should also be recognised that many models are now very sophisticated, and can represent complex physical, chemical and biological processes. Such models enable the trained user to advance their understanding of a particular regime quite significantly; likewise, they often help to identify processes which are not obvious, or even counter- intuitive.

Within the Solent, models have a long history of utilisation, from the early physical models established at Southampton University to present-day numerical modelling studies undertaken by the many agencies who have interests in a wide variety of applications (see below). This paper will focus upon those applications that relate to physical processes and, in particular, tides, waves, sediment transport and morphological change. Within this field of interest, there exists a host of different spatial and temporal scales. Tidal modelling necessarily ranges from the English Channel to exchanges within tidal creeks. Sediment exchanges can be considered over the basin as a whole, but must also address how sediment feeds the upper intertidal and saltmarshes around the Solent. Process time-scales of relevance vary from those of short period wind-waves (a few seconds) to geological time (1000's of years).

Examples from around the Solent provide numerous illustrations of important mechanisms, which give rise to this spatial and temporal variability. Such variability, in turn, points to the care required when applying models to complex situations and similarly, to the need to focus on the understanding offered, as a means of assessing any quantitative predictions. For general location and main sites referenced, refer to *Map A* and *Map B* in the *Preface*.

HISTORY OF MODELLING IN THE SOLENT

The Solent and Southampton Water, due to their importance for the local community, commercial and leisure interests, have been studied in detail with the aid of hydraulic and numerical models. In 1957, a large physical model of the Solent and Southampton Water was built at Southampton University (Wright & Leonard, 1959). The model was finally demolished in 1979, after being used for many studies, including those that looked into the proposed developments to the Port of Southampton. As part of a study to examine the best location for a sewage outfall (Watson & Watson, 1972), the Hydraulics Research Station (now known as HR Wallingford Ltd.) used a two-dimensional numerical model, with a grid resolution of 1 km, to examine tidal flows and pollutant dispersion between the East Solent and the Nab Tower. Compared with model and computer capabilities available at present, this model was of relatively coarse resolution and would have been slow to run. However, it would have enabled comparative testing of numerous discharge sites to have been undertaken more rapidly than with a physical model.

Present-day numerical models run much faster than in the past and, therefore, can represent far greater resolutions. There are, in existence, models of the Solent and Southampton Water, which can simulate two and three-dimensional hydrodynamics, sediment transport and waves, and a mixture of all of these to predict morphological change. These models are based upon regular grids, nested grids or meshes, with variable resolution. The resolution within these models varies over the model domain, such that sufficient resolution is applied to particular areas where it is needed; typically, it varies from a few hundred metres to a few tens of metres.

Numerical model results can provide valuable information about the natural environment where limited fieldwork has been undertaken, as long as the assumptions used are acknowledged. Quite often boundary conditions for models are derived from other numerical model results. This approach reduces the need for some fieldwork and can be reasonably 'cost-effective'.

TIDES

The tidal characteristics of the English Channel are the dominating factors driving the complex tidal features observed within the Solent. The time of low water at Penzance is approximately the same as the time of high water in the eastern reaches of the English Channel, with the natural period of the channel being about 10 hours. Consequently, the channel exhibits a significant amount of resonance (Webber, 1980). Tidal range is at a minimum within Poole Bay because of a degenerate amphidrome inland of Weymouth, centred remarkably close to Stonehenge. The location of the amphidrome in relation to the Solent means that there is a significant gradient in tidal range, from a mean of 1.2 m in Christchurch Bay in the west, to 3.0 m at Chichester Harbour in the east. Close to the nodal point in Christchurch Bay, the semi-diurnal harmonic constituent is relatively weak and, in combination with a strong shallow water tidal constituent and the bathymetry of Christchurch Bay, the impact on the phase relationship between these two harmonic components results in a double high water at Calshot. Farther to the east, near Spithead in the eastern Solent, the phase relationship alters slightly, resulting in an extended (rather than a double) high water (Webber, 1980).

Models encompassing the whole of the English Channel are now quite common and usually have a resolution of between 1 and 10 km, depending upon the application. The boundary conditions for such models are usually established using harmonic constituents obtained from the continental shelf model for the coast around the British Isles, developed by the Proudman Oceanographic Laboratory. The purchase of this data is much more cost-effective than deploying field instruments to measure water levels along the model boundaries. Model results from an English Channel model (calibrated to water levels at sites within the Channel) can be used, in turn, to provide boundary conditions for models of even finer resolution.

The tidal elevations (at Calshot) obtained from a nested numerical model of the Solent and English Channel (1200m resolution– English Channel, 400 m resolution– the Solent) are shown in Figure 1. The model was run with and without the Isle of Wight in place, purely as an exercise to examine the resonance of the English Channel in the vicinity of the Solent. It can be seen that the double high water and flood stand are still evident without the island, thus dispelling the myth that the island is the cause of the double high water. It is evident from the results presented in the figure, however, that the island does assist in the amplification of the tide, by constraining it within the West and East Solent. This technique shows a simple but effective method by which an hypothesis can be tested, relatively easily and inexpensively, with numerical models.

In Southampton Water, the tidal characteristics of the Solent are exaggerated further by some internal resonance within the estuary (Figure 2); these can be described in terms of a 'young flood stand' and a double high water period, with little change in water level (lasting for up to 3 hours). The stand on the flood is most pronounced on spring tides and can last for about 2 hours. The tidal profile is also asymmetrical, with the ebb phase of the tide taking less than 4 hours, compared to nearly 9 hours for the flood. This pattern means that Southampton Water is ebb dominant as the estuary has a shorter time over which to release its water on the ebbing tide, than it does to fill up on the flooding tide. Such a characteristic has the effect of creating higher current speeds on the ebb tide, providing a mechanism for the flushing of sediment and pollutants out of the estuary. Numerical predictions of peak velocities at Calshot, on a spring tide, indicate speeds of 1.0 m/s on the ebb compared to 0.7 m/s for the flood (ABP Research, 1995).

Present-day numerical models can not only predict the water speed, due to tidal movements, but can also combine this with additional effects due to the influence of wind and waves. Therefore, limitations in the predictive powers of areal models (two or three-dimensional), experienced in the past, when trying to reproduce specific events, can now be reduced. With this inter-linking of the various models, extreme events can be simulated through the utilisation of more of the physical processes that occur in nature.

However, there are certain physical processes that are not represented by the majority of the numerical tidal models in existence. Features on the bed, which have a frictional effect

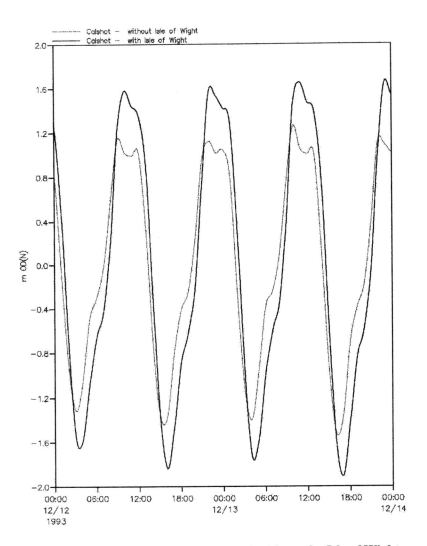

Figure 1 Tidal curve at Calshot, both with and without the Isle of Wight

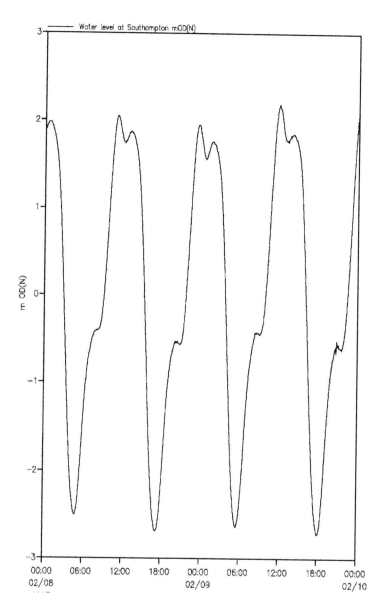

Figure 2 Typical tidal curve for Southampton Water

upon the flow of water, may vary with time. Sand ripples and other bed forms may develop in certain areas and will consist probably of an asymmetrical shape. Such a profile will produce a different frictional effect, depending upon the direction of flow of the overlying water. Most models will assume this bed roughness to be constant, throughout time. Similarly, different storm events may alter the composition of the bed. Biological features, such as large expanses of seaweed, can also influence the tidal flow; these may vary seasonally, but most models do not include this level of detail.

With the emergence of new technologies, the data produced by these models should be assessed for the benefits they may provide to the calibration procedure. For example, systems that can produce high-resolution data on bed levels could be used to provide additional information regarding bed forms. These results could be analysed, then fed back into numerical models as a bed roughness term, to aid in their calibration.

WAVES

The largest waves in the Solent are principally generated by winds blowing along the East and West Solent, as these are the two directions that have the greatest fetch lengths. The effects of swell waves in the West Solent and Southampton Water are likely to be minimal, due to the protection gained from Hurst Spit and Bramble Bank, respectively. Areal wave modelling provides a means of examining spatial variations in wave height and the influence of wave-driven currents. This effect is illustrated in Figure 3, which uses wave vectors and contours of wave height, describing the direction and magnitude of the waves, as they propagate through the Solent, from the west/southwest. The model used for this simulation (DHI, 1997) includes the effects of refraction, shoaling, bed friction, wind and wave breaking.

The results derived from the wave model were then incorporated into a hydrodynamic model to calculate the combined tidal and wave-driven currents. The results, for the vicinity of Hillhead on the flood tide, are shown in Figure 4. It can be seen that although the tidal flow is in a westerly direction, the flow of water due to the breaking waves in the nearshore zone is moving in the opposite direction; this is caused by the dominant wave-induced flow. This pattern suggests that under severe storm conditions, when the waves are propagating from the western Solent, there is likely to be an easterly movement of sediment along the shoreline at Hillhead. This interpretation is supported by field observations of the sediment distribution within groyne fields and has been reported elsewhere (Bray *et al.* 1995).

SEDIMENT TRANSPORT

The Solent

There is very little information available on the nature and movement of sediment within the Solent. Sediment distributions and transport pathways throughout the Solent have been presented by Dyer (1980); however, the sediment information is restricted to the type of material (sand/gravel/mud). There is some sediment type and size analysis data available within the Solent, but this is sparse.

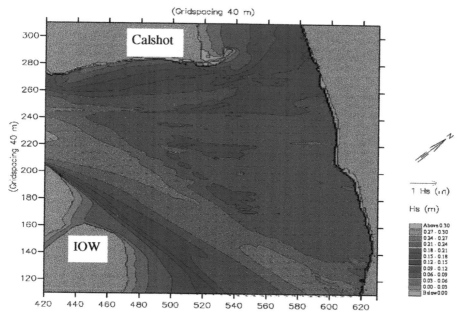

Figure 3 Significant wave height in mid-Solent

Using the results from the combined wave and hydrodynamic models (see above), potential sediment pathways have been predicted for a range of sediment sizes. Figure 5 shows the residual transport vectors around the mid-Solent area calculated for a spring tide. Offshore of the Medina Estuary the sediment transport is directed in an easterly direction (A); however, within the entrance there is some evidence of a gyre (B). Farther north of the easterly-moving sediment pathways, the transport direction is towards the west (C). This pattern highlights the different ebb- and flood-dominated transport directions that occur on either side of the Ryde Middle Bank. Another important feature that can be seen in this figure is that associated with the Bramble Bank, centred at the confluence of Southampton Water, the eastern and western arms of the Solent (D). The sediment transport pathways show this as a convergence zone, with sediment originating from a number of directions and meeting near to this point; this is illustrative of a potential area of (sediment) deposition; hence the reason for the existence of Bramble Bank. As mentioned in the previous Section regarding waves, there is also evidence of longshore littoral transport along the Hillhead shoreline. These features predicted by the model correlate well with reported pathways (Dyer, 1980; Bray et al. 1995) and the results of field work (Gao & Collins, 1997).

Due to a lack of spatial data detailing the sediment distribution on the bed of the Solent, together with only a few limited studies into transport pathways, model results provide a very good indication as to the probable movement of sediment. Such results from models could provide the basis for detailing field surveys in the future. There is also little information known about the sources of sediment into the Solent, especially that of fine-grained material. In order to understand the complex nature of the sediment regimes within the Solent estuaries, it is important that these sources are investigated.

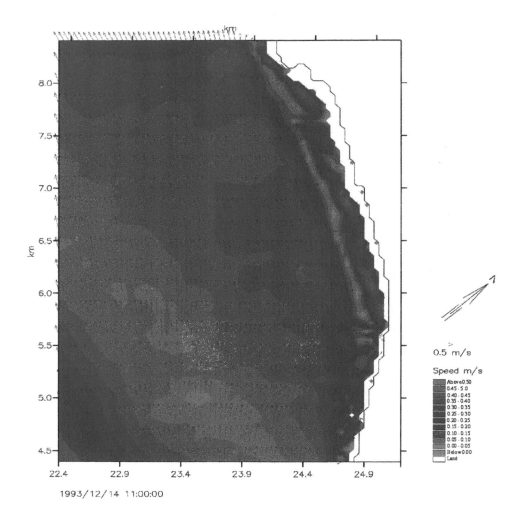

Figure 4　Tidal and wave driven currents near Hillhead

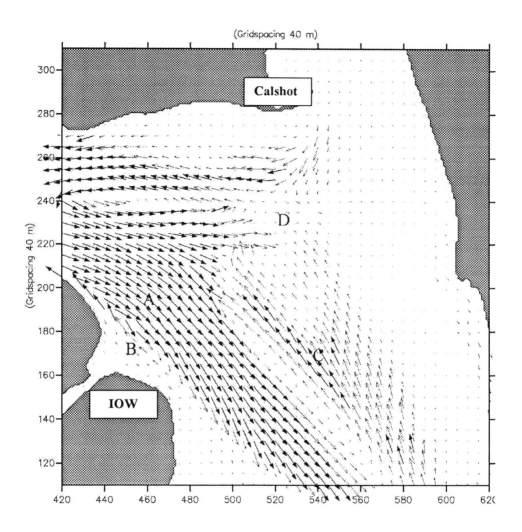

Figure 5 Sediment transport vectors in mid-Solent

Southampton Water

The unique tidal curve in Southampton Water influences sediment transport within the estuary, in a complex way. The asymmetrical nature of tidal propagation, with stronger ebb than flood currents, means that there is the potential for a net seaward movement of coarse bed-load sediments (particularly in the main channel).

Deposition of sediments within the estuary may be influenced by the length of the slack water period occurring prior to the ebb phase of the tide. The slack water period before the ebb is longer than before the flood phase. Therefore, fine-grained sediment transported into the estuary has a longer duration within which to settle out of suspension, before transport in a seaward direction occurs.

Although the asymmetry of the tidal wave is apparent throughout the length of the estuary, the ebb-dominance reduces in an upstream direction. This change may lead to a longitudinal variation in the intensity and direction of sediment transport. Physical evidence for this phenomenon comes from the spatial distribution of dredging requirements within the estuary. Although overall siltation is low, maintenance dredging takes place predominantly within the inner part of the estuary, upstream of Dock Head, and then, mainly in the berth pockets. Little or no dredging is needed in the outer part of the estuary.

Recent numerical investigations have helped to develop a better understanding of the complexities of the hydrodynamic regime and sediment transport pathways in Southampton Water. During the mid-flood stand and double high water period, when the rate of rise of the water level reduces, the volume of water within the estuary continues to move upstream. Since frictional effects are stronger on the intertidal zone and the water body has less momentum, the flow slows down faster than in the channel. In the channel, however, the water still continues to move upstream. This causes a water level gradient sloping towards the mouth of the estuary. The effect of this is for flow on the intertidal to move in the opposite direction to the flow in the main channel. Such movement continues until the water in the channel slows down, or the incoming tidal level starts to rise again. These opposing flows occur on the Hythe Marshes, the eastern foreshore between the Rivers Itchen and Hamble, and on the Dibden Bay foreshore. Tidal flow vectors during the period of the flood stand are shown in Figure 6.

This flow pattern has a complex effect on the local fine-grained sediment transport regimes. Between the opposing flow regimes (in effect, the centre of an elongated gyre), an area of deposition occurs where the tidal currents are low. After the flood stand, the tidal flows continue upstream and can re-erode this newly deposited fine-grained sediment. As the water floods the intertidal areas, the flow is directed onshore and can transport sediment higher up onto the intertidal regions and onto the saltmarshes, if water levels are high enough. Therefore, the flood stand (and, similarly, for the double high waters) creates flow patterns which enable suspended sediment to be transported laterally across the estuary, from the ebb-dominant regime of the main channel to the intertidal/saltmarsh areas. This reverse flow has been recorded using an Acoustic Doppler Current Profiler (ADCP). It has also been observed that the bottom water in the main channel undergoes a period of reverse

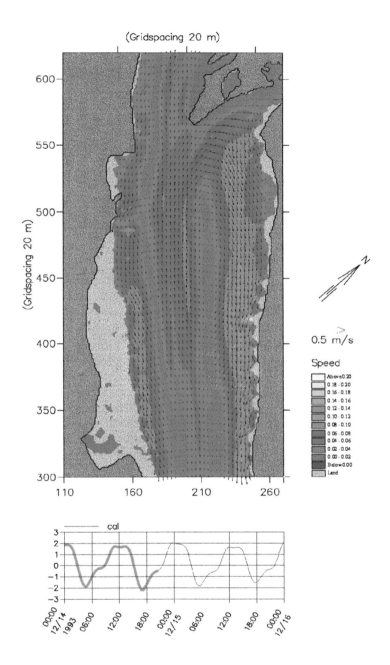

Figure 6 Depth-averaged velocity vectors, within Southampton Water, at the time of mid flood stand.

flow, suggesting that more detailed field work is required to investigate this phenomenon further.

These observations were the result of extensive fieldwork; conducted to support the detailed calibration and validation of the various models used. The added detail led to a much better understanding of the processes involved within the estuary, which may not have been identified had a coarser resolution or less well calibrated model been used. It is important to recognise, however, that the identification of the mechanism for feeding the intertidal areas with fine sediment was not a model output. Indeed, such recognition only came about following a detailed examination of what, at first, appeared to be unusual flow patterns generated by the model. This interpretation highlights the need for good data with which to prove the model; sufficient detail to ensure that processes are properly resolved; and an investigative approach to the interpretation of model outputs. Finally, the conclusions should be tested, within the context of how the system works as a whole.

ESTUARY MORPHOLOGY

One of the difficulties in predicting long-term change is the limited state of current knowledge on the way that estuarine systems evolve; likewise, how to represent this in some form of model. Detailed process models attempt to represent the way in which water movements erode, transport and deposit sediment to bring about changes in bathymetry. Running such a model for a long enough period could provide, in theory, suitable predictions. However, such an approach is likely to be limited to a few years (at most) because of:

- the error terms in predicting sediment transport;
- the non-linearities in the system response potentially leading to diverging states, over time, from closely similar starting conditions (deterministic chaos); and
- multiple 'most-probable' states, i.e. there may be more than one end state - such that it is only possible to predict a range of possible outcomes, or to define a set of possible states.

The alternative to the modelling approach outlined above is to prescribe the basis for dynamic equilibrium, or a steady-state condition, then use this to determine the changes in form, as the boundary conditions change (sea level, energy and sediment inputs etc.). This approach usually treats the system as an integral whole and seeks some form of mass or energy balance, or preserves a particular hydraulic/form relationship (usually referred to as a regime, hence the term regime model). A key difficulty with this type of approach is that the hydraulics and form within an estuary are inextricably linked and there is no clear cause-effect hierarchy.

In order to develop a more rigorous approach, the derivation of minimum entropy production in a river system was re-examined. For the evolution to a probable state in a system near to equilibrium, it has been suggested that the entropy production (per unit volume) will tend to evolve to a minimum, compatible with the conditions imposed on the system (Prigogine, 1955; Leopold & Langbein, 1962). Relating this to an estuary suggests that, in the long-term, a natural system will tend to evolve in an attempt to achieve the most probable distribution of tidal energy. However, the time taken to evolve to this state will be dependant upon constraints imposed upon the system (such as geological constraints and the supply of sediments). Such constraints may be significant enough to prevent the evolution to the most probable state in which entropy is maximised, or may induce a switch to some other steady state. Another complication is that the energy available to the system varies temporally over the evolutionary time-scale, in response to climatic changes, sea level rise, etc.

To establish any changes over time, a 2-D hydraulic model was established for the 1783, 1926 and 1996 bathymetry within Southampton Water. Water elevations and discharges were extracted from the model, at intervals along the length of the estuary; these were used to derive the energy transmission over a tidal cycle. The energy curves for each individual river, as a difference from the most probable state, are shown in Figure 7. In the River Test, there appears to have been little change between 1783 and 1926; followed by a substantial move towards the most probable state, in the period to 1996. This change is probably a consequence of the reduction in tidal volume, caused by the construction of Western Docks in Southampton, during this time. The differences in the Itchen are an order of magnitude smaller; these suggest a much earlier move towards the most probable state, followed by a localised move away in the reach 3 to 5 km from the head. Differences in the Hamble are again an order of magnitude smaller showing a small move away from the most probable state over time. This latter pattern may be due to dredging, associated with leisure developments in the river. Finally, the connecting reach shows a marked move away from the most probable state; this is most likely due to the channel deepening that has taken place, in stages, over the time interval studied (although notably not at the mouth, such that the boundary conditions used remain valid).

Generally, the bed within Southampton Water is remarkably stable and there has been little change in sub- and intertidal areas, other than as a direct consequence of dredging and reclamation. This pattern reflects the relatively non-erosive nature of the stratigraphy and the limited sediment supply to the system. Isolating the natural changes suggests that a 3% change in the tidal prism, in conjunction with an anthropogenic adjustment of some 20%, has brought about the observed changes in energy transmission (33% change in the Test). This pattern suggests that a further 35-45% siltation is required, to achieve the most probable state. Based on the rate of natural change observed, this could take some 400 to 800 years; this assumes that the boundary conditions and constraints remain constant. Given that there have been long periods with no observed change, together with the face that some components of the system have moved away from (rather than towards) the most probable state, it is more likely that the response time of the system is of the order of thousands, rather than hundreds, of years.

As noted at the outset, morphological modelling is presently a somewhat inexact science. Nonetheless, the approach, which has been briefly summarised, helps to develop an improved understanding of the functioning of the system. The role of the models is to add information, which enables the experienced user to test accepted hypotheses, in order to build a conceptual model for the system as a whole. As we have endeavoured to illustrate, this requires work on a range of spatial and temporal scales.

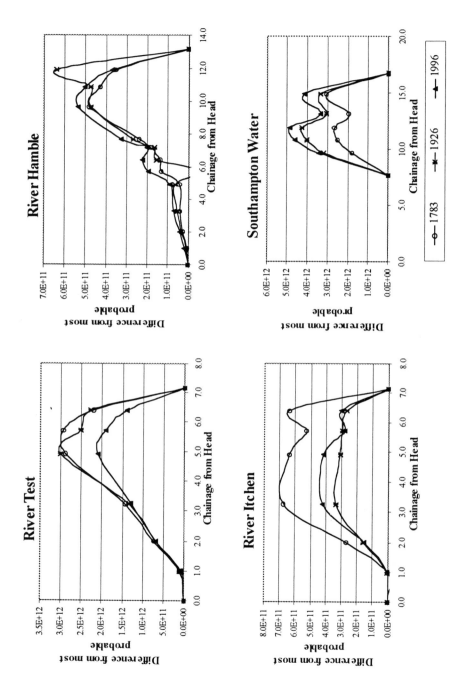

Figure 7 Changes in energy distribution in Southampton Water and the Test, Itchen and River Hamble.

CONCLUSIONS

Numerical models are now used widely for coastal and river engineering projects, as a mechanism for predicting changes to environmental conditions, whether from natural or anthropogenic causes. As such, they tend to be used for 'what if' scenarios by engineers, with output from studies being in the form of 'figures of change' produced as 'snapshots' at particular times. It is quite easy to accept a model as calibrated and then disregard all of the data it then produces because all that is of interest to the user are the changes it can predict. Very high-resolution data (both spatially and temporally) can be predicted by numerical models, which enables the user to examine very complicated flow structures and even fill in the gaps between measured data sets. This procedure can be of great assistance in developing an understanding of how a particular system works, providing a more robust basis for decision-making.

Estuary regimes can now be examined in detail, with the use of models; they are not 'complete' and are, inevitably, a simplification of the real world. This limitation should not, however, be used as a basis for dismissing models outright; they can provide valuable insights and help to develop understanding, if properly applied. The present scope of hydrodynamic, sediment transport and morphological models within the context of the Solent has been summarised. In particular, the importance of field data, to support the modelling effort and interpretation, and to make best use of model outputs, has been emphasised.

Future research will see the continued development of the various models and, progressively, a greater degree of integration between the models. Such integration is already taking place within the context of physical processes. Similarly, studies are beginning to highlight the potential for the greater integration of the physical, chemical and biological processes. In due course, this approach should provide the scope for more robust models of coastal and estuary systems. Within the Solent and Southampton Water, there is a general lack of data; specifically, long-term freshwater and tidal flows, suspended sediment, waves and sediment size/type. To achieve the goal of a more integrated systems approach, more of this type of information needs to be collected.

ACKNOWLEDGEMENTS

We are grateful to Bill Roberts at HR Wallingford, Janet Hooke at University of Portsmouth and Mike Collins at the University of Southampton, for providing information about modelling that they have undertaken in the Solent; this helped us to compile this overview.

REFERENCES

ABP Research. 1995. ABP Southampton - channel deepening, Environmental Assessment: assessment of dredging methods. Report R 493, Figures 6 and 7.

Bray, M.J., Carter D.J. & Hooke J.M. 1995. Littoral cell definition and budgets for Central Southern England. *Journal of Coastal Research*, **11**: 381-400.

DHI. 1997. MIKE 21, CD-ROM (v3) Danish Hydraulics Institute, www.dhi.dk.

Dyer, K.R. 1980. Sedimentation and sediment transport. In: *The Solent System – an Assessment of Present Knowledge*. N.E.R.C. Publications, Series C, No. **22**: 20-24.

Gao, S. & Collins, M.B. 1997. Changes in sediment transport rates caused by wave action and tidal flow time-asymmetry. *Journal of Coastal Research*, **13**: 198-201

Leopold, L.B. & Langbein, W.B. 1962. The concept of entropy in landscape evolution. *Geological Survey Professional Paper*, **500**-A, A1-A20

Prigogine, I. 1955. *Introduction to thermodynamics of irreversible processes*. John Wiley & Sons, London.

Watson, J.D. and Watson, D.M. 1972. Marine disposal of sewage and sewage sludge. Report to South Hampshire Plan Advisory Committee.

Webber, N.B. 1980. Hydrography and water circulation in the Solent. In: *The Solent System – an Assessment of Present Knowledge*. N.E.R.C. Publications, Series C, No. **22**: 25-35.

Wright, W. & Leonard, R.D. 1959. An investigation of the effects of a proposed dredging scheme in Southampton Water by means of a hydraulic model. *Proceedings of Institution of Civil Engineers*, **14:** 1-18.

Wisdom of Hindsight: Palaeo-Environmental and Archaeological Evidence of Long-Term Processual Changes and Coastline Sustainability

David Tomalin

County Archaeologist, Isle of Wight Council, 61 Clatterford Road, Newport, Isle of Wight, PO30 1NZ, U.K.

CHANGING PERCEPTIONS OF THE PALAEO-ENVIRONMENTAL RECORD

Since the Steers Report on Britain's coastline was commissioned in 1945 (Steers, 1946), there have emerged two divergent perceptions of the palaeo-environmental resources contained within the coastal zone. The most innovative view was that offered by Professor Steers. During the 1940's, Steers had been prompt to recognise that archaeological and palaeo-environmental deposits in the coastal zone offered the key to the calibration and interpretation of changes in the shoreline. This 'interrogative approach' had been pioneered in the Cambridge and Lincolnshire fens, where a multi-disciplinary team of archaeologists and environmental scientists had successfully traced a history of Holocene coastal changes. Episodes of marine transgression and retreat were mirrored by strategic movements in the human population. In these studies, the new science of palynology was found to be a powerful tool and the Cambridge team were soon to apply these techniques to the submerged margins of the Solent (Oakley, 1943; Godwin & Godwin, 1945).

In 1964, Professor Steers called for an integrated approach to the study and management of the coastline, a process in which he proposed a key role for archaeological and palaeo-environment scientists. By then, however, an approach which might be described as the 'portfolio principle' had muddied the water. For many years, the *Ancient Monuments Act* had focused the mind of Government on a schedule of those historic sites that might be deemed to be of national importance. This approach befitted the protection of standing historic monuments, but it was not so well-suited to the appraisal of prehistoric archaeological sites; at these, arrays of cultural material and associated environmental evidence often lay well-concealed below a blanket of soil or coastal sediment.

During the 1970s and 1980s, new attention was given to the development of 'environmental data-bases' and 'sites and monuments records' (SMRs). These records identified sites of all levels of importance, or potential; they fitted well within the structure of local authorities, when performing their roles as regulators of development and curators of the landscape (PPG16, 1990). Because these authorities were statutorily charged with development control and the constraint of environmental impacts, it now became common to consider the identification of archaeological and other palaeo-environmental sites, essentially, as a conservation issue. Thought was soon focused less upon the scientific significance of these sites, more upon the status they should claim in the planning process. This approach allowed them to be viewed alongside Ramsar sites, Sites of Special Scientific Interest (SSSIs) or Special Areas of Conservation (SACS) as places where 'problems' with management and protection should be anticipated.

Some notable exceptions to the portfolio approach arose in coastal counties such as Wight, Hampshire, East Sussex, Essex, Cleveland and Northumberland. Here data-collection was extended into the intertidal and sub-tidal zones. *Planning Policy Guidance on Coastal Planning* (PPG20), issued by Government in 1992, helpfully endorsed this vision, acknowledging that "information may be needed" on "off-shore processes, ... the current state of the environment and the nature, scale and pace of coastal change". This guidance rightly stressed that "earth science and ecological information should provide a firm base for development plans which seek to assess potential risks".

An unfortunate weakness in PPG20 arose from its singular address to Planning Authorities. This planning policy document omitted cross-reference to the work and research of Operating Authorities and Coastal Study Groups, who were charged with coastal protection and flood defence. Commentators were soon to add that the "sharing of information between local planning authorities, coastal defence groups, the National Rivers Authority [Environment Agency], coastal protection authorities and conservation interests" was particularly important and that "in the interests of economy and efficiency, local planning authorities should make full use of information being compiled as part of the preparation of Shoreline Management Plans". Moreover, this liaison "should be undertaken as a matter of course by local planning authorities before plans are prepared or reviewed" (Lee, 1996). Unfortunately, many plans still lacked clear scientific and philosophical perspectives, capable of ensuring consistency in their comprehension of the environment. Whilst points shown on a map might provide an illusion of site-specific knowledge in a Shoreline Management Plan, a common omission was a comprehension of the *in situ* physical evidence which each site could contribute to the understanding of sea-level rise and coastal change.

In 1992, the 'Yellow Manual' on 'the economics of coastal management' (Penning-Rowsell *et al.*, 1992) offered coastal managers a broader vision. This advice stressed that "designation should only be used as an initial indicator of site importance and the appropriate local expertise, whether from a statutory agency or a voluntary body should always be asked to advise on site characteristics". Key points in the manual proffered pertinent questions including "what proportion of the total of archaeological sites has so far been identified?" and "what is the future potential of each area for further discoveries?" The manual was less specific in linking these topics with coastal processes, but a further key point was precisely to the mark: "Uncertainty is another name for ignorance: ignorance over some areas can be reduced, but there are some areas where we do not even know that we are ignorant".

This concern was further emphasised by English Nature in its *Strategy for the sustainable use of England's estuaries* (EN 1993a). This document recognised the need to identify "the location of important sites and those of high potential". The topic of "archaeology/heritage" was then cited in the "matrix of conflicts", a vital component of this particular guide.

The companion volume, the *Co-ordinators guide to estuary management plans* (EN, 1993b) reinforced this issue, explaining that

> *"the integrated study of cultural and environmental evidence preserved in estuarine sediments can provide valuable insights into sea-level change and other coastal processes".*

It has been this last document which has re-focused our attention on the 'integrated approach,' first advocated by Steers. Solent fishermen, commercial aggregate prospectors and the Crown Estate have all made and reported significant discoveries during their commercial activities (Draper-Ali, 1996; RCHME, 1996; Tomalin, 1997), it is clear, therefore, that the call for integration should be cast as wide as possible. Such integration is particularly important for the Solent region. We are reminded by Government that "the Solent provides an example of a region working, incrementally, towards a common CZM [Coastal Zone Management] strategy". The region is one of the best researched in the UK and includes "considerable knowledge of maritime archaeology and coastal processes" (D.o.E., 1996). This timely reassurance leads us to consider the means by which wisdom of hindsight might now be drawn from past events and long-term trends that have led to the present changes which are occurring in our coastline.

EROSION, ENVIRONMENTAL CHANGE AND THE QUESTION OF TIME-SCALES

Despite the confidence expressed by earlier observers and scientists, the age of the Solent, as an open east-west seaway, still eludes us. Nevertheless, a number of coastal processes are proceeding within time-scales, which are certainly important to us. The floor of Hurst Narrows appears to be still actively scouring from a charted depth of 54 m, in 1781, to 60 m in 1988. Deep localised pits can be detected in the seabed off Yarmouth, where the inner threshold of the Hurst Channel appears to be scouring to depths exceeding 50 m.

Along the southern coastline of the West Solent, active cliff erosion can be observed at Bouldnor, Colwell Bay and Headon Warren (McInnes, 1994; Halcrow, 1997); for general location and main sites referenced refer to *Map A* and *Map B* in the *Preface*. On the northern shore of this seaway, the coastal salt marshes between Keyhaven and Sowley have been detectably retreating since the 18th century at least (Hooke & Riley, 1987). The long-term stability of Hurst Spit and its castle is, at best, uncertain (Bray *et al.*, 1992). Due to differential downwarping, local sea-level within the Solent region is demonstrably rising, amounting to 1-5 mm per year in the Western Solent and 5-8 mm per year in the Eastern Solent (Bray *et al.*, 1992; MAFF, 1995). In contrast with the Eastern Solent, the bed formations of the Western Solent are dominated by poorly-sorted gravels and sands (Hamblin & Harrison, 1989 (map 4); Solent Forum, 1996 (map 5)). All of this evidence indicates a high energy environment, in which the seaway appears to be undergoing geomorphological processes that are still relatively young. These stem from a severance event or 'tidal linkage,' which marks the base-line for many of the coastal processes we are now witnessing. Before future trends and scenarios are discussed, the chronological fixing of this event is a priority.

Wave-cut platforms, stacks and stumps

Notch lines and the gradients of wave-cut platforms are signatures left by advances and relative still-stands of the sea (Inman, 1983). This means that the Chalk wave-cut platforms and notched cliffs of the West Wight still have valuable evidence to impart on the long-term progress of coastal erosion. Other submerged platforms and rock outcrops requiring further underwater examination include the Owers and Mixon Rocks (Wallace, 1996) and the Bembridge Ledges.

At the foot of the Needles Lighthouse, a submerged wave-cut platform and notch-line can be found at an approximate depth of -2.6 m O.D. (Tomalin, *this volume*, Figure 3 location 8). Eastward, the height of the platform rises progressively over 1.6 km until it reaches the modern tidal notch line at Alum Bay beach (Tomalin, *this volume*, Figure 3, location 7). The gradient and width of this platform are parameters reflecting erosion and sea-level rise, since the sea stood at 2.6 m O.D. This height implies that wave-attack and the removal of land composed of Eocene sands and clays, within Alum Bay, may have occurred no earlier than the 1st millennium BC. The platform also offers an index of coastal erosion, influenced by the tidal linkage of the Western Solent. This event might be equated with the outermost and submerged stumps of the Needles stacks. Between the Needles and Compton Bay (Tomalin, *this volume*, Figure 3, location 9), a further platform offers a threshold for the measurement of sea advance over a distance of 7.5 km. Preliminary experiments undertaken at the Needles, with an Isis 1000 swath-bathymetry unit, shows highly successful results; these indicate that the topographical modelling of these platforms should be included in the future research agenda.

The coastal peats of the Solent

Since pollen was extracted from the Empress Dock peat in 1940, palynology has been highly successful in reconstructing past coastal environments. Recently, the combined efforts of Dr Robert Scaife (University of Southampton) and Dr Anthony Long (now, University of Durham) have identified a number of key sedimentary sequences; these have permitted the construction of an outline Holocene sea-level curve for the Solent (Long & Tooley, 1995; Long and Scaife, in preparation). The integrated study of diatoms from these sequences has extended the process of environmental reconstruction, to cover contemporary changes in salinity. If we are to interpret past changes in the Solent's coastal habitats, these integrated studies should now be pursued. This work should include the quest for past scenarios of environmental change, which may be used to model and test the events of the present.

A principal source of palaeo-environmental information has been the coastal peats, especially those at Fawley, Stansore Point, Wootton-Quarr, Newtown and Yarmouth (Hodson & West, 1972; Long & Tooley, 1995; Scaife, pers. comm. & forthcoming), for locations see *Map A* and *Map B* in the *Preface*. General observations suggest that the accretion of these peats ceased during the 2nd millennium BC, yet it must be emphasised that a full Solent-wide survey and assessment has yet to be made. Potential sites include the Medina, the Lower Itchen and the Chichester Channel, where there is clearly much intertidal evidence yet to be interrogated (Russell, pers. comm.; Cartwright, 1984; Wallace, 1996; Bingeman & Mack, pers. comm.). The sea-level curve for the Solent, particularly for the last two millennia, will remain imprecise until the sediments within the inner reaches of the undredged estuaries are fully examined.

The Solent's submerged forests and landscape

The need to interrogate submerged forests for coastal palaeo-environmental information has been emphasised recently by English Heritage (Fulford *et al.*, 1997). In the Solent, studies have focused, so far, on just two sites. In the Western Solent at Bouldnor (Tomalin, *this volume,* Figure 3, location 10) the 'Cross Forest', composed of oak and pine, lies some 300 m offshore at an approximate depth of 12 m O.D. Tree boles, root systems and recumbent trunks are present and have been dated at 6430-6120 Calendar year (cal) BC (GU-5420[1]). This forest might be tentatively equated with conditions contemporary with the persistence of the proposed land-bridge or 'Wight Umbilical' between Bouldnor and Pitts Deep (Tomalin, *this volume*, Figure 3, location 1).

At Wootton-Quarr 58 oak trees have been sampled on a submerged land-surface, which has so far been traced to 2.9m O.D. Sub-bottom images suggest that there may be a seaward extension of this land-surface, perhaps to as deep as 9 m O.D. The intertidal section of this forest has produced a tree-ring sequence, spanning the period 3463 to 2557 BC (Hillam, in preparation). These trees first show stress around 3200 BC, but growth recovers before 3000 BC. Subsequently, a progressive decline occurs perhaps leading to a final demise after the close of 3rd millennium BC, when just a few trees appear to have been surviving in the coastal peat at Wootton; these are dated at between 2200 and 1750 cal BC (UB-3271, UB-3272 & UB-3274). The diminution in forest growth on the coastal peat can be attributed tentatively to progressive saline invasion of the root systems. This is a process that may mirror the type of coastal change we are witnessing, at the close of 20th century. To secure a clearer understanding of past and present changes of this kind, the region requires new survey; within which datable timber and submerged land-surfaces should be identified and examined, on other tracts of the Solent's intertidal and sub-tidal margins.

Submerged archaeological structures in the Solent

Ancient structures erected on past strand-lines may serve to calibrate sea-level rise. The Wootton-Quarr archaeological survey has exemplified this approach, identifying no less than 160 ancient structures on 3 km of intertidal coast (Loader *et al.*, 1997). The structures include putative fish traps, which commence at 4040-3780 cal BC (GU-5251), and post alignments or fish weirs beginning at 840-600 cal BC (GU-5052). The construction of Neolithic intertidal trackways, at Newtown and Quarr, begins in the mid- 4th millennium BC and continues until 2900-2500 cal BC (GU-5341; GU-5582), when the coastal forest seems to have been succumbing to a rising sea. At the head of the Chichester Channel, submerged Roman structures identified by Cunliffe (1971, 1996) and Wallace (1997) provide evidence of post-Roman sea-level rise, worthy of further calibration. At Sinah Lake, in Langstone Harbour, a submerged circular wooden structure of Saxo-Norman date attests to sea-level rise since its construction in the period cal AD 980-1180 (GU-7275) (Allen & Gardiner, in preparation).

---

[1] Reference label for laboratory that carried out the carbon dating analysis.

Plate 1 Oak trunk sampled in the submerged Neolithic forest traced to -2.9 m OD at Wootton/Quarr, Isle of Wight. After 282 years of stressed growth, this tree died in the winter of the year 2777 BC.

Plate 2 At the head of Newtown Salt Marsh, marine advance has poisoned these ancient oaks. A rise of 0.4 m in the land surface permits the survival of neighbouring oaks.

Plate 3 Flint picks from the Solent seabed attest the presence of submerged land surfaces once occupied by man. These contexts are highly susceptible to unsustainable damage by dredging and harbour works.

Particularly disturbing is the exposure and destruction of *in-situ* archaeological remains, which have been formerly blanketed by intertidal muds bonded by *Spartina* grass (Tomalin, 1997). The current die-back of this species and exposure of previously entombed prehistoric structures provides incontestable evidence that certain long-term processes of sedimentation within the Solent are now being reversed. A study of *Spartina* die-back and intertidal habitat-loss should be firmly integrated, with an examination of datable archaeological structures and palaeo-environmental horizons.

The rias of the Solent and their sediment archives

The survival of deep 'sediment archives', attesting local or regional coastal change, has recently been demonstrated in the drowned creeks at Yarmouth, Newtown and Wootton (Loader *et al.*, 1997). The importance of such archives has been emphasised by Bell and Neumann (1997). It is evident that the scientific potential of these deposits can be readily threatened or destroyed, by modern demands for building development, mineral extraction or navigational dredging (Tomalin, 1997; 1998). A strategy is urgently required which will enable the ria and estuarine locations within the Solent to be assessed for their palaeo-environmental potential, before further losses are imposed by natural and human agencies. Such a strategy should enable primary evidence of early human activity and coastal change history to be retrieved and interpreted.

The spits

Since the early observations of Webster (1816) and Fox (1862), it has been evident that the age of Hurst Spit and Calshot Spit is central to the dating of the Western Solent seaway. At Hurst, the present spit overlies Pleistocene terrace gravels, a palaeochannel (Tomalin, *this volume,* Figure 3, location 12) and Holocene salt-marsh deposits. The genesis of the spit has been generally attributed to major hydrodynamic changes coincident with the drowning of the Western Solent and the subsequent formation of the 'tidal link' (Nicholls & Webber, 1987). Unfortunately, present knowledge of Hurst Spit cannot provide a proxy date for the opening of tidal linkage through the Western Solent seaway. A problem is posed by the apparent presence of a proto-spit, which is deemed to have occupied a position which could lie as much as 1 km southwest of the present spit (Nicholas & Webber 1987, Figure 8). An exploratory core examined by Nicholls & Clarke (1986), on the leeward side of Hurst Spit (Tomalin, *this volume,* Figure 3, location 13) attests the presence of datable Holocene peats and estuarine deposits. However, the position of this particular core invalidates its use as a means of dating the genesis of the spit. A new programme of coring and analysis is required, to secure palaeoenvironmental evidence from the west side of the spit.

At Calshot Spit, a cross section compiled by Hodson and West (1972) is more informative. The data suggest that Holocene deposits of sand and gravel, comparable with the composition of the present spit, could have begun to accumulate when sea-level stood at around 15 m O.D. This height does not accord with the levels of 5.5 and 6.8 m O.D., which have been tentatively attributed to the event of tidal linkage at Yarmouth and Newtown. The beach deposits at Calshot have protected a leeward series of accumulating saltmarsh and peat-forming environments (Hodson & West, 1972, Figure 3; Long & Tooley, 1995). There can be no guarantee that these are contemporary with the genesis of this spit, yet the sediment transport paths which feed the spit from this seaway (Bray *et al.*, 1991) seem unlikely to have developed until such a linkage was achieved.

Wrecks and portable artefacts as environmental indicators

The arrival of a wrecked ship or a discarded artefact on the seabed marks a chronological event, from which subsequent changes in the seabed environment might be measured. The entombment of the Mary Rose, by sediments accreting since AD 1545, is an example described recently by Justin Dix (pers. comm.). Elsewhere, 'artefact fields' or 'anchorage strews,' dating from Roman and later times have been identified on the floor of the Solent (e.g. Yarmouth Roads; Tomalin, *this volume*, Figure 3, location 14). These offer a longer period over which the movement of bedload, together with colonisation by benthic communities, might be measured (Tomalin, 1997). The medieval anchorages associated with the Port of Southampton have yet to be identified and there is a pressing need to map their seabed footprints and to assess their vulnerability to modern harbour works and navigational dredging activities. This is a responsibility of 'due regard,' which the port authorities have yet to achieve.

At Wootton Haven, the sedimentary entombment of Roman, Saxon and medieval metalwork has led to the formation of semi-stable patinas, on objects of silver and copper alloy. These patinas have shown evidence of rejuvenated chemical deterioration, where an environment of stable accretion had been interrupted by dredging, or a renewed erosion event. Such examples demonstrate that, when viewed in its depositional context, archaeological evidence can be used as an index of long-term environmental change on the Solent shoreline.

READING THE PAST, ANTICIPATING THE FUTURE

With increasing economic demands from two international ports and a large urban population, the environmental stresses on the Solent region are particularly severe. Conflicting needs arise from leisure, tourism and fishery industries, which are highly dependent upon the natural and historic resources of the coast. Into this volatile 'melting pot' has been thrown the element of uncertainty posed by natural and humanly-induced environmental change. For planners, developers, harbour authorities and coastal managers operating in this setting, there can be little room for poorly-informed decision-making; however, a clear and unified environmental strategy for data-collection, assimilation and policy-making has yet to be fully defined.

The Strategic Guidance for the Solent (Solent Forum 1996; 1997) has outlined six specific objectives for the historic environment; of these, the need to improve base-line information has been rightly identified as a component of the Strategic Priority (or 'Flagship' Project 1). The need to utilise geo-archaeological information for developing sustainable coastal defence policies has also been recognised. This present paper identifies seven means by which these improvements can be achieved. Some pertinent issues concerning long-term change, habitat loss and strategic options for wildlife resources have already been summarised elsewhere (Tubbs, 1991), yet databases for the natural and historic environment of the region are still fragmentary. There remains a need for a regional GIS, which is capable of assembling a unified overview. The absence of such a capability suggests that we may still be impeded by the 'portfolio perception,' whilst the integrated approach has yet to be achieved.

It is disquieting to find that the palaeo-environmental and geo-archaeological evidence that should guide our understanding of on-going changes in our coastline, can be so readily assigned to a box marked 'conservation issues' or hung with the title of 'built environment'. We should take heed that this former title was used to introduce the current topic, within the agenda of the Solent Science Conference [reorganised within these Proceedings -Eds.]. These aspects of the 'Historic Environment' have recently been discussed in an international forum assembled under the European L'Instrument Foncier pour l'Environnement (LIFE Programme). Here, present work in the Solent has been compared with parallel research on the estuarine coastlines of the Shannon (Ireland) and the Gironde (France). This forum has posed a number of pertinent questions, as outlined below.

1. Where and what are the sites that offer scientific information which will strengthen our understanding of environmental changes in terms of:
    - the nature, scale and pace of coastal processes;
    - global or regional climatic processes;
    - pollution; and
    - past and present human impacts of the environment?

2. Do we have sufficient archaeological and palaeo-environmental evidence to gain an understanding of the experiences and failures of past human communities, on this particular coastline?

3. Can present proposals for coastal management and protection demonstrate wisdom drawn from a long-term understanding of past natural and human events, at this particular location?

4. Will it be necessary to gather and analyse new palaeo-environmental and archaeological information, on past natural and human events, before wisdom of hindsight can be expressed in plans and predictions for the future?

The LIFE forum was agreed that, by addressing these questions, coastal planners, engineers and harbour authorities should be able to demonstrate that all reasonable precautions and 'due regard' had been taken in their decision-making processes. Equally, without the application of such a checklist, it should be apparent that reasonable precaution and regard cannot be demonstrated.

A FIELD AUDIT OF THE SOLENT'S PALAEO-ENVIRONMENTAL AND ARCHAEOLOGICAL RESOURCES

Within the Solent estuarine system, sufficient evidence has been glimpsed relating to the topics of wave-cutting, palynology, dendrochronology, marine seismology, coastal archaeology, sedimentology, coastal geomorphology and habitat change, to see that an integrated field audit of palaeo-environmental resources is now required. Such an audit should be 'seamless' in its remit, embracing the onshore, intertidal and sub-tidal zones; it should extend and enhance the absolute chronology, which is now being assembled by radiocarbon dating and dendrochronology. The first objective of this field audit should be to identify and assess those sites which serve as 'archives'; these are those which are best endowed with palaeo-environmental information, pertinent to our understanding of regional or global environmental change. The second objective should advance the interpretation of these sites; a third should ensure their long-term sustainability, in accordance with the aims of *European Agenda 21*.

A project design for such an audit will be an ambitious task, requiring close collaboration between the Universities of Southampton and Portsmouth, the Hampshire and Wight Trust for Maritime Archaeology and the Environment Agency. Local Authorities have a central role to play, as Coastal Operating Authorities and as Strategic Planning Authorities, with responsibilities for the recognition and curation of the natural and historic environment. Logic might argue that the Solent Forum is the natural locus to initiate a study such as this, whilst the 'Steers philosophy' might claim that this would be no more than a coming of age.

RECOMMENDATIONS

1. The Holocene chronology and development of the Solent should be researched and calibrated, to provide a base-line from which the progressive effects of tidal linkage, sea-level rise, coastal processes and heritage-loss can be accurately measured.

2. A *'Strategy for the Historic Environment of the Solent'* is required. This approach should allow palaeo-environmental and archaeological resources in the intertidal and sub-tidal zones to be assessed, for their scientific potential. Protection, intervention, or research should then be prioritised, before further losses are imposed by natural and human agencies. The strategy should enable local authorities, harbour authorities and offshore operators to plan with confidence and integrity, whilst fulfilling their obligation to give due regard to the natural and historic environment and to meet the requirements of *European Agenda 21*.

3. A comprehensive *'Field Audit'* of palaeo-environmental and archaeological resources should be executed, to inform the Strategy; its aim should embrace the LIFE objectives set out in this paper.

4. An *Advisory Group for the Historic Environment of the Solent* is needed: to develop the research strategy for this topic; to promote research-funding; to guide the field audit; to monitor the implementation of the strategy; and to encourage an integrated approach, to the study and understanding of long-term processual and environmental changes in the coastal zone. Expertise of the group should include palynology, coastal processes and geomorphology, sea-level change, GIS and data-base management, marine seismology, coastal and geo-archaeology. Early dialogue should be sought between this advisory group and the proposed specialist coastal research advisory groups, presently contemplated by MAFF.

ACKNOWLEDGEMENTS

This paper has been advanced by the Archaeology and Historic Environment Service of the Isle of Wight Council, with assistance from English Heritage and the Hampshire and Wight Trust for Maritime Archaeology. Scientific analyses, advice and field information has been provided by Dr Rob Scaife, Dr Justin Dix, Dr Adonis Velegrakis, Jean Dean, Daffydd Jones, John Cross, all of the University of Southampton, and Hume Wallace, John Bingeman and Arthur Mack. Dendrochronological data and diatom analyses have been provided, respectively, by Jennifer Hillam and Dr Nigel Cameron. I am especially grateful to my colleagues Frank Basford, Rebecca Loader, David Motkin, Robin McInnes and Nick Blake, for assistance in assembling the array of data used in this discussion. Radiocarbon dates quoted have been provided by the Ancient Monuments Laboratory and Dr Mike Allen of the Trust for Wessex Archaeology.

REFERENCES

Allen, M. & Gardiner, J. (in preparation). Our changing coast; an inter-disciplinary survey of Langstone Harbour, Hants, Wessex Archaeology.

Bell, M. & Neumann, H. 1997. Prehistory, intertidal archaeology and environment in the Severn Estuary, Wales. *World Archaeology*, **29**(1): 95-113.

Bingeman, J. & Mack, A. 1996. Personal communications, from John Bingeman and Arthur Mack, report intertidal wooden structures and prehistoric lithic scatters visible at low water in the Chichester Channel.

Bray, M.J., Carter, D.J. & Hooke, J.M. 1991. Coastal sediment transport study, report to SCOPAC (Vol 3). University of Portsmouth, 30 pp.

Bray, M.J., Carter, D.J. & Hooke, J.M. 1992. Sea level rise and global warming; scenarios, physical impacts and policies; report to SCOPAC. University of Portsmouth, 169 pp.

Cartwright, C. 1984. Field survey of Chichester Harbour. *Sussex Archaeological Collections*, **122**:23-27.

Cunliffe, B. 1971. Excavations at Fishbourne. *Society of Antiquaries Research Report*, No. 26. (2 vols.), London, 620 pp.

Cunliffe, B., Down, A. & Rudkin, C. 1996. *Chichester excavations 9; excavations at Fishbourne 1969-1988*. Chichester District Council, 246 pp.

Dean, J.M. 1995. Holocene palaeo-environmental reconstruction for the nearshore Newtown area, Isle of Wight. University of Southampton, Department of Oceanography. Unpubl. M.Sc. Research Project Report, 110 pp.

D.o.E. 1996. *Coastal zone management; towards best practice*. Nicholas Pearson Associates for the Dept of the Environment, 74 pp.

Draper-Ali, S. 1996. *Marine archaeology and geophysical survey: a review of commercial survey practice and its contribution to archaeological prospection*. Hampshire & Wight Trust for Maritime Archaeology. Southampton, 104 pp.

English Heritage & Royal Commission on the Historical Monuments of England. 1996. England's coastal heritage; a policy statement, 16 pp.

English Nature 1993a. *Strategy for the sustainable use of England's estuaries*. Peterborough, 43 pp.

English Nature 1993b. *Estuary management plans: a co-ordinators guide*. Peterborough, 88 pp.
Fox, W.D. 1862. How and when was the Isle of Wight separated from the mainland. *The Geologist*: 452-452.
Fulford, M.G., Champion, T.C., & Long, A.J. 1997. *England's coastal heritage: a survey for English Heritage and the Royal Commission on Historical Monuments in England*. English Heritage Archaeological Reports No. 15, 268 pp.
Godwin, H. & Godwin, M.E. 1945. Submerged peat at Southampton; data for the study of Postglacial history. *New Phytologist*, **39**: 303-307.
Halcrow, W. & Partners 1997. *Isle of Wight coast shoreline management plan Vols. 1 & 2*. Sir William Halcrow & partners, Swindon, for Isle of Wight Council. Limited edition.
Hamblin, R.J.O. & Harrison, D.J. 1989. *Marine aggregate Survey Phase 2; South Coast*. British Geological Survey Marine Report 88/31. Keyworth, Nottingham, 30 pp and maps.
Hodson, F & West, I.M. 1972. Holocene deposits at Fawley, Hampshire and the development of Southampton Water. *Proceedings of the Geologist Association*, **83**: 421-441.
Hooke, J. & Riley, R. 1987. *Historic changes in the Hampshire coast 1870-1965*. A report for Hampshire.
Inman, D.L. 1983. Reconstructing palaeocoastlines. In: *Quaternary coastlines and marine archaeology; towards the prehistory of land bridges and continental shelves*, Masters, P.M. & Flemming, N.C. (Eds.): Academic Press, London etc., 1-49.
Lee, E.M. 1996. Earth science information in support of coastal planning: the role of shoreline management plans. In: *Coastal management: putting policy into practice*. Flemming, C.A. (Ed.), Telford, London, 54-65.
Loader, R., Westmore, I. & Tomalin, D.J. 1997. *Time and Tide: an archaeological survey of the Wootton-Quarr coast*. Isle of Wight Council. Newport. 32 pp.
Long, A.J. & Tooley, M.J. 1995. Holocene sea-level and crustal movements in Hampshire and southeast England, United Kingdom. *Journal of Coastal Research*, Special Issue **17**, Holocene cycles: climate, sea levels and sedimentation: 299-310.
McInnes, R. 1994. *A management strategy for the coastal zone*. South Wight Borough Council.
MAFF 1995. *Shoreline management plans; a guide for operating authorities*. MAFF publications PB 1471, 221 pp.
Nicholls, R.J. 1987. The evolution of the upper reaches of the Solent River and the formation of Poole and Christchurch Bays. In: *Wessex and the Isle of Wight field Guide*. Barber, K. (Ed.). Quaternary Research Association, Special Publications, Cambridge, 99-114.
Nicholls, R.J. & Clarke, M.J. 1986. Flandrian deposits at Hurst Castle Spit. *Proceedings of the Hampshire Field Club*, **42**:15-21.
Nicholls, R.J. & Webber, N.B. 1987. The past present and future evolution of Hurst Castle Spit, Hampshire. *Progress in Oceanography* **18**: 119-137.
Oakley, K.P. 1943. A note on the post-glacial submergence of the Solent margin. *Proceedings of the Prehistoric Society*, **9**: 56-59.

Penning-Rowsell, E.C, Green, C.H., Thompson, P.M., Coker, A.M., Tunstall, S.M., Richards C. & Parker D.J. 1992. *The economics of coastal management; a manual of benefit assessment techniques.* Bellhaven, London & Florida, 380 pp.

PPG 16. 1990. *Planning Policy Guidance 16: Planning and Archaeology.* London, HMSO, 20 pp.

PPG 20. 1992. *Planning Policy Guidance 20: Coastal Planning.* London, 20 pp.

RCHME 1996. *The national inventory of maritime archaeology for England.* Royal Commission on the Historical Monuments of England, Swindon, 36 pp.

Solent Forum 1996. *Towards strategic guidance for the Solent. Solent Forum*, Hampshire County Council, Winchester, 133 pp.

Solent Forum, 1997. *Strategic Guidance for the Solent. Solent Forum.* Hampshire County Council, Winchester, 200 pp.

Steers, J.A. 1946 & 1964. *The coastline of England and Wales* (1st & 2nd editions). Cambridge, 750 pp.

Tomalin, D.J. 1996. *Towards a new strategy for curating the Bronze Age landscape of Hampshire and the Solent region. Archaeology in Hampshire: a framework for the future.* Hampshire County Council, Winchester, 13-25.

Tomalin, D.J. 1997. Bargaining with nature; considering the sustainability of archaeological sites in the dynamic environment of the intertidal zone. Preserving archaeological remains in situ. In: *Proceedings of the conference of 1st-3rd April 1996 at the Museum of London.* Museum of London/University of Bradford, 144-158.

Tomalin, D.J. 1998. Sustaining a non-renewable resource. In: *Wessex before words; some new research directions for prehistoric Wessex.* Woodward, A. & Gardiner, J. (Eds.) Wessex Archaeology for the Council for British Archaeology, Wessex and the Forum for Archaeology in Wessex. Salisbury, 10-11.

Tomalin, D.J. (this volume). Geomorphological evolution of the Solent seaway and the severance of the Wight: A review. In: *Solent Science - A Review.* Collins, M.B. & Ansell, K. (Eds.), Proceedings in Marine Science Series, Elsevier, Amsterdam.

Tubbs, C. 1991. *The Solent: a changing wildlife heritage.* Hampshire and Isle of Wight Wildlife Trust. Romsey, 40 pp.

Wallace, H. 1996 & 1997. The barrier beach across Bracklesham Bay and the lands it protected: Rumbrug, Medberry and the lost half of Hayling Island: A Roman decorated spring and a medieval mill beneath the mill-pond at Fishbourne, West Sussex. In: *Sea-level and Shoreline between Portsmouth and Pagham for the past 2500 years.* A discussion document, privately circulated.

Webster, T. 1816. Letter VI, (dated Yarmouth June 12, 1811). In: *A description of the principal picturesque beauties, antiquities and geological phenomena of the Isle of Wight.* Englefield, H.C (Ed.), London, 157-163.

# Shoreline Management Plans – A Science or an Art?

Jonathan McCue

Halcrow Group Ltd., Burderop Park, Swindon, Wiltshire, SN4 0QD, U.K.

INTRODUCTION

Within certain scientific circles, Shoreline Management Plans (SMPs) are not commonly brought into serious discussion. It is perceived that they do not really deliver much in terms of new science or findings and do not utilise what science has uncovered to its fullest potential. There is speculation as to how far science used in SMPs has advanced our understanding. There is also speculation as to whether new scientific information is being used effectively in SMPs, or that scientific understanding has been manipulated to merely prop up existing management practices. Overall, sceptics are of the view that SMPs successfully present an artistic view of 'old science', presenting information that has not been contested in recent years or, at the very best, is in need of update; in reality, this is not true. However, as expected, the level of 'new science' or the type of information contained within SMPs is ultimately dependent upon the budgets made available.

SHORELINE MANAGEMENT PLANS

What is their Purpose?

For centuries, the coastline has been a focus for a variety of activities including industry, agriculture, recreation, and fisheries. These are the nation's economic assets and have developed and flourished despite constant changes in the physical characteristics of the coast. The coastline is a national heritage and, in order to sustain it for future generations, proper management of coastal development, including the planning of its defences, is essential. Consequently, there has been a need over recent years to look at the management of the coast from a more strategic perspective.

SMPs are 'non-statutory', as they are not required by law. Nevertheless, for the purpose of any non-statutory document, such as an SMP, suggestions and policy strategies for the future must consider the existing statutory planning system. It is also important that the SMP focuses upon all the issues relative to the shoreline, within the geographical area concerned.

Through the SMP development process that has taken place to date, it is now understood widely that the goal of developing sustainable shoreline management policies (through close dialogue and consultation) is dependent upon achieving a successful balance between the requirements of all the coastal interests in an area. The advent of SMPs mark an important step in achieving this objective, through the elimination of conflicts over the way in which the coastline is protected.

What do they provide?

For the purpose of assigning and implementing shoreline strategies or 'policies', the MAFF guidelines (1995) suggest that the coastline should be divided into 'management units'. These units are sections of coastline that are sensibly consistent, in terms of the coastal processes acting upon them and, in terms of land use and existing features. Each management unit should also have a coastal defence policy, together with preliminary guidelines for future implementation. For each unit, a range of generic options are appraised: 'do-nothing'; 'advance the line'; 'retreat the line'; and 'hold the line'.

Before SMPs are adopted, each coastal defence option is considered in relation to its impact (both positive and negative) upon the various factors that are influenced by, or are influential upon, the condition of the coastline. It is reviewed based on: compatibility with natural processes; implications for the human and natural environments; technical soundness and sustainability; economic viability; and the wider impacts on adjacent shorelines and economies.

What are the Needs of SMP Users?

SMPs address the interests and issues of all the identified authorities and organisations active on the coast. This wealth of information may be of potential use to scientists, planners, engineers, and the public alike (Reeve & McCue, 1997). SMP formats, however, have been criticised as being not 'user friendly' in the way that the various points are put across. Everyday terms used by coastal practitioners, in academia or local authorities, are usually not the same terms used, or understood, by most stakeholders (i.e. interested parties). Therefore, SMPs need to present a common language, free of specific jargon so that all parties involved can comprehend.

In terms of defining key SMP user groups, the main three are scientists, planners, and engineers. 'Scientists' include those studying 'non-anthropogenic' disciplines i.e. biological and physical sciences; 'planners' are those who are ultimately concerned, as Orbach (1996) states with "the governance of human behaviour"; whereas 'engineers' include specialists concerned with the management, design, construction and maintenance of the built environment.

All of these groups play a part in achieving the requirements and needs of the SMP. From this perspective, one main criterion to establish is how SMP users presently communicate with each other. The role of the scientist and the engineer needs to be made clear as, amongst other roles, they need to: inform planners of the likely morphodynamic consequences of policy decisions; and to define appropriate methods of mitigating any adverse event. In practice, such clear-cut lines of responsibility are rarely achieved. A rather more pragmatic view is to consider the role of scientists as providing the decision-makers with the necessary information to ensure that 'appropriate shoreline management' is applied.

What Does the SMP Need?

SMPs need information; this may originate from a desk study of existing scientific data, or through the initiation of new fieldwork. Clear scientific understanding is paramount to the success of the SMP process. One common problem is related to the availability of scientific information. Where SMPs have undertaken new work to understand more clearly certain features (e.g. sediment movements), the additional work has obviously been readily available, and has proven useful when the results have been interpreted. However, SMP budgets commonly do not stretch to undertake new fieldwork, as unfortunately, this is not the normal procedure. Instead, most SMPs have to be reliant on existing data, or are expected to purchase time-series data on a commercial basis. This financial barrier to attaining scientific information has created problems for some of the SMPs that have so far been produced.

A key component in the creation of many SMPs, including the West Solent to Southampton Water Plan, has been the use of Geographical Information System (GIS) i.e. a computer system capable of merging, manipulating and analysing digital data that is held spatially. Through the goodwill generated by the bodies represented on SMP Steering Groups, considerable quantities of data held, much of it in digital format, is being released for use. In the case of the Environment Agency, their holdings of OS raster tiles, extensive archive of aerial photographs and associated beach profile interpretation is invaluable when producing SMPs. Other sources of data from county councils (their database of historic sites, archive of historic and agricultural classification maps) is also desirable, whilst English Nature are able to supply digital conservation site boundary data. An assimilation of these data sets, in conjunction with data already held by the private sector, has enabled such data to be combined and presented at a range of scales and in different formats.

THE PROBLEMS

The Availability of Data

SMPs are reliant on reviewing 'existing' information. As already stated, budgets rarely allow for new research or fieldwork to be undertaken unless specifically requested in the Terms of Reference for the study. Consequently, an initial 'trawl' of scientific data, through contacting a wide range of consultees, is undertaken early on in the SMP process. Whilst replies may be returned stating that relevant information may exist, or research is currently being undertaken, it does not necessarily mean that information will be readily available for use in the SMP. Copyright rules, confidentiality clauses, or financial barriers to data use can arise.

There are also technological and institutional barriers to the delivery of coastal information to those who most need it. Attempting to obtain scientific information with restricted access within institutions or privately funded confidential research, has occasionally been a problem. Where strategic studies, such as SMPs, have not been budgeted to carry out their own new fieldwork or research, this data availability issue is likely to remain a problem. The SMPs prepared to date, within the Solent, have been fortunate enough to benefit from a number of high quality studies carried out, for example, by academic institutions. However, it is of interest to note that most of the SMPs produced are still recommending more specific field studies to be undertaken, to improve knowledge on certain aspects of the coastal environment. If this is the case, perhaps the SMP process should have been delayed until the field studies have been completed. In reality, the process needs to start somewhere; hence, it is argued that research to improve our understanding of the coastline needs to be driven by this process.

Regardless of this dilemma, more information has become available as a result of the setting up of Coastal Groups around the country. As such, genuine attempts have been made to break down barriers between local and national government agencies and commercial agencies, so that pertinent information is made available.

Has 'Science' Alone Helped to Manage the Coast?

In order to assist management decisions, the West Solent and Southampton Water SMP introduced the concept of 'Process Units'. These are sub-units of the study coastline, defined by physical processes and characteristics. The concept of being able to divide up the coast is one that is easily grasped by various user groups and enables stakeholders to focus on aspects of the SMP that have significant geographical relevance to them i.e. where they live, or where their issue of concern lies. Using a scientific philosophy to establish their derivation, the Process Unit approach provides also a logical means by which information can be divided, ordered, and presented. Of greater importance is the basis they provide for the development of sustainable coastal defence strategies.

To make implementation manageable, the Process Units are subdivided into smaller *management units* - each consisting of a stretch of coastline with similar coastal process and land-use characteristics. The coastline of the Western Solent and Southampton Water has been subsequently divided into 51 separate Management Units. By considering data and ultimately managing the shoreline within the framework of Process Units, this advocates that science and coastal processes, in particular, are providing the background against which sustainable strategies for the coast should be developed.

A science theme ultimately runs throughout the preparation of SMPs. This approach was certainly that adopted in the West Solent and Southampton Water SMP, although it is not the only component required in their successful implementation. Commitment from stakeholders is required; this was pursued early in the SMPs preparation, by ensuring that all their concerns and issues were raised and evaluated in addition to the more technical ones posed from original research programmes. In order to achieve this, improved communication using 'jargon-free' messages is needed.

Getting the Point Across

Presentation of Data

Whilst much of SMP development has been prescribed by MAFF guidance and by individual commission 'Term's of Reference', format and presentation have tended to be left to the discretion of Consultants. No official standard format has been prescribed for SMP production; however, it is inevitably the case that, commonly, budget availability dictates the usability and graphical quality of the end-product.

Where Authorities have made it clear that they would not be excessively influenced by price, when assessing SMP tenders, it can be argued that freedom in interpretation has probably been beneficial to the presentation of the end-products. Aspects such as OS Royalties and colour-copying are somewhat expensive; however, they do add interest and clarity to documents, particularly when scientific information is being put across to the public or non scientist. Therefore, improved communication should be an underlying objective in future SMP productions.

Public Consultation

Consultation can either be carried out by pro-active, but time-consuming methods (such as setting-up meetings with key individuals or groups), or by less pro-active methods (such as standard mailshot methods) arguably, as effective. Its success is dependent upon the objectives of the exercise, the types of stakeholders involved and the enthusiasm put in by the Project Team.

The question of consultation is always difficult when dealing with strategic documents. In terms of the West Solent SMP, it was decided to limit consultation on Stage 1 because it simply reported on the data collection phase of the study. Instead more emphasis was placed upon the second and final stage; these set out the proposed strategic defence options. In this way, the Plan's time-scale is tightly controlled, whilst allowing both key consultees and the public a reasonable period to comment.

The success of the consultation exercise for the West Solent and Southampton Water SMP, unfortunately, ranks very closely to other SMPs that have taken place elsewhere. Lukewarm responses appear to be the normal reaction, regardless of the amount of effort. A variety of consultation methods was attempted, including radio and newspaper advertisements. 2000 copies of a glossy Consultation Summary were issued to the Coastal Group members for distribution ahead of a series of consultation meetings held at Calshot Spit Activity Centre (in July 1997). The content of these pamphlets was accessible and well written; they avoided being too 'academic' or full of 'government speak' and the balance of academic information to informative illustration seemed to work well, according to responses received. Undoubtedly, to date this type of approach has proved by far to be the most successful.

The 'art' of conflict resolution

The 'art' of conflict resolution and knowing how to effectively involve local communities, in the overall process of coastal and shoreline management, is an important concept to evaluate; without this, the long-term success and sustainability of the policies proposed is brought into question. Issues such as urban encroachment within the coastal zone and conservation requirements often result in inevitable 'multiple-use conflict'. Concentrating upon the conservation of natural assets here, a few examples are described to exemplify how science and stakeholder co-operation is needed to resolve specific issues (or conflicts) that have arisen.

Bray & Hooke (1995) have discussed the need for the adoption of special approaches to conserve dynamic coastal landforms and their associated habitats, along with soft geological features. It is accepted generally that this approach is preferred over the long-term, but in the meantime, conflicts of interest will undoubtedly arise. For example, it is rare for geological conservation objectives (relevant to both hard and soft coastal sites) to be isolated from other issues of relevance to the coast, such as archaeological preservation or natural habitat protection. Hence, SMPs must seek to address all issues when assigning policies and strategies for lengths of the coastline (McCue, 1998).

Interesting examples of important geological conservation sites on the south coast include, amongst many others, those at Barton-on-Sea and at Hengistbury Head (refer to *Maps A and B* in the *Preface*, for general location and main sites referenced). The cliffs around Barton-on-Sea in Hampshire are renowned for their Tertiary fossils and are still visited by many tourists and amateur fossil hunters each year. The problem at Barton is the need to protect land that has experienced an accelerated erosion rate, as a consequence of piece-meal coastal engineering schemes down-drift, whilst seeking to preserve the geological conservation interest in the site.

A similar classic 'multi-user' management problem occurs to the west at Hengistbury Head; this is an exposed site, experiencing erosion and cliff retreat. The Head affords valuable protection to Christchurch Harbour and the surrounding lowlands, as well as being important for its environmental, archaeological and geological qualities. Through continual consultation with the key stakeholders at Hengistbury Head, an approach has been agreed to balance the various needs. A dynamic beach management approach was introduced to help resolve the problems at the site; this involves maintaining the beach at a critical width, such that marine erosion of the cliff toe is controlled (but not completely halted).

In addition to the need for latest scientific information, conflicts of interest such as those outlined above will not be resolved in the SMP unless improved co-ordination, communication and co-operation techniques are installed amongst the key stakeholders involved. The 'art' of improved communication goes beyond the realm of pure science yet plays arguably just as important a role.

LESSONS LEARNT FROM THE SMP PROCESS.

Learning from the experienced gained from each others problems is vital in the future, so much so that adjacent SMPs within the Solent should be listed as obligatory reference documents; this would ensure compatibility, in any subsequent revisions to the strategic options (Beech & Nunn, 1996). Important lessons have been learned from the SMP exercise, as outlined below.

Data Presentation and Dissemination

Improved data presentation and dissemination is a key aspect in the success of future shoreline management planning. However, together with improved data transfer and availability (new information technology and a wider understanding of the SMP process), there is some uncertainty over the amount of data required; this, in turn, makes it difficult for managers and decision-makers to judge the quality and relevance of observations.

With the SMP's undertaken to date, disparate authorities and bodies have pooled information. By presenting and handling such data in digital form, there are benefits to be had from being flexible in the way that data is presented. Therefore, it is recommended that Operating Authorities be encouraged to establish information systems that achieve their objectives and those of more regional groups (such as the Standing Conference on Problems Associated with the Coastline (SCOPAC)). For example, in conjunction with other departments, Local Authorities may well be able to justify the adoption of GIS. Much of the information being gathered is potentially relevant to wider interests, than purely coastal defence, such as emergency planning, pollution control and nature conservation.

Whilst the GIS option may prove an excessive burden for some, continually improving PC-based data handling systems are becoming a preferred option. Data handling systems, such as Halcrow's Shoreline and Nearshore Data System (SANDS) have the capacity to hold and analyse the new and existing data, as well as carry much of the broader data generated by SMP's. The potential for storage of SMP data, and ultimately the full SMP, on such systems is likely to benefit not only the SMP process in the future, but also Operating Authorities as changes arise, whether through the availability of new data or review of preferred strategies. This approach would enhance the status of the SMP as a 'living' document; as such, this would move away from the need for major reviews of the Plan, together with the associated cost.

Improving Coastal Education

Improving communication and raising awareness is seen as an important objective of SMPs that is arguably not paid enough attention. Planners, engineers and scientists all play a role in SMP generation either directly or indirectly, though with all, the art of communication (to each other and to the wider public to improve coastal understanding) is a commonly-overlooked issue. Whilst SMPs do not have a remit to improve coastal education *per se*, there would be overall benefits if information and interpretation is improved for coastal stakeholders and the public alike. Terms such as 'managed retreat' are not clear in most peoples' minds; such uncertainty needs rectifying if such strategic options are to be widely accepted by the public.

Facilitating Discussion and Communication

Various non-statutory plans have been and/or are being prepared for the Solent coastal area; all these seek to manage its many assets, in an integrated and sustainable manner. The scope and methods adopted for the production of these plans are often similar, containing several areas of overlap. Given this overlap, there have to be benefits in attempting to ensure co-ordination between the plans, to avoid duplication and conflict. In this way, confusion between consultees should be minimised; likewise, it should promote existing partnerships with statutory and non-statutory bodies. The formation of the Solent Forum, in 1992, has proved to be very successful in breaking down cross-sectoral barriers in the area. With a membership of over 50 organisations and bodies (Inder, 1996), the Solent has undoubtedly benefited as a result of this initiative.

Prioritise Research Requirements

One of the main recommendations arising from SMPs has been the need to initiate a suite of coastal research programmes. With reference to the West Solent and Southampton Water SMP, the key recommendation for research relates to the need for a regional Saltmarsh Management Strategy for the SCOPAC Region. It is recognised that areas of saltmarsh require careful management, if they are to be retained as beneficial conservation areas and for their natural defence capability. Until more scientific information is available, on the reasons behind saltmarsh loss and its role in sustainable shoreline management, the implementation of coastal defence options will remain uncertain along stretches of the coastline.

CONCLUSIONS

In conclusion, science plays an important role (but is by no means the only criterion) in SMP production. The science must be presented and integrated in such a way that is immediately accessible to decision-makers along the coast. Together with improved scientific knowledge is the need to improve coastal education, conflict resolution techniques, and a better understanding of 'social' science considerations. Without this SMPs and more multi-disciplinary integrated coastal management plans will not be as effective.

Therefore, SMP success is dependent equally upon improving communication between the public, conservationists, and engineers, as well as influencing coastal decision-makers in authority by forging working partnerships, disseminating information and promoting positive management. Only then will sustainable management objectives be implemented successfully and attained. It is with anticipation that, following review of the first round of SMPs for the English coastline, an appreciation of present limitations are acknowledged and rectified. If this philosophy is embedded into future SMPs for the SCOPAC region, then we will have gone a considerable way to achieving our goals.

REFERENCES

Beech, N.W. & Nunn, R. 1996. Shoreline Management Plans – the next generation - In *Partnership in Coastal Zone Management*. Taussik, J. & Mitchell, J. (Eds.), Samara Publishing Ltd., 345-353.

Bray, M.J. & Hooke, J.M. 1995. Strategies for Conserving Dynamic Coastal Landforms. In: *Directions in European Coastal Management* Healy, M.G & Doody, J.P (Eds.), Cardigan: Samara Publishing Ltd., 275-290.

Inder, A. 1996. Partnership in Planning and Management of the Solent – In: *Partnership in Coastal Zone Management.* Taussik, J. & Mitchell, J. (Eds.), Samara Publishing Ltd., 405-413.

Ministry of Agriculture Fisheries and Food (MAFF). 1995. *Shoreline Management Plans - a Guide for Coast Defence Authorities,* London.

McCue, J.W. 1998. "Sense and sustainability" – achieving geological conservation objectives as part of the present Shoreline Management Plan process. In *Issues in Environmental Geology: a British Perspective.* Bennett, M.R. & Doyle, P. (Eds.), 381-399.

Orbach, M.K. 1996. Science and policy in the coastal zone. In: *Partnership in Coastal Zone Management.* Taussik, J. & Mitchell, J. (Eds.), Samara Publishing Ltd., 7-15.

Reeve, D.E. & McCue, J.W. 1997. - Shifting Sands off East Anglia – Understanding their importance for Appropriate Future Shoreline Management. In: *Tidal 96 – Interactive Symposium for Practising Engineers* – 12 November 1996, Brighton University.

SECTION 2

Short Contributions

Late Pleistocene/Holocene Evolution of the Upstream Section of the Solent River

A.F. Velegrakis, J.K. Dix and M.B. Collins

School of Ocean and Earth Science, Southampton Oceanography Centre, University of Southampton, European Way, Southampton, SO14 3ZH, U.K.

INTRODUCTION

During the Pleistocene lowstands, a major river (the "Solent River") was present in the area of the modern Solent (West, 1980). This river flowed along a west-east trending watershed incised on Tertiary sediments and surrounded by high Chalk country. Most of this watershed was inundated during the Flandrian sea-level rise and only parts of its tributary rivers are still intact, forming the modern drainage network of the area (Figure 1).

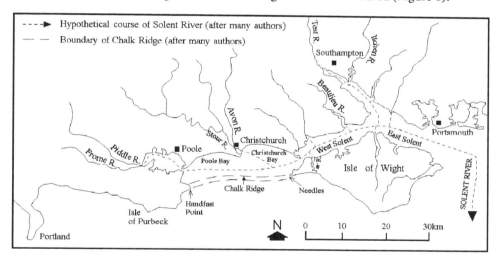

Figure 1 Hypothetical offshore course of the Solent River and the offshore trace of Purbeck-Wight Chalk Ridge.

DATA ACQUISITION AND ANALYSIS

Shallow seismic and echo-sounder profiles (560 line km) were collected in Poole and Christchurch Bays, in 1990 and 1991. The morphology and elevation of the upper bedrock erosional surface have been estimated, by combining the overall thickness of the sediments with the water depth measured at each the same position (tidally-corrected and reduced to OD Newlyn).

RESULTS AND DISCUSSION

The sediment thickness of the area was found to be generally limited; sediment deposits of significant thickness are associated only with the modern tidal deltas, offshore sand ridges and buried valleys. These sediments rest upon a bedrock erosional surface of complex relief. Seven palaeovalleys were identified (Figure 2). In Poole Bay, Palaeovalleys I, II, and III appear to cut through the Purbeck-Wight Ridge. In contrast, Palaeovalleys IV, V, VI and VII in Christchurch Bay do not appear to cut through the Ridge. Valley-infilling sediments of significant thickness are found only within Palaeovalleys I and II; in the remainder of the palaeovalleys, such sediments are either missing, or have small thickness. Estuarine transgressive sediments are observed only within Palaeovalleys I and II (Velegrakis *et al.*, 1999).

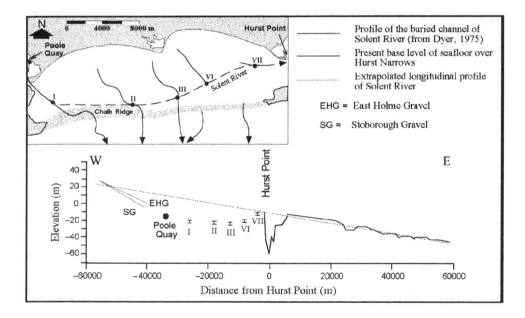

Figure 2 The upstream extrapolation of the longitudinal profile of the Solent River System (after Dyer, 1975), the elevation of the Poole Quay deposits (after Nicholls, 1987), the elevation of the late Devensian terraces of the Frome River, as shown by East Holme Gravel and Stoborough Gravel (after Allen and Gibbard, 1993), and the elevation of the palaeovalley thalwegs. The elevation of the channels has been taken at their junction with the reconstruction of the Solent trunk valley (see inset).

Allen and Gibbard (1993) identified pre-Flandrian gravel terraces in the Frome and Avon-Stour valley systems. The elevations of the two youngest of the Frome terraces (the East Holme and the Stoborough Gravels) appear to correlate well with both the elevation of the base level of the Poole Quay buried channel (Nicholls, 1987) and the base level of the trunk valley of Palaeovalley I (Figure 2). Therefore, it is likely that the age of the Palaeovalley I is similar to that of these terraces. The nature of the preserved sedimentary infill of the Palaeovalleys I and II indicates a relatively calm transition, from fluvial to marine sedimentary environments, for Poole Bay during the late stages of the Flandrian Transgression. Conversely, the nature and infilling sediments of the palaeovalleys observed in Christchurch Bay show evidence of an abrupt transition. The exact date of the embayment inundation (and the separation of the Isle of Wight from the mainland) cannot be concluded from the available data. However, if the elevation of the bedrock erosional surface over eastern Christchurch Bay is compared with local sea-level curves, then that event must have taken place not later than 7,000 to 7,500 BP (Velegrakis *et al.*, 1999).

CONCLUSIONS

The upper reaches of the Solent River system were irreversibly disrupted before the Flandrian Transgression, as a result of the breaching of the southern barrier of the system (the Purbeck-Wight Chalk Ridge) by three southerly-flowing rivers. Poole Bay was the first to be submerged. The submergence of Christchurch Bay took place at a later time *c.* 7,000-7,500 BP.

REFERENCES

Allen, L.C. & Gibbard, P.L. 1993. Pleistocene evolution of the Solent River of Southern England. *Quaternary Science Reviews,* **12**: 503-528.

Dyer, K.R. 1975. The buried channels of the 'Solent River', Southern England. *Proceedings of Geological Association of London,* **86**: 239-245.

Nicholls, R.J. 1987. The evolution of the upper reaches of the Solent River and the formation of Poole and Christchurch Bays. In: *Wessex and the Isle of Wight: Field Guide*, Barber K.E. (Editor). Quaternary Research Association, Cambridge, Special Publication: 99-114.

Velegrakis, A.F., Dix, J.K. & Collins M.B. 1999. Late Quaternary evolution of the upper reaches of the Solent River, Southern England, based upon marine geophysical evidence. *Journal of the Geological Society of London,* **156**: 73-87.

West, I.M. 1980. Geology of the Solent Estuarine system. *In: The Solent Estuarine System: An Assessment of the Present Knowledge.* N.E.R.C. Publications, Series C, No **22**: 6-19.

Sea-Level Rise in the Solent Region

M.J. Bray, J.M. Hooke and D.J. Carter

Department of Geography, University of Portsmouth, Buckingham Building, Lion Terrace, Portsmouth, PO1 3HE, U.K.

INTRODUCTION

Sea-level has been a major factor in the evolution of the Solent. Rapid post-glacial sea-level recovery, between 15,000 and 5,000 years BP, inundated the system and continuing rising sea-levels control contemporary biogeomorphological environments; similarly, they pose a potential threat to human occupation and uses. Concerns relating to the future effects of climate change and sea-level rise have led to several studies specific to the Solent region (Ball *et al.*, 1991; Bray *et al.*, 1992, 1994, 1997).

SEA-LEVEL RISE

Local rates of relative sea-level rise (RSLR) have been developed from a variety of sources, including tide gauge data analyses and dated sedimentological and archaeological sea-level indicators summarised by Bray *et al.* (1994 and 1997). The latter reveal mean RSLR of 1-3 mm a^{-1} over the past 6,000 years, due to slow crustal subsidence. Analysed local tide gauge data are sparse, cover short periods and indicate a range of 2-5 mm a^{-1}. Reliable analyses from long-established gauges located well outside of the Solent suggest 2 mm a^{-1}. The latest updated sea-level trends, from the UK national network of tide gauges, are available on the Internet[1]. Measurements of recent marsh sedimentation indicate RSLR of 4-5 mm a^{-1}, assuming equilibrium between accretion and RSLR (Cundy & Croudace, 1996). RSLR is predicted to accelerate significantly in the future, due to the effects of global warming. Latest medium-high scenario modelled estimates indicate a rise of 34 cm, or 6.5 mm a^{-1}, by 2050 (two to four times more rapid than at present) assuming a conservative estimate for local subsidence. Other estimated impacts of climate change for the UK include increased winter gales and rainfall (Hulme & Jenkins, 1998). Significant uncertainties remain attached to these estimates, although an increasing consensus has developed between scientists over the past eight years.

MAIN IMPACTS

The main impacts of RSLR are likely to be an increased frequency of extremely high sea-levels and potentially increased flooding of low-lying land and erosion of beaches (Bray *et al.*, 1997). Saltmarshes and wetlands are very sensitive, especially where they are backed by defences and vulnerable to "coastal squeeze," i.e. becoming trapped as narrowing margins between rising sea-levels and inflexible defences. Marshes around many of the estuaries and harbours of the Solent are backed by defences and are already suffering marsh erosion

---

[1] http://www.nbi.ac.uk/psmsl/psmsl.info.html

following 'die back' of *Spartina* vegetation (Haynes and Coulson, 1982). Many relict or marginally stable cliffs around the Solent could become unstable, due to increased preparatory toe erosion and higher triggering winter rainfall. Cliff erosion sediment yields are likely to increase (Bray & Hooke, 1997), providing valuable contributions to beaches in adjoining areas.

IMPLICATIONS FOR FUTURE PLANNING

Climate change and RSLR dictates that continued priority should to be given to addressing flooding and erosion problems, the development of monitoring and warning systems and the planning of future land uses to avoid hazardous areas (Bray *et al.*, 1997). Where developed lands are threatened, it should remain feasible to provide reliable protection by improving defences (Ball *et al.*, 1991). Full protection of less densely populated agricultural areas may not be sustainable, or desirable, and alternative policies of managed retreat or realignment are applicable in some areas to conserve habitats and facilitate 'natural' adjustments to RSLR. Regional research should aim to detect local impacts, through long-term monitoring of tidal levels, waves, and shorelines. Results should be reviewed periodically to differentiate trends from natural variations. The improving precision of regional climate change estimates, which are likely to become available in the future, should provide opportunities for enhanced prediction of impacts.

REFERENCES

Ball, J.H., Clark, M.J., Collins, M.B., Gao, S., Ingham, A. & Ulph, A. 1991. *The Economic Consequences of Sea-level Rise on the central South Coast of England*, Report to MAFF. 2 Volumes. Geodata Institute, University of Southampton: 140 pp.

Bray, M.J., Carter, D.J. & Hooke, J.M. 1992. *Sea-Level Rise and Global Warming: Scenarios, Physical Impacts and Policies.* Portsmouth Polytechnic. Report to SCOPAC, 205 pp.

Bray, M.J., Hooke, J.M. & Carter, D.J. 1994. *Tidal Information: Improving the Understanding of Relative Sea-level Rise on the South Coast of England.* University of Portsmouth. Report to SCOPAC, 86 pp.

Bray, M.J., Hooke, J.M. & Carter, D.J. 1997 Planning for sea-level rise on the south coast of England: advising the decision-makers. *Transactions of the Institute of British Geographers*, NS, **22**: 13-30.

Bray, M.J. & Hooke, J.M. 1997 Coastal cliff prediction with accelerating sea-level rise. *Journal of Coastal Research,* **13** (2): 453-467.

Cundy, A. & Croudace, I. 1996. Sediment accretion and recent sea-level rise in the Solent, southern England. *Estuarine, Coastal and Shelf Science*, **43**: 449-467.

Haynes, F.N. & Coulson, M.G., 1982. The decline of *Spartina* in Langstone Harbour, Hampshire. *Proceedings of the Hampshire Field Club, Archaeological Society*, **38**: 5-18.

Hulme, M. & Jenkins, G. 1998. *Climate Change Scenarios for the UK*. UKCIP Technical Report No.1, 80 pp.

Littoral Sediment Transport Pathways, Cells and Budgets within the Solent

M. J. Bray, J. M. Hooke, D. J. Carter and J. Clifton

Department of Geography, University of Portsmouth, Buckingham Building, Lion Terrace, Portsmouth, PO1 3HE, U.K.

INTRODUCTION

This paper summarises a study in which present knowledge of the pathways of coarse sediment transport around the Solent shoreline has been compiled (Bray *et al.*, 1998). The net-long term directions of sand and shingle drift, within littoral cells and sub-cells between transport boundaries, have been identified using the methods of Bray *et al* (1995). Hence, an understanding of the budgets of these materials and of the operation of the coast as a sediment system has been developed. The analysis demonstrates, to managers, the value of a regional context within which to develop their work and identifies the uncertainties requiring further research.

METHODS

The major part of this synthesis is based upon work completed for the regional group of local coastal defence authorities SCOPAC (Standing Conference on Problems Associated with the Coastline). The study involved the use of their sedimentation database to identify relevant information; this was then extracted, analysed for reliability, and mapped (Bray *et al* 1991). This work has now been updated for the Solent, following review of the results of the three recently completed Shoreline Management Plans and other studies concerning sediment inputs to the coast (Posford Duvivier, 1997). The overall work is based upon the results of numerous research and consulting studies, demonstrating the extent of understanding available within the Solent. Results are presented as maps, accompanied by data tables identifying budget elements e.g. cliff erosion inputs, storage volumes etc. Analysis involves evaluation of the overall budget, especially the importance of contemporary sources, and the effects of management activities. For general location and main sites referenced, refer to *Map A* and *Map B* in the *Preface*.

CLIFF EROSION SEDIMENT INPUTS

Marine erosion and coastal landsliding result in detachment of geological materials from cliffs and delivery to the shore, where sediments may contribute to the beach, be transported alongshore (coarse gravels and sands), or be removed offshore in suspension (fine sands and clays). Sediment inputs may be quantified from knowledge of the cliff elevation, its rate of recession (historical map and/or air photography comparisons) and knowledge of the geological materials outcropping. Such studies reveal that overall, cliff inputs are substantial from the exposed Christchurch Bay and Isle of Wight coasts, but much reduced within the sheltered waters of the Solent, where the erosion is less intense.

An exception is the Isle of Wight coast of the Western Solent, where the clayey Tertiary strata form high cliffs that degrade rapidly by landsliding. The modest wave action observed is sufficient to remove clay debris from the foreshore, thus maintaining slope instability, recession and sediment inputs. The clays supplied are likely to be important sediment sources for the Solent mudflats and saltmarshes. The significance for the Solent, of the large sediment yields from the open coasts, remains uncertain.

BEACH NOURISHMENT (INPUTS TO SHORELINE SYSTEM)

Over the past 20 years, a growing number of coastal defence schemes have involved imports of replenishment materials to stabilise eroding beaches. Marine gravels are preferred, due to their durability. This process introduces new material to the shoreline sediment transport system and contributes positively to the budget. The Lee-on-the-Solent and Hurst Spit schemes utilised local material, which would otherwise have been unavailable to the shoreline. Measures are usually required to control losses and to recycle, or periodically top-up, the beaches as material, is transported. A potential limitation of this technique is the need to find suitable gravel deposits that can be dredged economically, without resulting in adverse impacts elsewhere.

PRINCIPAL COARSE SEDIMENT STORES AND SINKS

The characteristics and size of the principal sediment stores and sinks have been summarised (Bray *et al.*, 1998). Knowledge is based upon the results of numerous studies which have applied techniques such as historical map and chart analysis, sediment sampling, hydrographic and side-scan survey, boreholes and vibracoring. The major sinks within the Solent are thought to result from accumulations which began some 6500-8400 years ago, when the Solent was inundated by rising sea-levels and the present coastal regime was initiated. Many stores, such as those around the East Solent and the Harbours, are more recent. These are thought to result from extensive reworking of sediments from the floor of the English Channel and Solent approaches, then driven into the Solent as barrier beaches by rising sea-levels over the past 10,000 years. Overall, well in excess of 100 million m^3 of potentially mobile shingle and sand has become stored within the Solent, making it a major sink area on the South coast of England. Major accumulations such as the Brambles Bank, Ryde Sands and the East Winner Bank remain unquantified. Some 15-25 million m^3 of material have been dredged from these inshore deposits, to maintain navigable channels and provide materials for reclamation and the aggregates industry.

LITTORAL CELLS AND SHORELINE MANAGEMENT PLANS ALONG THE SOUTH COAST OF ENGLAND

Littoral sediment cells and sub-cells comprise relatively self-contained units, within which coarse sediment circulates have been mapped according to the continuity of shoreline sediment transport. Major cell boundaries are identified where transport is permanently (50-100 yrs) and completely interrupted *e.g.* major hard rock headlands and inlets. Sub-cell boundaries are identified, where the effects are partial and/or time dependent. Boundaries separate those parts of the coast that are linked from those that are independent and where interruption of transport should not have a significant effect upon neighbouring cells. They form valuable units, within which sediment budgets may be compiled and management be carried out. Shoreline Management Plans (SMPs) are developed within the major cells, with slight adjustment of boundaries to accommodate estuaries and fine sediments. Sub-cells define units of interlinked processes and form discrete process units, within the SMPs. Specific management objectives are formulated for each unit, based upon knowledge of its characteristics. Future management options developed within SMPs are checked for their compliance with these processes, operating within the relevant process unit. In this manner, links are developed between the understanding of the physical environment and the requirements of management. The method appears successful for the management of beaches within the Solent region, but less so in the case of mudflats and saltmarshes sustained by the cycling of fine sediments; these are somewhat unaffected by shoreline cell boundaries and require more regional approaches.

SEDIMENT TRANSPORT PATHWAYS WITHIN THE SOLENT AREA

Principal sediment transport pathways derived from local area studies, *e.g.* Christchurch Bay, have been mapped to reveal the net long-term sources, transport routes, storage areas and final destinations (sinks) of shoreline sediments within the Solent. Cliff erosion and beach replenishment are identified as important inputs of sediment that sustain beaches along transport pathways and ultimately contribute to stores and sinks. The pathways mapped reveal that sediments are transported westward into the Eastern Solent, whereupon they are cycled within beach, spit and tidal delta systems associated with the harbour inlets before becoming deposited in major sinks. In contrast, relatively little coarse materials enter the West Solent from the open coast, with the major sink (Shingles Bank) being located outside - a function of the presence of Hurst Spit and ebb tidal flow dominance at the Hurst Narrows inlet. Many beaches throughout the region have become depleted, due to the effects of coastal defences in stabilising eroding cliff sediment sources and intercepting sediment transport pathways.

CONCLUSIONS

Complex circulations of coarse sediment, involving reversals of drift direction and exchanges with nearshore stores, are identified within the Solent. Primary controls relate to the availability of materials (Holocene inheritance and local erosion), the shelter afforded by the Isle of Wight, and dominant ebb tidal flows at inlets to the harbours. Mobile, coarse, open coast sediments have penetrated into the East Solent, to a much greater extent than they have the Western Solent. Local cliff erosion remains important, in sustaining some low flux transport pathways within the West Solent and inner parts of the East Solent. However, human activities including inshore dredging and coastal stabilisation have significantly affected some pathways and have contributed to problems of beach erosion. In a growing number of cases, restoration has been achieved by replenishment schemes. The cells and pathways identified have established a physical basis for the development of SMPs, to deliver sustainable long-term management of shorelines throughout the Solent.

FURTHER RESEARCH

The uncertainties outlined below are identified as requiring further research:
1. Exchanges between the shoreline and the inshore deposits.
2. Beach volume changes (indicative of changes in transport rate or sediment supply).
3. Quantification of the volumes of all the significant stores and sinks.
4. Evaluation of the sources, sinks and transport pathways for fine sediments.
5. The long-term and wider effects of SMP-recommended strategic coastal defence options.
6. Separation of short-term variability in behaviour, from long-term trends, especially within the context of future climate change.

REFERENCES

Bray M.J., Carter, D.J. & Hooke, J.M. 1991. *Coastal Sediment Transport Study* (5 Volumes). Department of Geography, Portsmouth Polytechnic. Report to SCOPAC, 535 pp.

Bray, M.J., Carter, D.J. & Hooke, J.M. 1995. Littoral cell definition arid budgets for central southern England. *Journal of Coastal Research,* **11**(2): 381-400.

Posford Duvivier. 1997. *Sediment Inputs Research Project. Phase 2: Cliff Erosion.* Posford Duvivier, Consulting Engineers, Report to SCOPAC, 141 pp.

Bray, M.J., Hooke, J.M., Carter, D.J. & Clifton, J. 1998. *Sediment Transport Pathways, Cells and Budgets within the Solent.* Poster presented at Solent Science Conference, Southampton Oceanography Centre, 21-22 September, 1998 (available upon request from the author).

My Collins and K. Ansell (editors)
© 2000 Elsevier Science B.V. All rights reserved.

Residual Circulation and Associated Sediment Dynamics in the Eastern Approaches to the Solent

D. Paphitis, A.F. Velegrakis and M.B. Collins

School of Ocean and Earth Science, Southampton Oceanography Centre, University of Southampton, Southampton, SO14 3ZH, U.K.

INTRODUCTION

In tidal coastal environments, non-linear interactions of the tidal flow with the physiographic features of the coastline and/or the seabed give rise to complex residual flows. Wakes and eddies, having scales of several kilometres, have been observed around islands and headlands (Pingree and Maddock, 1985; Pattiaratchi, *et al.*, 1986; Hatayama *et al.*, 1996). Residual eddies have been found, elsewhere, to be associated with extensive sand deposits on the seabed, often referred to as convergence zones (e.g. Ferentinos & Collins, 1979). This contribution is an attempt to study the spatial distribution of unconsolidated sediments and associated transport pathways in the Eastern Approaches to the Solent, a coastal environment where tidally-induced secondary circulation has been observed.

DATA ACQUISITION AND ANALYSIS

Data obtained from two deployments (1995 and 1996) of the Ocean Surface Current Radar (OSCR), together with a two-dimensional (2-D) mathematical model (Salomon & Breton, 1993), were used to study the spatial and temporal distribution of the surface and depth-averaged tidal currents over the area, respectively. A data set consisting of 127 surficial sediment samples and side-scan sonar observations (collated/collected) was also analysed, to provide information on the seabed morphology and the prevailing dynamics. Using the sand-size fraction of the bed sediments of the area (medium to fine-grained and moderately to well-sorted), a grain size trend analysis (based on the Gao & Collins (1992) technique) was undertaken.

RESULTS AND DISCUSSION

The output from the 2-D hydrodynamic model, together with the harmonic analysis of the results of the OSCR deployments, reveal the presence of a tidally-induced residual circulation (eddy) in the vicinity of Selsey Bill. This eddy is (probably) a headland-associated eddy, resulting from vorticity generated by frictional effects due to the presence of the Selsey Bill headland, which protrudes into the tidal flow: such conditions favour flow separation, resulting in the formation of the residual eddy.

A complex spatial distribution of lithological types (mainly gravels, sandy gravels and sands) characterises the area under investigation (Figure 1). A significant sand deposit (Medmerry Bank, see Figure 2) is found at the eastern margin of the area, located below the observed eddy; this may have provided the necessary hydrodynamic conditions, favouring the initial accumulation of sand i.e. formation at the small sandbank here. As the Medmerry Bank developed, its interaction with the tidal flow may have further enhanced (maintained) the eddy. The proposed mechanism for the formation and maintenance of the Medmerry Bank is similar to that proposed by Pingree (1978).

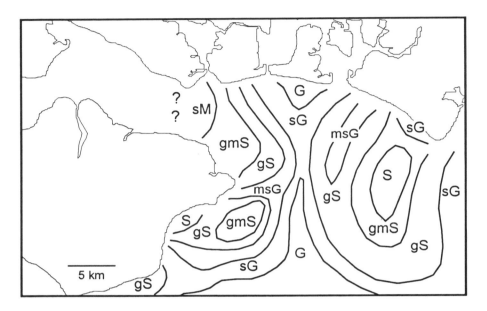

Figure 1 The spatial distribution of lithological types in the area under investigation.

Two independent approaches based on grain-size trends and bedform asymmetries have provided information on the general sand (bedload) transport pathways of the area (Figure 2). Sediment convergence was identified over two regions; these two depositional centres are associated with the Medmerry Bank and Princessa Shoal. An offshore (southerly) transport, exporting material out of the tidal harbours (i.e. Portsmouth, Langstone and Chichester) was identified; this is probably the result of the strong ebb tidal flow at the harbour entrances. Material from the East Solent appears to be transported offshore along the eastern coast of the Isle of Wight. Along the eastern margin of the area, easterly transport was also identified.

CONCLUSIONS

The dominant hydrodynamic feature identified in the area is an eddy located over its eastern margin; this appears to be a headland-associated eddy. This eddy may be related with the formation and maintenance of a small sandbank (Medmerry Bank). Bedload sand transport pathways appear to be converging towards this particular area.

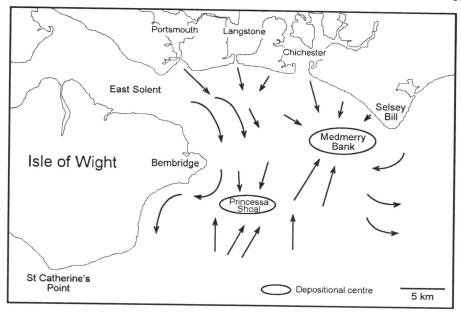

Figure 2 Bedload transport pathways based on grain-size trends and bedform asymmetries (cf. Figure 1).

REFERENCES

Ferentinos, G. & Collins, M.B. 1979. Tidally induced secondary circulations and their associated sedimentation processes. *Journal of the Oceanographical Society of Japan*, **35**: 65-74.

Gao, S.& Collins, M.B. 1992. Net sediment transport patterns from grain-size trends, based upon definition of "transport vector". *Sedimentary Geology*, **80**: 47-60.

Hatayama, T., Awaji, T. & Akitomo, K. 1996. Tidal currents in the Indonesian Seas and their effect on transport and mixing. *Journal of Geophysical Research-Oceans*, **101** (C5): 12353-12373.

Pattiaratchi, C., James, A., & Collins, M. 1986. Island wakes and headland eddies: a comparision between remotely sensed data and laboratory experiments. *Journal of Geophysical Research*, **92** (C1): 783-794.

Pingree, R.D. & Maddock, L. 1985. Rotary currents and residual circulation around banks and islands. *Deep Sea Research*, **32** (8A): 929-947.

Pingree, R.D. 1978. The formation of The Shambles and other banks by tidal stirring of the sea. *Journal of Marine Biological Association U.K.*, **58**: 211-226.

Salomon, J.C. & Breton, M. 1993. An atlas of long-term currents in the Channel, *Oceanologica Acta*, **16** (5-6): 439-448.

Seabed Mobility Studies in the Solent Region

A.F. Velegrakis[*], A.H Brampton[+], C.D.R Evans[++] and M.B. Collins[*]

[*] School of Ocean and Earth Science, Southampton Oceanography Centre, University of Southampton, European Way, Southampton, SO14 3ZH, U.K.

[+] HR Wallingford Ltd, Howbery Park, Wallingford, OX10 8BA, U.K.

[++] British Geological Survey, Keyworth, Nottingham, NG12 5GG, U.K.

INTRODUCTION

The nearshore environment is of considerable interest to Local Authorities, as well as the shipping, fishing, recreation and marine aggregate industries; thus, conflicts of interest may occur. One of the major concerns is related to the extraction of marine aggregates from the nearshore areas, which has increased steadily over the past years (Humphreys et al., 1996). Thus, a non-statutory procedure, the Government View, has been developed to address these concerns (Hughes, 1996). An important part of the procedure is related to the assessment of the impacts of the offshore dredging on the prevailing hydrodynamic regime, the sediment transport and the morphodynamics of the adjacent coastlines. In 1992, a SCOPAC[1]/HR Wallingford funded study on the mobility of the offshore sediments was carried out for the area to the east of the Isle of Wight (Figure 1, area A). In 1996, a similar study, funded by BMAPA[2], CIRIA[3], DETR[4], English Nature, MAFF[5], SCOPAC and the Westminster Dredging Co., was undertaken for the area to the west of the Isle of Wight (Figure 2, area B). This contribution summarises the results of these studies.

METHODS

During both studies, the research approach included: (i) analysis of both previously-available and purposely-collected information, on the seabed morphology and sediment distribution and the hydraulic regime; (ii) tidal and wave modelling; and (iii) estimations of the seabed mobility, based upon the results of the hydrodynamic modelling and theoretical, laboratory and field estimations of certain parameters, e.g. the bed roughness and the critical shear stress for the initiation of sediment movement. As wave/current interaction may result in significant changes of the seabed mobility, its effect was also studied. A different methodology was used in the two studies. For area A, wave/current interaction was investigated using the probability distributions of the tidal velocity (generated using the FLUXMANCHE tidal numerical model (Salomon & Breton, 1991) and the seabed wave

---

[1] Standing Conference on Problems Associated with the Coastline
[2] British Marine Aggregates Producers Association
[3] Construction Industry Research and Information Association
[4] Department of the Environment, Transport and the Regions
[5] Ministry of Agriculture, Fisheries and Food

orbital velocity (estimated from the wave climate). These distributions were combined to produce an exceedance curve of the bed shear stress (Brampton, 1993). To estimate the sediment transport rates for area B, the SANDFLOW-2D model (Brampton *et al.*, 1998) was used. The model utilised, as input, the tidal currents simulated by the TELEMAC-2D model (Hervouet, 1991). The HR TELURAY model was used for the inshore transformation of the waves.

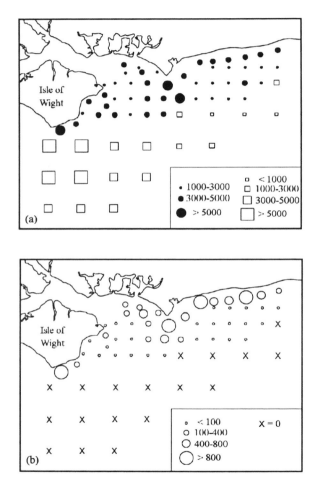

Figure 1 Mobility (in hours/year) of (a) sand under tidal action (□) and waves and currents (•); and (b) gravel under waves and currents in Area A (east of the Isle of Wight).

RESULTS

In area A, sand appears to be mobile for considerable periods of time over the whole of the area. In contrast, gravel is mobile only over the shallow inshore areas (Figure 1), where wave action enhances the near-bed flow.

For area B, the modelling shows that sand transport is relatively weak during calm weather conditions, particularly in the inshore areas (Figure 2). These transport patterns in the inshore areas appear to agree with those inferred from the bedform asymmetry, grain-size trends and *in situ* current measurements (Brampton *et al.*, 1998). The superimposition of waves alters significantly the sediment transport patterns and fluxes (Figure 2), particularly in the inshore areas where significant transport may occur under severe storm conditions (see also, Velegrakis *et al.*, 1994). Finally, modelling of the effects of potential dredging activities shows that, although gravel dredging in the offshore areas (in water depths in excess of 18 m) is not likely to have an effect on the coastline, sand dredging in the nearshore areas should be viewed with caution.

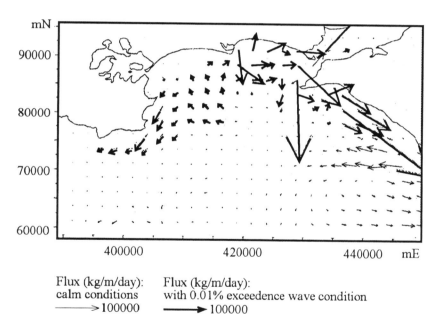

Figure 2 Wave effects on sediment transport in Area B (west of the Isle of Wight).

CONCLUSIONS

These studies have shown that sediment mobility and transport patterns are complex in the approaches to the Solent. Moreover, wave-current interaction appears to have a significant effect on the sediment mobility and fluxes. Finally, reliable sediment mobility studies require the combination of various research approaches.

REFERENCES

Brampton, A.H., Evans, C.D.R. & Velegrakis, A.F. 1998. *Seabed Mobility Study (West of Isle Wight):* CIRIA Report 65, CIRIA, London, 218 pp.

Brampton, A.H. 1993. *South Coast Seabed Mobility Study (East of Isle of Wight): Summary Report.* HR Wallingford Report EX 2795, 24 pp.

Hervouet, M. 1991. *TELEMAC, a Fully Vectorised Finite Element Software for Shallow Water Equations.* Electricite de France, Report HE43/9: 1-7

Hughes, D. 1996. *Environmental Law* (3rd Edition). Butterworths, London. 628pp.

Humphreys, B., Coates, T., Watkiss, M. & Harisson, D. 1996. *Beach Recharge Materials: Demands and Resources.* CIRIA Technical Report 154. London, 171 pp.

Salomon, J.C. & Breton, M. 1991. FLUXMANCHE's first results in modelling hydrodynamics through the Channel and Dover Strait. In: *FLUXMANCHE: Hydrodynamics and Biogeochemichal Fluxes in the Eastern Channel.* First Annual Report to EC: 9-14.

Velegrakis, A.F., Voulgaris, G. & Collins M.B. 1994. The role of waves and currents in the maintenance of a nearshore gravel bank. *Abstracts of the 2nd International Conference on the Geology of Siliciclastic Shelf Seas, Ghent,* Belgium: 124-126.

Lee-on-the-Solent Coast Protection Scheme

L. Banyard and R. Fowler

Halcrow Maritime, Burderop Park, Swindon, Wiltshire, SN4 0QD, U.K.

INTRODUCTION

The 2 km shingle beach at Lee-on-the-Solent is part of the North-Eastern Solent Coastal Sub-Cell, which extends from the Hamble estuary to Portsmouth Harbour entrance (for general location and main sites referenced, refer to *Map A* and *Map B* in the *Preface)*. The beach is backed by a vertical concrete seawall, promenade and low, graded Pleistocene gravel cliffs. Underlying the shingle beach are the silty clays of the Bracklesham Beds, which are exposed on low spring tides. These exposures are classified as Sites of Special Scientific Interest (SSSI).

The undermining and collapse of sections of the seawall at Lee-on-the-Solent, during the winter of 1991/92, highlighted the need to take action to mitigate the effects of the coastal erosion, if serious loss of the town's frontage was to be avoided.

METHODOLOGY

A sequence of mathematical models was run, to study the behaviour of this stretch of coastline (Figure 1). The modelling enabled selection of the most suitable solutions to the erosion problem.

The wave conditions at Lee-on-the-Solent were established by combining remotely-generated offshore waves, with locally-generated waves from within the Solent (Figure 2). The remotely-generated waves were propagated into the Solent using Halcrow's wave propagation model. The locally-generated waves were hindcast from wind records[1].

The wave conditions were used as input to Halcrow's Beach Plan Shape Model, which showed how the beach would develop if no action was taken (Figure 3). At the east end of the frontage, the shoreline would retreat by approximately one metre per year.

---

[1] Source: HMS Daedalus, Lee-on-the-Solent (Met Office Station 5661), 50 deg 49 min N, 01 deg 13 min W; Coastguard Station, Lee-on –the-Solent (Met Office Station 5662), 50 deg 48 min N, 01 deg 13 min W.

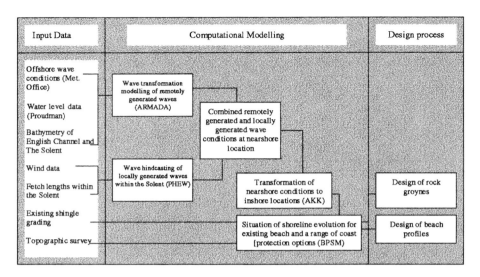

Figure 1 Sequence of mathematical modelling.

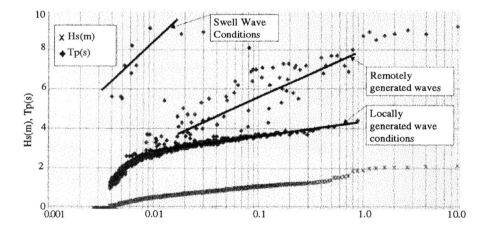

Figure 2 Wave height exceedance curve with associated wave periods.

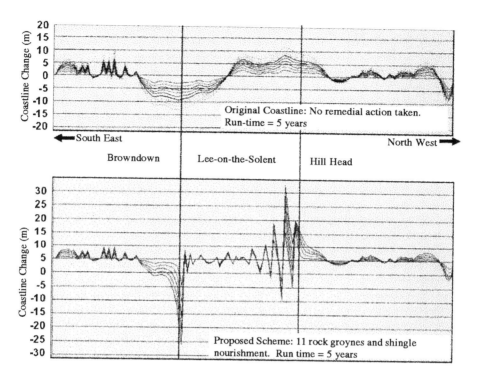

Figure 3 Beach plan shape modelling.

RESULTS

By the introduction of a series of rock groynes, this progressive erosion could be halted; however, the modelling highlighted the potential for downdrift erosion. Halcrow proposed that the replenishment of the beach would address this problem.

CONSTRUCTION

During the winter of 1996/97, eleven rock groynes were constructed at Lee-on-the-Solent. In spring 1997, 300,000 m^3 of shingle dredged from Southampton Water was pumped onto the beach and profiled.

The improvement in coastal protection at Lee-on-the-Solent was demonstrated clearly during the storm of 2nd January 1998. Prior to the scheme, significant quantities of water and shingle would have overtopped the seawall, under such conditions. The promenade would have been impassable at the eastern end. Most importantly, the foundations of the seawall would have been at risk from wave-induced scour. With the scheme in place, the highest wave run-up level was located at least 30 m to seaward of the promenade.

SECTION 2

Workshop Findings

Findings of the Coastal Processes Workshops

Session Leader: Michael Collins (*SOES, University of Southampton*)

INTRODUCTION

Five workshops were held to discuss the various coastal processes in the Solent region, as outlined below[1]:

1. Shoreline Management
2. Sea-level Rise
3. Saltmarsh Erosion
4. Navigational Dredging
5. Design and Monitoring of Coastal Defences

SHORELINE MANAGEMENT

Rapporteur: Doug Wright (*New Forest District Council*)

Main Issues

- The choice of strategic defence options for each management unit is influenced too heavily by land use, economic and social considerations, at the expense of strategic science. There is a need to refer back to coastal cell processes more thoroughly, before selecting the most appropriate defence option. It was recognised that social and economic factors are important in evaluating the appropriate option, but it was felt that the balance was not right in the first series of Shoreline Management Plans (SMP's).

- The four specified generic coastal defence options ('do nothing', 'hold the line', 'advance the line' and 'managed retreat') were felt to be a constraint on evaluating the most appropriate defence option; this has led to their meanings becoming confused, when they are 'extended' to deal with unusual situations. For example, managed retreat has been used to describe the limiting of soft cliff erosion rates, by recharge of the adjacent beach. Whilst sensible, in terms of explaining the management process, it does not accord with the accepted definition of 'managed retreat'. The solution seems to be to expand the number of generic defence options and clarify their exact meaning, in time for the first SMP review; in most cases, this is four or five years away.

- The implications of managed retreat are not clear at the moment. Consequently, it would be useful to utilise the time before the SMP's are reviewed to determine more clearly the likely short- and long-term implications of this defence option. It was felt that the provision of this information 'up front' would cause much less of a reaction to managed retreat, than has so far been experienced. Besides, in the potential managed

---

[1] Note: Discussion on archaeology is included in the Biodiversity and Conservation Workshop findings.

retreat areas, where economic loss creates opposition to the idea, there are areas where retreating the existing line would create, or improve, one type of wildlife habitat at the expense of another. Presently, environmental organisations are not certain which is preferable in the long-term.

- When referring to the science used in SMP's, it is important to include social science. One of the greatest failures of the SMP process, so far, is the inability to engage the interest and participation of the public and politicians, except where they have a direct vested interest in the management of the coast. A form of communication is required that makes the subject appealing to a wide audience, without losing the essential scientific underpinnings of the process. It was suggested that the USA could offer examples of proven and successful techniques.

- It was considered that the current SMP's are not clear enough about the time-scale over which the proposed strategic defence options remain valid. In many cases, a 'hold the line' or 'do nothing' option may be appropriate and/or politically expedient, in the short- to medium-term; however, factors like sea level rise may make it impossible, or economically unrealistic, to continue in the long-term. The next series of SMP's should propose defence options for the short- medium- and long-term, where appropriate. It was suggested that, linked to this idea, the historic, present and predicted future costs of implementing the chosen strategy should be included in SMP's.

- In general, it was considered that the first series of SMP's are credible documents, based upon good existing scientific data and a rational and well presented analysis of the data. However, cost and time constraints have led to little new knowledge being revealed and the possibility that not all the existing knowledge has been properly exploited. Despite this limitation, it was considered that the development of strategy plans (as the 'stepping stone' between SMP's and scheme implementation), together with the regular SMP review timetable, provided enough opportunity to identify and supply information to fill the gaps in scientific knowledge that presently exist. The plans could also develop a better means of engaging the interest of the wider public, in the SMP process.

Research/Management Priorities

Managed Retreat

- Clarify the long-term ecological implications of this strategy.
- Link back these implications into the planning system (land-banking).
- Under this heading, move forward on the development of a saltmarsh management strategy for the wider Solent (or SCOPAC area).
- Resolve and make public the issue of compensation for economic losses.

Communication/Consultation

- Investigate other country's procedures for consulting the public and disseminating information, particularly USA practice.
- Consider the composition of SMP Steering Groups and the possible need for some form of political input, or a political superstructure.

SMP Outputs

- Clarification of the exact output of SMP's, in terms of future action, implementation and research needs.
- Setting targets and time-scales for further information gathering. Specifying the level of scientific data required and how it should be fed back into the SMP/strategy plan process.
- Regional evaluation of research and information needs identified in SMP's and an agreed programme of implementation; this would require agreed criteria for evaluating the priority of each proposal. SCOPAC are presently looking at this, through a study being undertaken by Portsmouth University - *'Critique of the Past, a Strategy for the Future'*.

SEA LEVEL RISE (SLR)

Rapporteur: Malcolm Bray (*University of Portsmouth*)

Main Issues

Short-term

- Lack of monitoring i.e. establishment of a long time-series of observations.
- Biodiversity and ecological functions.
- Funding of research and compensation for managed retreat sites.

Medium-term

- Need to generate regional climate model results.
- Managed retreat - how?

Long-term

- Regional strategy, to deal with long-term issues.
- Planning issues: links with coastal plans.
- Reactivation of coastal landslides.

Research Priorities

Short-term

- Monitoring.[2]
- Habitat creation.
- Managed retreat and saltmarshes.

Long-term

- Regional climate model outcomes.
- Wave climate and storm variability.

Collaborative Research Priorities/Opportunities

Short-term

- National or regional co-ordination of monitoring, by MAFF or SCOPAC[3].
- Variability of processes.[4]
- Easier access to baseline and monitoring data.

Management Priorities

Short-term

- Address sea level rise within SMP's.
- Education of public and managers.
- Link SMP's and development plans.

Medium-term

- Monitoring.
- Public participation.

---

[2]Sea-levels, extreme water levels, waves, beaches, saltmarshes, rainfall, coastal winds. (Note: regional standardisation and central archiving required.)

[3]Consistency is important at all scales.

[4]Funding becomes an issue, within this particular context, as there are few sources available for monitoring.

Long-term

- Long-term strategic planning needed - well beyond political time-scales.
- Must maintain or enhance habitats.

Other

Short-term

- Valuation of the environment and habitats.
- Political decision-making needs to transcend 5 year intervals.

Medium-term

- Develop alternatives to cost-benefit and contingent valuation methods.

Long-term

- Traditional analyses.

Important issues raised

- Sea Level Rise (SLR) emerged as a very important issue, to all concerned.
- Not that much discussion on science of SLR - all accepted SLR as fact.
- Main issue was how we should respond. All agreed that planning and monitoring needed to start *now*, especially with respect to winds, wave climates, beaches and saltmarshes.
- Monitoring to detect changes is vital, but inadequate, at present. There is a need to establish baselines.
- General consensus that key habitats are at risk and much more guidance was needed on habitat conservation.
- Developed areas should be protected relatively easily, as long as funds continue to be available.
- More guidance needed on how to implement managed retreat and plan for the future mobility of coastal landforms - better links between coastal defence and development planning.

Issues emerging from the discussion

- Scale of the problem – multidimensional, with important considerations at each scale.
- Global - the science of climate change and the ability to provide reliable regional estimates of change.

- National – funding, research and monitoring? Capability to organise effectively.
- Local - level at which impacts actually felt and communities actually affected (Note: overall, the discussion concluded that more should be done nationally and regionally and that a 'Solent-limited' perspective was not sufficient).

Sea Level Rise

SMP's - SLR focuses interest on the need to identify strategic and sustainable coastal defence options. SLR is one of the major pressures forcing the changes in attitude toward coastal defences.

SLR has funding implications including:

- cost of defences;
- costs of implementing retreat i.e. to permit the necessary mobility of shorelines;
- costs of implementing habitat replacement; and
- a long-term issue, but work *now* needs to be started.

Cost Considerations

Who pays, who gains? In terms of the balance between local and national interests - the locals gain most, so should the whole nation pay? – Or, are our coasts a national resource for all?

- Political process - based on no more than 5 years ahead (term of office); this makes it difficult to tackle long-term strategic issues, such as SLR.
- Biodiversity - losses, even if natural processes are allowed to go unhindered.
- Education is needed, if the public is to participate in decisions relating to the type of coast that they wish to see in the future and what they may be prepared to pay.
- Monitoring is a key issue, but who pays? Who ensures the standardisation of methods and the archiving to common formats?
- Changes in coastal and shoreline management options are clearly needed, but how should these be implemented? *Enforcement* - new laws to force retreat? *Inducements* - e.g. compensation? The view of the Group was that both a "carrot" and "stick" approach would be necessary.
- Valuation of the environment and habitats – a more deliberate and inclusive means of managing the value conflicts that will arise from SLR should be explored, as an alternative to existing (but flawed) cost-benefit and contingent valuation methods.

SALTMARSH EROSION

Rapporteur: David Johnson *(Maritime Faculty, Southampton Institute)*

Main Issues

- More information on coastal processes is needed e.g. sediment budgets, rates of erosion.
- Scope for change - importance of the "overall picture" (underpinning knowledge).
- A definitive map is required of saltmarsh state, given that presently there is no strategic view or quantification of the distribution of vegetation types.
- The collation of existing studies.
- Alternative methods of assisting accretion and the re-creation of intertidal flats/ saltmarshes.
- Stability of sediment i.e. geochemical considerations (the release of pollutants) and the importance in terms of their sinks of nitrogen.
- Role of maintenance dredging, to retain sediment within the system.
- Value of saltmarshes - wider monetary value in economic terms (Note: on-going SCOPAC study).
- Vigour of vegetation – understanding the reasons for the die-back cycles.
- Modelling of erosion, related to 'coastal squeeze'.
- Flora and fauna inventories.
- Erosional processes – the balance between sheet and cliff erosion.
- The potential use of Fawley as a Case Study.
- The importance of (long-term) tidal cycles (reaching a 19 year peak, within the tidal curve).
- The macro/micro effects of changing wind/wave climate (associated with, at the same time as, SLR).

Research Priorities

- Review/collation of existing Solent 'site specific' studies e.g. extensive studies have been undertaken in relation to Langstone Harbour. The production of a definitive map of saltmarsh state, providing area, height, stability (associated with a review of data quality).
- Rate/causes of erosion, scope for modelling/prediction of future trends.
- Underlying coastal processes and morphology (incorporating studies of physical properties).

- Appropriate protection and enhancement techniques.
- Establishment of the position of saltmarshes within a regional pattern i.e. within coastal systems, together with their conservation in comparison to other habitats.

Further, based on the observations of the Session Chairman, there is a need to:

- establish regional sediment transport pathways and budgets; and
- investigate prevailing hydrodynamics, sediments, and intertidal and sub-tidal communities.

NAVIGATIONAL DREDGING

Rapporteur: Mike Thorn *(HR Wallingford)*

[Note: the Group comprised of representatives of Local Authorities and Agencies - *not* consultants, dredging contractors or port/marine operators]

Main Issues

Spoil disposal i.e. the discharge of 'contaminated' muds onto intertidal flats, requires knowledge of:

- the degree, if any, of contamination.
- the complexity of gaining permission for the re-use of dredged sediments.
- the timing of dredging, to avoid environmental "damage" e.g. during salmon runs;
- dredging activities changing the dispersion of discharges (both legal and illegal).
- dredging changing the sediment type, by creating a sediment sink and disrupting the sediment balance.
- the whereabouts and accessibility of Solent data - sediment type, geochemistry, and contaminant distribution.

Actions proposed

- All data/information needs to be collected together, maintained and made available (Note: SCOPAC might facilitate this particular requirement).
- A dredging Strategy Plan is needed for the co-ordinating and/or timing of dredging operations - for different areas and at different times, according to operational and environmental constraints, with the possibility of beneficial use. Could this be integrated within the coastal defence strategy?
- The whole process of approval for dredging needs to be clearer and simpler, in operation.
- New dredging technologies (e.g. jet dispersion) should be considered.

- Further research should be undertaken into the underlying causes of sedimentation, and the need for possible reductions in dredging quantities.
- What needs to be known?
- Where can dredged sediment be placed?
- Where are the data and are they commercially-restricted?
- What is contained within the dredged sediment especially originating from marinas?
- Is it true that sediment present alongside the western side of the (Solent) system is associated with cadmium and copper contamination?
- What are the boundaries or limits of acceptance for dredging/disposal?

Additionally, the Session Chairman added a number of observations (see below).

- Climate change, in general, needs to be considered.
- Research undertaken at the SOC indicates that wave heights in the NE Atlantic are increasing, based upon buoy and satellite data.
- There is a predicted increase in storminess with sea level rise, together with directional changes in wave approach.
- Research is required into extreme sea levels, including storm events and surges.

DESIGN AND MONITORING OF COASTAL DEFENCES

Rapporteurs: Andrew Bradbury *(New Forest District Council)*,
Alan Allinson and Tim Kermode *(Environment Agency)*

Main Issues

- Monitoring.
- Analysis of large datasets.
- Learning from mistakes.
- Establishing a database.
- Research.
- Priority of resources and money available.
- Free exchange of data between organisations.
- Economics.

Research Priorities

- Monitoring of engineering works, past and future predictions.

- Sea level rise - effect on structures and new works (increase in wave height and storminess).
- Research on environmental parameters - physical, chemical, and biological.

Collaborative Research /Priorities/Opportunities

- Collaboration between organisations, for exchange of present and future datasets. (Note: time required analysing large data sets).
- Adopt an holistic approach to sea defences i.e. consider the 'knock on' effect.
- Calibration and validation of past, present and future data sets and projects, through monitoring.

Management Priorities

- Using research for Cost-Benefit Analysis – gaining grants, etc.
- Flooding, strategic plans and SMP's.
- Cost analysis – how to put a price on human needs, compared to those of the environment.

Other

- Public understanding.
- Who will pay?
- Planning process – adopt an holistic approach i.e. how does the planning process affect the private individual?

Additional Information Requirements Identified by Session Speakers

A. Velegrakis (SOES, University of Southampton)

- Offshore geology.
- Pleistocene terrace gravels.
- Sediment distribution - temporal/seasonal and extreme variability.
- Bedforms, as indicators of transport pathways.

J. Sharples (SOES, University of Southampton)

- Availability and analysis of tidal data.
- Transport processes, incorporating the analysis of ebb- and flood-dominated areas of the system.

- Lateral variability in flow characteristics, including the presence of local eddies.
- Water column structure.
- Long time-series requirement.

I. Townend (ABP Research)

- Insufficient field measurements.
- Calibration and validation of instrumentation and measurements, in terms of potential errors.
- Wave/current interaction.
- Net (fine-grained) sediment transport onto intertidal flats: coarse-grained transport out of estuary.
- Balances - geological, historical, 10 year cycles.
- 'Most probable' state - response times.

J. McCue (Sir William Halcrow & Partners)

- More science needed, in general.
- Extreme water levels.
- Need for additional biological information.
- Information on saltmarsh management and processes requested - suspended sediment transport and plants (as indicators).
- Need for consultation processes.
- Production of 'Strategy Plans', incorporating research requirements.

Additional requirements:

- Bed form stability, historical/geological.
- Cycles of equilibrium of sediments in estuaries (short-and long-term).
- Long-term estuarine evolution.
- Climate change - the establishment of regional sediment transport pathways and budgets.
- Public awareness.
- Extreme sea levels, including storm events and surges – including: (a) increase in wave heights in the NE Atlantic; and (b) the predicted increase in 'storminess,' associated with a rise in sea level (SLR).

SECTION 3

Water Quality and Chemistry

Nutrients in the Solent

David Hydes

NERC, Challenger Division, Southampton Oceanography Centre, University of Southampton, European Way, Southampton, SO14 3ZH, U.K.

INTRODUCTION

On a world-wide basis, there is much debate about the possible deleterious effects of increased concentrations of nutrients in surface waters. In the Solent, an important area of scientific concern is the stability of the extensive mats of macro algae (*Ulva* and *Enteromorpha*) in the natural harbours of the region (Portsmouth, Langstone and Chichester) - the reasons for their existence, the impact they have on the sediment ecosystem and on populations that feed on them (such as Brent Geese). In Southampton Water, understanding the relationship between the high inputs of nutrients to the system and the intensity and variability that has been observed in phytoplankton blooms, is of importance. A wide range of chemical elements (nutrients) in different forms is required for organisms to grow. Within surface waters, the growth of plants (biological primary production) tends to be limited by the availability of nitrogen (N) compounds and phosphorus (P). Increased inputs of these nutrients from rivers and atmosphere, to the seas of the European shelf are well-documented (Howarth *et al.*, 1996). Nutrient inputs can originate from both point and non-point sources. Point sources include waste water treatment plants (human and industrial waste); non-point sources include inputs from the atmosphere and agriculture (both land drainage and animal wastes). Between 1930 and 1980, there was a six fold increase in the use of nitrogen fertilisers in the United Kingdom (Anon., 1983). Nitrogen and phosphorus losses from farm land increase both in proportion to the addition of fertiliser and to the intensity of land use (Harper, 1992). In the Hampshire region, increases in agricultural inputs have followed this general pattern. Inputs from sewage sources have increased, in line with the population growth (in Hampshire) which has been one of the fastest growing areas of the UK in the 1980s and '90s.

Excess plant growth can be a consequence of nutrient enrichment and problems such as anoxic events and blooms of nuisance algae may result (Lancelot, 1990). A fundamental feature of the behaviour of marine phytoplankton in temperate latitudes is a period of rapid population increase, referred to as a "bloom" (Vollenweider, 1992). An increase in nutrient concentrations within a body of water should be termed 'hypernutrification', while the term 'eutrophic' has come to be applied to those waters where the increase is considered likely to produce deleterious changes in the ecosystem. Two important pieces of European legislation are now in place, with the purpose of curtailing potential damage from the discharge of excess amounts of biologically-available nitrogen and phosphorus; these are the Urban Wastewater Treatment (UWT) Directive (EEC, 1991a) and the Nitrates Directive (EEC, 1991b). These Directives require a better understanding of the problem and that remedial action be taken where necessary. The UWT requires the assessment of areas, as being sensitive or less sensitive to anthropogenic discharges of nutrients. The criteria that have been established in the UK for this assessment include: winter concentrations of nutrients; concentrations of plankton; the duration of large plankton populations; effects on dissolved oxygen concentrations; changes in the fauna; changes in macrophyte growth; occurrence of paralytic shell fish poisoning; and the formation of algal scums.

A specific recipe controls the growth of phytoplankton - 3.3 kg of phytoplankton requires 95 g phosphate, 832 g nitrate, 1.6 kg water, 4.6 kg carbon dioxide and 310 kJ of sunlight. Silicon is also required by some species, particularly diatoms. Phytoplankton growth is limited when any one ingredient is in short supply. At the temperate latitude of the UK, the limited amount of light energy available in late autumn and winter imposes an annual cycle of growth and decay on phytoplankton populations. In the last decade, we have begun to understand some of the complexities that this recipe imposes on the system. The requirement of light energy for photosynthesis has been shown to be particularly critical, so that simply higher nutrient levels in the water do not mean higher biomass unless the physical conditions in the water are correct (Tett & Walne, 1995). Blooms in surface waters occur when the net rate of biomass production is greater than the loss rates due to respiration, grazing, sinking, horizontal transport by advection and turbulent diffusive processes (Cloern, 1996). Estuaries such as Southampton Water have high nutrient availability, so the initial formation of a bloom tends to be a function of available light and physical processes; these control the dispersion of phytoplankton communities (Fichez *et al.*, 1992). Chlorophyll-*a* (Chl-*a*) concentrations exceeding 10 µg/l have often been observed, whilst levels of up to 40 µg/l have been recorded (Purdie, 1996; Crawford *et al.*, 1997). The persistent occurrence of blooms in excess of 10 µg/l Chl-*a* is one definition of a system that has become eutrophic (Anon., 1983). This has not been the case in Southampton Water, where blooms usually only exist for about two weeks. It has been documented that tidal conditions play a major role in the incidence of blooms within Southampton Water and other shallow coastal environments (e.g. Cloern, 1996).

Recent developments arising from the Southern Nutrients Study (SONUS) are considered here; this has been a project funded by the Department of the Environment, Transport and the Regions (DETR), since 1995. The aims of the study have been: to investigate why blooms are not continuous in a hypernutrified estuary, like Southampton Water; and to assess the contribution that the Solent system makes to nutrient levels in the English Channel (and its possible contribution to the eutrophication of the whole North Sea system). There have been three parts to the SONUS project: a data base has been established of available river discharge data for UK south coast rivers; and of observations of nutrient concentrations and pertinent hydrographic data from the English Channel[1]: regular surveys of Southampton Water (undertaken at Southampton Oceanography Centre); and the development of new methods for improving the collection of information.

---

[1] Data base is available from the British Oceanographic Data Centre (previously held at the Plymouth Marine Laboratory).

DESCRIPTION OF STUDY AREA

Pertinent background information on the Solent region is contained within the NERC publication "The Solent Estuarine System: An Assessment of Present Knowledge," edited by Dennis Burton (Burton, 1980). Important general ecosystem processes have been described by Tubbs (1980), sediment transport by Dyer (1980) and hydrodynamic processes by Webber (1980). Phillips (1980) has collated the existing information on nutrients and other contaminants.

Southampton Water is an estuary fed by two main rivers, the Test and the Itchen with a combined average annual discharge equivalent to 1.54×10^6 m^3/day. Generally, estuarine water flows into the English Channel, to the north of the Isle of Wight, through the Solent. The estuary is described as 'partially mixed', and 'macrotidal'. The maximum tidal range is around 4.5 m and the tidal excursion is 2.5 km. It is an approximately linear body of water, which is 2 km wide, and extends for 10 km; it is highly urbanised, with large industrial complexes along its length. In addition to river flow, Southampton Water receives a (consented) sewage discharge of about 0.1×10^6 m^3/day. Waste water can contribute up to 25% of the flow during periods of low river discharge. Salinities in the main body of the estuary vary between 30 psu and 33 psu. Using the approach of Officer (1976) and data from Webber (1980), flushing rates for spring tides of about 26 hours, and 76 hours for neap tides have been estimated (Wright *et al.*, 1997).

SUPPLY OF NUTRIENTS TO SOUTHAMPTON WATER FROM THE RIVERS TEST AND ITCHEN FROM 1974 TO 1998

The Solent is one of the major sources of freshwater and suspended particulate matter to the English Channel from the English coast. The rivers Itchen, Test, Hamble, and to a lesser extent the Beaulieu, all contribute dissolved nutrients to Southampton Waters and hence the Solent (for general location and main sites referenced, refer to *Map A* and *Map B* in the *Preface*). At present, it is still unclear as to what extent Portsmouth Harbour, Langstone Harbour and Chichester Harbour are likely to act as sources of phosphorus and nitrogen to the eastern Solent. The sum of the extensive macroalgal growth and associated denitrification in the sediments of the Harbours may be such that they are a net sink for nitrogen.

The rivers Test and Itchen are predominantly fed by high quality Chalk streams from catchment areas of 1260 km^2 and 400 km^2 respectively: the water is clear, hard and alkaline. On the River Test, consumptive use represents 2% of the annual average flow and 80% of water abstracted for domestic use is returned after treatment. The catchment of the Test is predominantly rural with the population spread across towns in the north, or concentrated in conurbations along Southampton Water. Above the tidal limit of the Test there are two substantial discharges of treated sewage: Fullerton (Andover) and Romsey with (consented) discharges of 16,000 m^3/d and 6,410 m^3/d, respectively. Other smaller works at Chilbolton, Kings Sombourne and Stockbridge, have discharges of less then 500 m^3/d. In addition, there are small works on the rivers Blackwater, Dever and Dun. The largest discharge to the River Test is in the upper estuary at Millbrook and Slowhill, amounting to 55,000 m^3/d. For the Itchen, treated water from Winchester is filtered though the Chalk back to the river

(9000 m³/d); above Winchester, the works at Headbourne Worthy have a discharge of 4100 m³/d. Increases in population have required the installation of three large sewage works, discharging directly to the lower reaches of the river above the tidal limit at Chicknell (Eastleigh 30,000 m³/d) and into the estuary at Portswood (27,700 m³/d) and Woolston (15,000 m³/d).

Since the establishment of the Water Authorities in 1974, data are available from the Environment Agency's Data Centre. At approximately weekly intervals, samples have been taken above the tidal limit of both rivers, for the determination of nutrient concentrations. The data from both rivers, for concentrations of nitrate and phosphate, and the corresponding flow data are summarised in Table 1. Over this period, flow in the Test has tended to be higher than in the Itchen. When data averaged for the individual decades are compared, it appears that the flow in the Itchen has remained constant while the flow in the Test has been more affected by the drought in the 1990s, prior to 1998. The concentrations of nitrate in both rivers are similar and show similar seasonal variations. Maximum concentrations of nitrate tend to occur in winter, during periods of high flow. A steady and progressive increase in concentration has taken place over the last 24 years, from 342 μM and 308 μM (1974-1979) to 422μM and 393 μM (1990-1997) in the Test and Itchen, respectively. Concentrations of the two other forms of nitrogen capable of assimilation by algae, ammonia and nitrite, are similar and low in both rivers where mean concentrations are 7 μM and 4 μM, for ammonia and nitrite, respectively.

Concentrations of phosphate over the period are plotted in Figure 1. In the Itchen, phosphate concentrations have been consistently higher than those measured in the Test. The seasonal variation in both rivers is the reverse of the nitrate pattern in that high concentrations tend to be associated with periods of low flow in summer. This pattern has two causes: one is the lower rate of solublisation of phosphorus from soils, which results in concentrations tending to be diluted at high discharges; the other is that sewage inputs make a higher contribution to river loads than is the case for nitrate these remain relatively constant throughout the year, whilst the base river flow changes with the amount of rainfall. The impact of a sewage source is the probable reason for the higher concentrations in the Itchen, relative to the Test. Concentrations of phosphate were at a maximum in the late 1980s. Concentrations decreased between 1990 and 1994; this may be due to the impact of measures to control discharges of phosphorus. Between 1995 and 1997, the concentration of phosphate increased in both rivers. However, calculation of the daily discharge load (by combining the concentration and flow data) suggests that the load of phosphorus has tended to fall during the 1990s; likewise that the increase in the concentration of phosphate is a function of reduced flows in the rivers.

Table 1 Summary of Environment Agency data for the Rivers Test and Itchen

	River Test			River Itchen		
	Flow (m3/s)	NO_3 (mM)	PO_4 (mM)	Flow (m3/s)	NO_3 (mM)	PO_4 (mM)
1974-1997						
average	10.2	385.9	5.3	5.6	362.1	10.7
maximum	30.4	578.6	16.1	15.7	695.7	39.7
minimum	1.5	94.3	0.3	1.6	136.4	0.3
stdev	4.6	66.1	1.9	2.2	62.0	5.0
1974-1979						
average	10.3	342.4	4.1	5.1	308.2	7.4
maximum	24.3	568.6	12.6	11.1	445.7	39.7
minimum	3.3	211.4	0.7	2.2	202.9	0.6
stdev	4.1	55.5	1.5	1.9	44.7	3.6
1980-1989						
average	11.1	380.3	6.1	5.9	365.1	13.1
maximum	25.3	578.6	11.9	10.6	695.7	29.4
minimum	5.1	94.3	0.3	2.7	136.4	0.3
stdev	3.7	61.2	2.0	1.8	57.5	5.7
1990-1997						
average	8.8	422.3	5.3	5.9	393.5	10.9
maximum	30.4	557.1	16.1	15.7	685.7	32.6
minimum	1.5	164.3	0.3	1.6	214.3	0.3
stdev	5.9	57.9	1.7	2.7	51.1	4.0

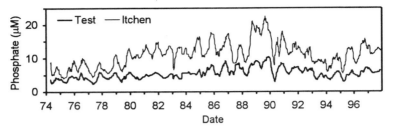

Figure 1 Environment Agency data for concentrations of phosphate in the rivers Test (–) and Itchen (–).

SEASONAL CYCLES AND BLOOMS

The first phase of the SONUS programme included 24 surveys of Southampton Water and the estuarine arms of the rivers Test and Itchen up to their navigable limits (between March 1996 and February 1997). These were all undertaken during spring tides, to make use of the highest navigable range.

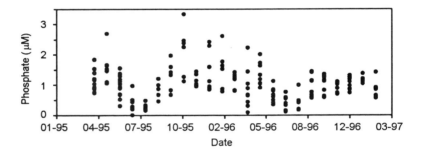

Figure 2 SONUS survey data for concentrations of phosphate (μM) in Southampton Water.

The impact of phytoplankton growth is to produce an annual cycle in nutrient concentrations in Southampton Water. Concentrations of both nitrogen and phosphorus are lower during summer, although in the case of phosphate (Figures 1 and 2), the concentration in the input waters is higher. The scatter in the data at each time point in Figure 2 is related to the range of salinity encountered on each survey. The lowest concentrations of nitrate and phosphate in Southampton Water were observed in July 1995. In Figure 3, the data for nitrate and phytoplankton chlorophyll from the July 1995 survey are plotted against salinity; this shows an almost complete removal of nitrate into the phytoplankton biomass at higher salinities. The nitrate deficit at each point in the survey can be calculated relative to theoretical conservative mixing of freshwater and the most saline sample. When the calculated deficit is converted to an equivalent amount of chlorophyll, assuming a carbon to chlorophyll ratio of 100, the shape of the estimated and observed chlorophyll distributions are similar

(Figure 3). This relationship suggests (a) that the observed chlorophyll is the product of *in-situ* production in Southampton Water; and (b) that although the rivers provide a constant supply of new nutrients, and they are present in sufficient quantities to support growth throughout Southampton Water, significant production takes place only in the outer region. Current research is being undertaken better understand the interactions between growing plankton populations and their physical environment. Factors such as water depth, stratification, turbulence and available light are known to determine the locations in which plankton populations develop (Wright *et al.*, 1997).

Observations made in 1996, from moored instruments (see below), show that in some years the levels of production are much less than others; this is related to the coincidence of unfavourable weather and hydrodynamic conditions. In June (1998), Southampton Water was surveyed during both spring and neap tidal conditions. A more intense bloom (50 μg Chl-*a*/l) was detected during the less turbulent conditions of a neap tide, relative to a spring tide (15 μg Chl-*a*/l). Although Southampton Water is a relatively well-sampled estuary, this year is the first that such a comparison has been done. The SONUS and other observations indicate that this spring-neap cycle in production may be a regular feature of the Solent system; the constant supply of nutrients from the rivers, together with the spring-neap tidal cycle providing suitable conditions for growth during the neap part of the cycle. Monbet (1992) has suggested, from an analysis of the available data from estuaries round the world, that macrotidal estuaries with mean tidal ranges >2 m exhibit a greater tolerance, than estuaries with smaller tidal ranges, to the likelihood of pollution from nitrogen discharges enhancing phytoplankton growth. On the other side of the Channel, off Roscoff (France), where the tidal range is greater, Sournia *et al.* (1987) saw little evidence for changes in productivity through the spring-neap cycle. The Solent system appears to be on the borderline in terms of the differentiation of systems, based upon their tidal range (as suggested by Monbet (1992)); and crosses over from being a low productivity to a high productivity system, through the spring-neap tidal cycle.

The contribution of sewage sources to the load of nutrients exported from Southampton Water

An important question relating to all estuaries is the extent to which both natural and anthropogenic processes modify the load of nutrients contributed to the receiving waters. In Southampton Water, for example, it is critical to assess the scale of the impact of direct sewage discharges, on the system. A maximum impact can be calculated from published population equivalents for the sewage works and *per capita* estimates of potential human discharges (3.3 kg/y of nitrogen and 0.4 kg/y of phosphorus). The potential magnitude of the direct discharges into the estuary, if the sewage were untreated are 1211 tonnes nitrogen/y and 147 tonnes phosphorous/y; these can be compared to those based upon observations. The SONUS survey data can be used to assess export loads from the estuary, in comparison with that which can be calculated from the river input flow and the concentration data; this is undertaken by extrapolation of the data collected in the outer estuary to zero salinity, to estimate the effective input concentration. Plots of nitrate within the estuary are characteristically linear throughout the estuary (Figure 4). Even in the summer months, where removal takes place at higher salinities (Figure 3) over the greater part of the salinity range, the linear trend in the data suggests that within the estuary inputs of nitrate-nitrogen are not significant. Consequently, this trend in the data should accurately define the effective concentration of nitrate in the river water sources to the estuary. When the calculated input and observed river concentrations are compared, the calculated input is generally lower than the river values. This pattern suggests that some loss of nitrate-nitrogen occurs between the Environment Agency sampling point and the head of the estuary; this may be due to denitrification occurring in the river's sediments and marshes (Nedwell & Trimmer 1996). The loss was equivalent to 27% (579 tonnes) of the river supply in 1995 and 10% (189 tonnes) in 1996.

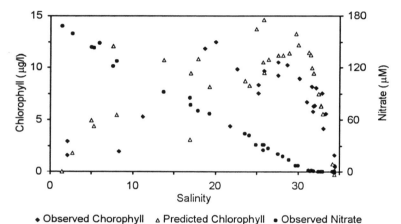

Figure 3 SONUS survey data for nitrate (μM) and Chlorophyll-a (μg/l) in Southampton Water, July 1995.

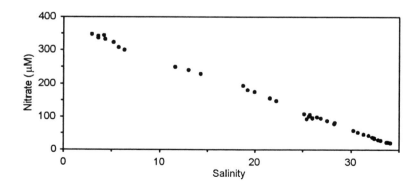

Figure 4 SONUS survey data for nitrate (μM) in Southampton Water, December 1996.

The data obtained from estuarine surveys of phosphate and ammonia are more complex; these show clearly concentrations in the mid-estuary which are well above those in the rivers Test and Itchen; typical plots of winter distributions (December 1996) are shown in Figures 5 and 6. For most surveys, the scatter in the data collected in Southampton Water, with salinities around 30 and above, is relatively small. Concentrations of phosphate and ammonia change gradually with changes in salinity; these data can be used to estimate the effective input concentration. The estimates suggest that the sewage load directly discharged to the estuary was 44 tonnes of phosphate, in 1995, and 19 tonnes in 1996; together with 3 tonnes of ammonia in 1995, and 18 tonnes in 1996. The increase in total nitrogen (due to ammonia discharge) is small compared to the river load; it appears to be less than the amounts of nitrate lost from the upper estuary.

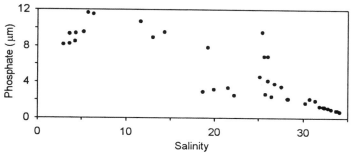

Figure 5 SONUS survey data for phosphate (μM) in Southampton Water, December 1996.

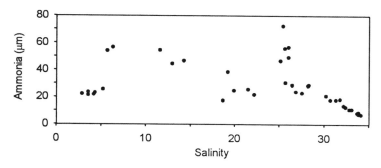

Figure 6 SONUS survey data for ammonia (μM) in Southampton Water, December 1996.

However, the increase in phosphate load is significant and is equivalent to 71% of the river load in 1995 and 26 % in 1996. Comparison with the above estimate of the potential load from untreated sewage suggests that treatment is effectively removing a large proportion of the potential nitrogen load, but that a third of the potential phosphate load was discharged to the estuary in 1995.

THE SONUS-DATA BUOY

An objective of the Southern Nutrient Study (SONUS) fieldwork programme was to use an instrumented buoy, moored within Southampton Water, to continually sample the estuary at a high temporal resolution (Wright *et al.*, 1997). It was hoped that interpretation of the data would consolidate concepts arising from a number of earlier observational studies (discussed above). To date, four deployments of moored instruments have been carried out. Data from the second deployment (in spring, 1996), as discussed here, covers the major bloom event of that year. The SONUS buoy was moored in the protected environment of the BP Jetty, at Hamble, on 4th April 1996, Julian Day (J.D.) 95; refer to *Map B* in *Preface,* for location site. A nearly-continuous record was achieved, over approximately 100 days, during the main period of phytoplankton production in spring and summer; curtailed when the logger circuitry was damaged by a severe thunderstorm over the site. Measurements made from the buoy were chlorophyll fluorescence, photosynthetically-available radiation light, water clarity (transmission), salinity, temperature, dissolved oxygen and pressure. The measurement rate was one reading every 15 minutes. Each set of measurements was transmitted back to SOC automatically using a cellular telephone based system.

The observed values for temperature and chlorophyll, as measured by the SONUS buoy are plotted together with the predicted tidal range, in Figure 7. The data set is incomplete during the bloom event in summer, due to an error in restarting the logger (during a service visit, carried out in poor weather conditions). Patterns within the data are discernible on three different time-scales; these correspond to the daily tidal cycle, the spring neap tidal cycle and longer-term changes, corresponding to changes in the weather and the development of a phytoplankton bloom. The temperature record proved to be a useful 'integrator' of weather conditions. Water temperature increased steadily, from 6°C to 11°C on J.D.118. During this time, the prevailing weather conditions were calm, with mostly sunny days. The temperature then remained relatively constant up until J.D.140, during two weeks of stormy and cloudy weather. The weather conditions then improved and temperatures increased steadily to about 18°C, by J.D.170. Water temperature then fell due to another period of un-seasonably stormy weather. After J.D.165, chlorophyll values increased to a maximum concentration of 3.6 µg/l, on J.D.179. Information from other investigations suggests the peak of the bloom was equivalent to 8 µg Chl-*a*/l. The composition of the bloom was dominated by the diatom *Skeletonema costatum*, and the dinoflagellate, *Peridinium* spp.

Data from the preceding ten years have suggested that both the summer and spring blooms tend to coincide with periods following spring tides, with tidal ranges less than 4 m: the first major bloom often occurring in May (Crawford *et al.*, 1997). This observation would suggest that a bloom should have occurred around J.D.138; however, no bloom occurred at this time. A bloom was observed around J.D.172; it coincided with a period of lower tidal energy; the magnitude of this was somewhat less than observed in previous years. For a bloom to occur there must be adequate levels of nutrients and light; likewise, the water column must be stable enough for the phytoplankton to absorb sufficient light energy. Water column stability appears to be important in bloom formation, especially in tidally-dynamic areas, such as Southampton Water. The results presented in Figure 7 suggest that the bloom can be linked to a period of lower tidal energy; however, non occurrence of the bloom, in May, shows that it is not the only factor.

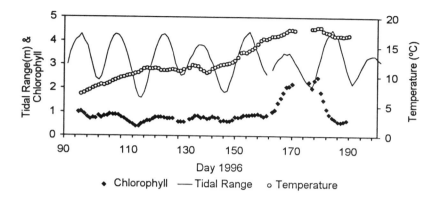

Figure 7 SONUS data buoy results for water temperature (°C) and Chlorophyll-a (μg/l) and Tidal range (m).

The pattern of temperature increase may provide a good record of overall weather conditions. Changes in temperature of the water depend upon the balance between heat input from solar radiation and cooling which can be enhanced by increased turbulence. The 'levelling off' of temperature during May corresponds to Meteorological Office total radiation data for the area. This showed that this period experienced only three relatively sunny days; these reach a total radiation level of c.7000 W/m²/day, then fall rapidly to less than 2000 W/m²/day. When the bloom did occur, the solar radiation levels were not only higher throughout the majority of this period, but there was a sustained level of relatively high radiation; five consecutive days having a total daily radiation in excess of 8000 W/m²/day.

New techniques

Monitoring schemes in which only a single sample is collected from each river, often no more frequently than weekly, can fail to identify transients, such as storm events, which can pass in a matter of hours. If such transient signals are not detected, estimates of the total discharge load are severely underestimated. Models of nutrient uptake by algae need to be validated against data sets collected at temporal resolutions that are difficult to match using conventional sampling techniques. In order to eliminate these problems in the monitoring of nutrients, the need exists for autonomous, *in-situ*, techniques to measure nutrients at a high temporal resolution that could be left within the environment for a period of time, giving continuous, longer-term (weeks to months) data. As part of the SONUS fieldwork project, evaluation and deployments have been undertaken of newly-developed chemical analysers designed for such *in-situ* use for the determination of concentrations of nitrate, phosphate and silicate (Hydes *et al.*, in press; Finch *et al.*, 1998).

The "Ferry-Box" name comes from an initiative of GOOS the International Global Ocean Observing System programme. The idea is to use new technologies in the creation of robust measuring equipment, that provide reliable data without human intervention when installed on commercial ships. Commercial ships run consistent routes, on a regular basis; this makes for the collection of systematic data. Systematic and regularly-collected data are what is required to detect changes in the "weather" and "climate" of surface waters. The approach is equally valid in smaller systems, such as the Solent. Here, the use of robust low maintenance sensors on a ferry will enable six surveys a day to be made in Southampton Water and across the Solent, rather than the one or two per month feasible at the moment. The unit will be fitted on the Red Funnel ferry the "Red Falcon," in December 1998. Measurements will be made of chlorophyll-fluorescence, nitrate, turbidity, salinity, temperature, ships position and time. The data rate will be 1 Hz, which will provide a spatial resolution of about 7m when the ferry is a full speed.

CONCLUSIONS

An understanding is developing of the conditions under which major bloom events occur, in conjunction with favourable weather conditions and neap tides. However, there is a lack of understanding and, therefore, an inability to predict the occurrence of blooms of specific algal types. A particular example is the red/brown tides that were common in the estuary in the 1980s, but were absent in recent years and returned this year (1998). This may be associated with relatively high fresh water discharges this year. Within the terms of the Urban Wastewater Treatment Directive, Southampton Water itself is not a eutrophic environment. Levels of nutrients in the rivers are moderate, for ones draining populated areas with intensive agriculture. Although sewage discharges of phosphate and ammonia are greater than the river water inputs to the estuary, dilution appears to be too rapid for significant de-oxygenation due to the nitrification of this ammonia to occur. This is a similar situation to that in the Ythan estuary, in Scotland (Balls *et al.*, 1995). Clear symptoms of eutrophication are present in the harbours (Langstone, Chichester and Portsmouth) in the form of the extensive beds of macro-algae, overlying anoxic sediments. Past research programmes have focused either on the Harbours or on Southampton Water. Work in the Harbours has not found any significant connection between the extent of algal growth and the direct sewage inputs to the Harbours (Soulsby *et al.*, 1985). In contrast in the Acachon Lagoon (southwestern France), the growth of weed beds does appear to be associated with particular river discharges (Rimmelin *et al.*, 1998). The link to the Southampton Water outflow, which is a probable source of a substantial part of the freshwater (with associated nutrients) entering the Harbours, has not been investigated.

REFERENCES

Anon. 1983. *The nitrogen cycle of the United Kingdom*. Report of a Royal Society Study Group, The Royal Society, London, 264 pp.

Balls, P.W., MacDonald, A., Pugh, K., & Edwards, A.C. 1995. Long term nutrient enrichment of an estuarine system; Ythan, Scotland (1958-1993). *Environmental Pollution*, **90**: 311-321.

Burton, J.D. (Ed.). 1980. *The Solent Estuarine System: An Assessment of Present Knowledge.* NERC Publication Series C.

Cloern, J.E. 1996. Phytoplankton bloom dynamics in coastal ecosystems: A review with some general lessons from a sustained investigation of San Francisco Bay, California. *Reviews of Geophysics,* 34: 127-168.

Crawford, D.W., Purdie, D.A., Lockwood, A.P.M. & Weissman, P. 1997. Recurrent red tides in the Southampton Water estuary caused by the phototrophic ciliate *Mesodinium rubrum. Estuarine and Coastal Shelf Science,* 45: 799-812.

Dyer, K.R. 1980. Sedimentation and sediment transport. In: *The Solent Estuarine System: An Assessment of Present Knowledge.* Burton, J.D. (Ed.), NERC Publication Series C, 20-24.

E.E.C. 1991a. The urban wastewater treatment directive. *Official Journal of the European Communities,* 91/271/EEC, 13 pp.

E.E.C. 1991b. Nitrates directive. *Official Journal of the European Communities,* 91/676/EEC, 8 pp.

Fichez, R., Jickells, T.D. & Edmunds, H.M. 1992. Algal blooms in high turbidity, a result of the conflicting consequences of turbulence on nutrient cycling in a shallow water estuary. *Estuarine and Coastal Shelf Science,* 35: 577-592.

Finch, M.S., Hydes, D.J., Clayson, C.H., Gwillam, P., Weigl, B. & Dakin, J. 1998. A low power ultra-violet spectrometer for the measurement of nitrate in seawater. Introduction, calibration and initial sea trials. *Analytica Chimica Acta,* 377: 167-177.

Harper, D. 1992. *Eutrophication of fresh waters - principles, problems and restoration.* Chapman and Hall, London. 327 pp.

Howarth, R.W., Billen, G., Swaney, D., Townsend, A., Jaworski, N., Lajtha, K., Doowning, J.A., Elmgren, R., Caraco, N., Jordan, T., Berendse, F., Freney, J., Kudeyarov, V., Murdoch, P. & Zhao-Liang, Z. 1996. Regional nitrogen budgets and riverine N & P fluxes for the drainages to the North Atlantic Ocean: Natural and human influences. *Biogeochemistry,* 35: 75-139.

Hydes, D.J., Wright, P.N. & Rawlinson, M.B. *(in press).* Use of a wet chemical analyser for the *in-situ* monitoring of nitrate. In: *Chemical Sensors in Oceanography.* Varney, M. (Ed.). To be published 2000, by Gordon and Breach, pp 95-105.

Lancelot, C. 1990. *Phaeocystis* blooms in the continental coastal area of the Channel and the North Sea. In: *Eutrophication and algal blooms in the North Sea coastal zones, the Baltic and adjacent areas: Prediction and assessment of preventive actions.* Lancelot, C., Billen, G. & Barth, H. (Eds.), Brussels. C.E.C, 27-54.

Monbet, Y. 1992. Control of phytoplankton biomass in estuaries: A comparative analysis of microtidal and macrotidal estuaries. *Estuaries,* 15: 563-571.

Nedwell, D. B., & Trimmer, M. 1996. Nitrogen fluxes through the upper estuary of the Great Ouse, England: the role of the bottom sediments. *Marine Ecology Progress Series,* 142, 273-286.

Officer, C.B. 1976. *Physical oceanography of estuaries (and associated coastal waters).* John Wiley and Sons Inc. New York. 477 pp.

Phillips, A.J. 1980. Distribution of Chemical Species. In: *The Solent Estuarine System: An Assessment of Present Knowledge*. Burton, J.D. (Ed.), NERC Publication Series C, 44-59.

Purdie, D.A. 1996. Marine Phytoplankton Blooms. In: *Oceanography- An illustrated guide*. Summerhayes, C.P. and Thorpe, S.A. (Eds.), Manson. London, 89-95.

Rimmelin, P., Dumon, J-C., Mnaeux, E., & Goncalves, A. 1998. Study of annual and season dissolved inorganic nutrient inputs in the Arcachon Lagoon Atlantic Coast (France). *Estuarine and Coastal Shelf Science*, **47**: 649-659.

Soulsby, S.B., Lowthion, D., Houston, M. & Mongomery, H.A.C. 1985. The role of sewage effluent in the accumulation of macroalgal mats on intertidal mudflats in two basins in southern England. *Netherlands Journal of Sea Research*, **19**: 257-263.

Sournia, A., Birrien, J.l., Douville, J.L., Klein, B. & Viollier, M. 1987. A daily study of the diatom spring bloom at Roscoff (France) in 1985: 1, The spring bloom within the annual cycle. *Estuarine and Coastal Shelf Science*, **25**: 355-367.

Tett, P. & Walne, A. 1995. Observations and simulations of hydrography, nutrients and plankton in the southern North Sea. *Ophelia*, **42**, 371-416.

Tubbs, C.R. 1980: Processes and Impacts in the Solent. In: *The Solent Estuarine System: An Assessment of Present Knowledge*. Burton, J.D. (Ed.), NERC Publication Series C, 1-5.

Vollenweider, R.A. 1992. Coastal marine eutrophication: principles and control. *Science of the Total Environment*, Supplement 1992: 1-20.

Webber, N. 1980. Hydrography and water circulation in The Solent. In: *The Solent Estuarine System: An Assessment of Present Knowledge*. Burton, J.D. (Ed.), NERC Publication Series C, 25-35.

Wright, P.N., Hydes, D.J., Lauria, M-L., Sharples, J. & Purdie, D. 1997. Data buoy measurements of phytoplankton dynamics in Southampton Water, UK, a temperate latitude estuary with high nutrient inputs. *Deutsches Hydrographisches Zeitschrift*, **49**, 201-210.

Trace Metals in Waters, Sediments and Biota of the Solent System: A Synopsis of Existing Information

Peter J. Statham

School of Ocean and Earth Science, Southampton Oceanography Centre, University of Southampton, European Way, Southampton, SO14 3ZH, U.K.

INTRODUCTION

There have been a significant number of studies on trace metals[1] in the Southampton Water and the Solent region, since the review of Phillips (1980). These new data are particularly important in relation to dissolved trace metals, where a new generation of techniques has resulted recently in accurate baseline measurements being made for several elements. This contribution summarises some of the available information on the concentrations and distributions of trace metals found in the waters, organisms, and sediments of the Solent system. Comparison of these data to environmental quality standards (EQSs), together with a consideration of the biogeochemical implications of the information, forms a logical basis for the identification of important topics for future studies. The information provided is based on apparently reliable data, which have been abstracted principally from the scientific literature, agency reports, or postgraduate theses and dissertations.

METALS IN WATERS

Most of the data reported in the literature utilises a conventional, but arbitrary, size distinction between metals in dissolved and particulate phases (at about 0.45 µm); this will be applied here, unless specifically indicated otherwise. For some elements, such as Fe and Pb, colloidal forms (approximately 1- 400 nm particle size) can be important. Additionally, the molecular or atomic form in which a metal occurs in solution (i.e. its speciation) can be important in terms of biological availability and geochemical reactivity. Very few studies on speciation have been undertaken for the Solent region. The behaviour of metals in estuarine systems can be complicated by processes other than simple mixing of river and seawater (conservative mixing), such as interaction with particles, and inputs from industrial and sedimentary sources (Millward & Turner, 1995). All the preceding factors should be considered when reviewing the summarised data. For general location and main sites referenced, refer to *Map A* and *Map B* in the *Preface*.

Whilst problems with contamination have invalidated many of the early measurements of dissolved trace metals, for some elements the techniques used were adequate to obtain accurate data. Molybdenum concentrations in the range of 4.9 to 14 µg/L, within a salinity

---

[1] The term "heavy metal" is used often to describe a set of mainly transition metals, having a high density in the metallic state. Because these elements are not normally present in metallic forms in natural waters, and typically only occur at low concentrations, the term trace metal will be used throughout to include this group of elements.

range of 14.3 to 34.6, have been reported for Southampton Water (Head & Burton, 1970), whilst selenium concentrations in the range 250-370 ng/L have been reported for the Test and Itchen estuaries (Measures & Burton, 1978).

Tankere (1992) made some preliminary measurements of trace metals in Southampton Water, using reliable methods, identifying the following concentration ranges: Cd, 0.12-1.34 nmol/L; Cu, 13-44 nmol/L; Mn, 95-440 nmol/L; Ni, 10-16 nmol/L. These metals showed non-conservative behaviour, and Cd, Cu and Ni (in particular) showed a considerable variation with salinity. Scatter in the data was attributed to point sources in the estuary such as industrial and sewage inputs. The most complete baseline data set of concentrations of Cd, Co, Cu, Fe, Mn, Ni, Pb, and Zn, for Southampton Water and the Itchen and Test estuaries, is provided in a study by Fang (1995). A summary of these data is given in Table 1.

Table 1 Average dissolved metal concentrations for summer and winter conditions, in Southampton Water and the Itchen and Test estuaries (Fang, 1995). The full data set provides information for 3 additional cruises, during other seasonal conditions. Metal concentrations in nmol/L units.

	07-Jul-1993	08-Feb-1994	07-Jul-1993	08-Feb-1994	07-Jul-1993	08-Feb-1994
	Estuary Test		Itchen Estuary		Southampton Estuary	
Salinity range	25.3-32.1	0.28-27.11	0.41-32.4	16.55-22.09	32.1-34.0	23.33-27.23
Cd	0.16	0.16	0.19	0.17	0.11	0.14
Co	2.96	3.13	2.23	3.07	1.81	2.59
Cu	16.3	19.5	22.7	21.5	16.1	22.9
Fe	114	147	189	101	41.6	122
Mn	110	138	104	163	46.8	126
Ni	11.5	17.9	15.2	18.8	10.2	15.4
Pb	0.45	0.24	1.56	0.18	0.3	0.24
Zn	31.1	45.8	47.4	54.1	20.1	45.6

In the Itchen and Test estuaries, no clear conservative mixing trends were obvious for any of the metals. Dissolved Cd and Co were clearly non-conservative, whilst Cu, Mn, Ni, and Zn remained relatively constant over a wide salinity range. Dissolved Pb in the Test estuary was also relatively constant, whilst values for the Itchen were more variable with time, possibly reflecting industrial and /or sewage inputs; concentrations were significantly higher than in the Beaulieu estuary. In Southampton Water, there were also no clear relationships of dissolved metals with salinity.

These complex and variable distributions of dissolved metals in the Southampton Water and Itchen – Test systems, represent significant localised anthropogenic inputs, such as industrial discharges from the Fawley refinery and chemical plants on the western side of the estuary; these combine with other sources, which include potentially-important diffuse benthic inputs. A significant fraction of the total dissolved copper, within this system was found to be in an organically associated form.

There have been several studies undertaken of trace metals in the Beaulieu estuary. This estuary provides a contrast to the Southampton Water system, as there is virtually no industrial activity and the river drains the rural New Forest area. Holliday and Liss (1976) observed, for this estuary, the large-scale removal of dissolved (i.e.<0.4 µm) iron at low salinities and, essentially, conservative behaviour for Mn and Zn over the full salinity range. A similar behaviour for Zn was reported by Dolamore-Frank (1984). Moore et al. (1979) studied the behaviour of dissolved organic carbon (DOC), iron and manganese during estuarine mixing in the Beaulieu estuary using ultra-filtration. The manganese behaved essentially conservatively in the estuary, whilst there was large-scale removal of iron through destabilisation of colloidal matter and flocculation of iron in the low salinity zone. Fang (1995) studied a range of metals (Cd, Co, Cu, Fe, Mn, Ni, Pb, and Zn) in dissolved (see Table 2), colloidal and particulate phases in the Beaulieu estuary.

Observed distributions and estuarine behaviour for dissolved Mn, Zn and Fe broadly followed patterns observed in previous work. Dissolved iron and lead are present in the river waters predominantly in colloidal forms; however, the significance of this size fraction diminishes rapidly, on mixing with seawater. Cadmium shows a mid-estuary increase related to release of Cd from particles; Co distributions are similar to those of Mn. Dissolved Ni showed close to conservative behaviour, but for Cu there was a maximum in concentration in the salinity range 20-25; this appears to be caused by release of Cu from antifouling paints on boats in this part of the estuary. Concentrations of metals in particles in the Beaulieu estuary, which were released by a 25% (v/v) acetic acid leach (Fang, 1995) are: Cd, 0.02-1.79 µg/g; Co, 1.6-70 µg/g; Cu, 5.0-24.8 µg/g; Mn, 75-2910 µg/g; Ni, 1.0-20.3 µg/g; Pb, 0.9-65 µg/g; Zn, 21-454 µg/g, and Fe, 0.3-14.5 %. Howard et al. (1984) reported concentrations of 0.03 –0.26 µg/L of inorganic As (III + V oxidation states) in the salinity range 10.5-32.7; within this range, near-conservative behaviour was found. However, laboratory experiments indicated substantial removal of As during low salinity mixing experiments; this infers removal of As during iron flocculation. Total dissolved selenium concentrations, ranging from about 0.14 µg/L at salinity 15 to about 0.09 µg/L at salinity 30, have been reported (Measures & Burton, 1978).

Concentrations of dissolved metals in the Solent and adjacent coastal areas are low; these are in line with other values reported for coastal and shelf sea systems (Fang, 1995; Tappin et al., 1993); representative data are given Table 3. The National Rivers Authority (NRA; superseded by the Environment Agency, EA) carried out a series of coastal surveys of water quality (1998), which included measurement of dissolved trace metals; none of the metals measured exceeded EQS values, in the Solent region.

Table 2 Dissolved metal data (<0.4 μm) for the waters of the Beaulieu estuary (Fang, 1995). All the data are in nmol/L units. Note: the range in the values represents consistent changes with salinity for most metals (see text).

	Cd	Co	Cu	Fe	Mn	Ni	Pb	Zn
17-Mar-94								
Average	0.29	10.50	13.0	4050	551	17.50	0.10	39.50
max	0.51	23.20	25.30	2750	1720	38.20	0.29	69.60
min	0.15	0.58	4.56	12.60	6	3.73	0.04	11.20
13-May-94								
Average	0.16	6.65	19.80	707	855	33.50	0.16	17.00
max	0.28	10.70	39.90	2120	1880	68.8	0.41	37.10
min	0.10	1.23	8.92	18	24	7.15	0.03	4.20
22-Sep-94								
Average	0.17	4.90	16.40	389	929	27.40	0.08	47.8
max	0.31	10.40	32.90	2030	2040	51.80	0.28	99.7
min	0.08	0.57	6.28	14.7	14.2	9.68	0.02	5.50

Mercury is regarded as a potentially important contaminant in marine waters; it is a List I substance under the UK Dangerous Substances Directive. Whilst there have been no recent detailed studies undertaken on mercury in Solent area waters, it has been monitored by the Environment Agency. Data from three stations in the Solent region (Dock Head, Calshot and East Brambles) for the period January 1994 to May 1998, show a maximum value of 0.05 μg/L Hg, with 84% of the total data points (n =111) being below the detection limit of the method (0.008-0.03 μg/L). As the EQS for mercury is 0.3 μg/L, and these reported concentrations are typically at least an order of magnitude below this value, mercury falls well below this quality threshold in the Solent system. Present UK EQS values, relating to the EC Directive on Dangerous Substances (76/923/EEC), are given in Table 4.

Other EQS values (with higher concentrations, relative to the Directive on Dangerous substances) exist for the Shellfish Waters Directive (National Rivers Authority, 1994). It is clear that for all the metals discussed above, the measured values lie below (frequently, orders of magnitude below), these EQSs; this infers that the Solent region is a relatively clean environment, from the point of view of the dissolved forms of the above trace metals. However, it should be noted that adjacent to metal sources the concentrations may be significantly higher, and that organisms may bio-accumulate metals, though routes other than the water (see later, for Cu). The special case of tributyltin is considered below.

Table 3 Ranges of concentrations of dissolved trace metals (nmol/L) in the waters at the mouths of the Beaulieu estuary, Southampton Water (Fang, 1995), and the Solent and Central English Channel (Tappin et al., 1993). Note: N.D. = no data.

	Beaulieu	Southampton Water	Solent	Central English Channel
Salinity range	32.99-34.08	32.1-34.40	34.52-34.74	34.86-34.93
Cd	0.08-0.15	0.10-0.54	0.20-0.26	0.18-0.31
Co	0.57-1.23	0.36-2.80	0.64-1.70	< 0.06
Cu	4.56-10.76	9.6-22.5	6.0-11.4	2.7-4.0
Fe	12.6-21.9	8.0-109	N.D.	N.D.
Mn	6.0-24.0	8.7-75.0	6.4-17.4	1.0-1.5
Ni	3.98-10.3	6.7-28.6	7.2-10.2	3.3-4.0
Pb	0.021-0.089	0.11-0.66	0.14-0.21	0.16-0.20
Zn	5.5-11.2	9.4-48.0	11.8-21.7	6.5-10.5

METALS IN SEDIMENTS

The varying composition of estuarine and coastal sediments, combined with the use of differing metal measurement techniques, can make difficult the direct comparison of published trace metal data. Estimates of the total content of a trace metal in sediments are frequently given; at the same time the analytical reliability of these data is becoming clearer, as certified reference materials are more frequently being used to demonstrate the accuracy of the measurements. Many sediment studies are concerned with the biologically or environmentally available fraction of the metals present (Luoma, 1995). As mineral-bound metals are only released over geological time-scales, the use of a range of partial leaches to study metal association with the more labile phases present, have developed. A consequence of using a range of techniques for the measurement of metals in sediments is that care is required in the interpretation and comparison of the available data.

Table 4 United Kingdom National Environmental Quality Standards (EQSs) for dissolved trace metals in saline waters. Data are taken from the Department of the Environment (now DETR) circular for List I (potentially most toxic) and List II substances, which relate to the Dangerous Substances Directive. All the figures presented are annual means of dissolved concentrations, in µg/L.

List I Metals					
	Cadmium	Mercury			
Estuarine	2.5	0.3			
Marine	2.5	0.3			

List II Metals					
	Arsenic	Copper	Chromium	Lead	Nickel	Zinc
Salt water	25	5	15	25	30	25

In the past, the Fawley Refinery has been a major source of dissolved Cu to Southampton Water; this has been reflected in the contamination of sediments adjacent to the site on the western side of the estuary. Armannsson et al. (1985) found high concentrations of copper in cores taken from the Fawley Oil Jetty (depth range 0-72 cm, 88-362 µg/g dry weight) and the nearby North Refinery (depth range 0-18 cm, 23-237 µg/g dry weight). The elevated copper concentrations coincided, to a substantial extent, with elevated hydrocarbon concentrations in these sediments. A strong correlation, between concentrations of copper and hydrocarbons, has also been reported by Oyenekan (1980). Inputs of copper to the estuary from Fawley have decreased dramatically since the early 1970s, when measures were initiated to decrease effluent inputs (Dicks & Levell, 1989). This decreasing input is well demonstrated by the vertical trends in the concentration of copper, in sediment cores from salt marshes in the estuary (Hart, 1997).

Sediments can act as useful recorders of metal inputs to estuarine and coastal sediments, as long as 1) diagentic changes (subsequent to deposition) do not change the concentrations present, 2) a good chronology is available, and 3) there has been no major physical disturbance of the sediment. Saltmarsh cores from the Hamble, Itchen, and Beaulieu estuaries (Cundy et al., 1997) have provided good records of Cu, Cs-137 and Co-60 inputs; these have been related to temporal or spatial point-inputs. The copper source has been discussed above, whilst the Cs-137 inputs were caused by fall-out from atmospheric nuclear weapon tests (maximum inputs in 1963) and a smaller input from the Chernobyl accident (1986). Cobalt-60 originated from discharges of the AEE Winfrith nuclear power station, in Dorset; this produced low-level discharges in the period 1970-1990, with maxima in 1975 and 1980/81. Interpretation of the sedimentary lead record was difficult, because of the complex mixed input of atmospheric and marine origins.

Algan (1993) carried out a study on the sedimentology and trace metal geochemistry of fine-grained sediments of the Solent area. As well as high metal concentrations in the sediments adjacent to the industrialised western shore of Southampton Water, anomalously high concentrations of Cu, Pb and Zn in some surface sediments of the Itchen, Test and Hamble estuaries were identified. These latter elevated concentrations were attributed to inputs from sewage discharges and outfalls in these areas. The high concentrations of copper in sediments close to the industrialised western shore has also been observed by Savari (1988) (see Table 7, below). Measurements of metals in sediments in Portsmouth Harbour were undertaken by Soulsby et al. (1978). Other sediment metal measurements have routinely been made by the NRA, and now by the EA, in Langstone Harbour and other sites in the Solent region.

METALS IN ORGANISMS

Very few extensive studies have been made of metals in organisms in the Solent system. Mostafa & Collins (1995) reported data on concentrations of cadmium, copper, lead, nickel and zinc in the sea urchin *Psammechinus militaris* in Southampton Water (Table 5). Whilst the concentrations of metals in most tissue components were similar to those in urchins taken from Egypt and western Ireland, copper and zinc concentrations in the digestive tract and connective tissue were much higher.

Weir (1995) measured cadmium, copper, manganese and zinc in the American slipper limpet (*Crepidula fornicata*), in the Solent region (Table 6). There were no statistically significant variations in metal contents, between the 9 sites sampled along the eastern side of Southampton Water and the northern shore of the western Solent. However, there was a trend of increasing dry weight metal concentration with decreasing mass of dry organism. This pattern was particularly evident for cadmium, where a strong logarithmic relationship was observed.

Table 5 Concentrations of trace metals in tissues of the sea urchin *Psammechinus militaris* from Southampton Water (Mostafa & Collins, 1995). All the data presented are in µg/g dry weight, whilst the range given is ± 1 standard deviation of the mean for the 15 individuals analysed.

Tissue	Cadmium	Copper	Lead	Nickel	Zinc
Gonad	0.35 ± 0.23	3.81 ± 2.24	0.80 ± 0.97	1.10 ± 1.00	77.6 ± 109
Lantern	7.55 ± 0.33	6.03 ± 0.26	47.4 ± 1.39	33.9 ± 2.21	1.2 ± 2.0
Connective tissue and digestive tract	1.95 ± 0.49	33.1 ± .88	6.80 ± 1.46	6.51 ± 2.16	328 ± 120

Table 6 Concentrations of trace metals within the tissues of the slipper limpet *Crepidula fornicata*, from the Solent and Southampton Water (Weir, 1995). All the data are in µg/g dry weight, whilst the range given is ± 1 standard deviation of the mean for the 116 individuals analysed.

Total tissue	Cadmium	Copper	Manganese	Zinc
Mean concentration	7.44	87.6	43.3	43.7
Range	6.85	51.5	32.6	20.4

A comprehensive study of trace metals in the common cockle, *Cerastoderma edule*, and the surrounding sediments was undertaken by Savari (1988) for the period 1985-86.

A trend of higher concentrations of Cu, Cd and other metals in tissue from organisms close to the industrialised western part of Southampton Water (Fawley, Tucker Pile, Bird Pile and Marchwood) relative to the east (Woolston, Dibden Bay and Netley), was evident (refer to Table 7). The range of concentrations of Cu and Cd in cockles collected close to the industrial sites was much greater than at the other sites: the smaller individuals tended to contain the greatest concentrations of these metals. A paucity of species in the intertidal parts of the estuary, particularly close to the industrialised western side, has been ascribed to the chemically-induced stress on organisms in this part of the estuary (Houston *et al.*, 1983). Some seasonal variations in concentrations of metals were also noted (Savari *et al.*, 1991). Other data for trace metals in shellfish tissue are available from the Environment Agency.

SPECIAL CASE OF TRIBUTYLTIN

It was realised in the early 1980s that tributyltin (TBT), which was being leached from anti-fouling paints on ships and boats, was having a major impact on non-target organisms in some coastal and estuarine environments. Effects included imposex and sterility in the dog-whelk, *Nucella lapillus* (Bryan *et al.*, 1986; Bryan *et al.*, 1987), at concentrations of only a few nanogrammes per litre of TBT; this led to the elimination of this species from the Solent, by the mid 1980s. The only remaining populations are on the southern coast of the Isle of Wight. As the source of TBT to the Solent is primarily from Southampton Water and associated estuaries, this impact indicates the extent to which the contamination extended; a similar situation occurred with TBT from Poole Harbour, extending into Poole Bay. Other impacts of TBT include shell deformation in species of oysters, at similar extremely low concentrations (Bryan & Gibbs, 1991). Legislation was enacted in 1987, which prohibited the use of TBT based paints on vessels of less than 25 m in length. The Southampton Water area and Poole Harbour were particularly susceptible to the impact of TBT paints, because of the high density of small boats, port status with vessels of over 25 m still using TBT based paints (particularly, for Southampton), and the relatively poor flushing of the waters (Poole).

Table 7 Seasonally averaged metal data for whole tissue of *Cerastoderma edule*, together with the surrounding sediments (concentrated nitric acid attack), in Southampton Water. All the data presented are in µg/g dry weight. Information on mean total petroleum hydrocarbon content (TPC) and the silt-clay fraction (SCF) of the sediments are presented, to complete the information. Data abstracted from Savari (1988).

	Fawley	Tucker Pile	Marchwood	Bird Pile	Woolston	Dibden Bay	Netley
Cerastoderma edule tissue							
Cu	10.4	8	9.6	6.9	5.7	5	5.4
Cd	1.2	1.1	1.4	1.2	1.1	1	1.1
Fe	809	567	645	551	446	442	539
Pb	11.2	8.7	10	8.4	8.4	8.2	9
Ni	33.1	37.9	58.3	45.7	44	42.4	53.9
Zn	98	95	118	85	85	77	89
Sediments							
Cu	99	72	67	94	52	50	33
Cd	1.34	1.33	1.62	1.33	1.43	1.29	1.33
Fe	13300	13400	13900	13500	11400	13100	13600
Pb	48.1	49	66.7	53.1	63.4	52.3	57.3
Ni	23	23.1	25.6	22.7	21	23.5	21.9
Zn	73	79	123	86	99	91	75
TPH	3800	2840	590	540	224	606	1860
%SCF	88	84	87	78	46	56	76

Studies were undertaken to determine if the TBT paint ban had a significant impact on the concentrations of TBT in the water column in the Southampton Water area (Langston *et al.*, 1994; Waldock *et al.*, 1993). Concentrations of TBT in the waters showed a logarithmic decline in the Itchen and Hamble rivers, over the period 1986-1992, with a half-time of removal of about 3 years (Langston *et al.*, 1994), and more recent EA data indicates a continued decrease to close to the EQS value for water (0.8 ng Sn/L). However, concentrations in the waters of the Test are still high and variable; these seem principally to reflect the significant number of large vessels using the port facilities, many of which will still use TBT on their hulls. Thus, whilst TBT concentrations are gradually decreasing in areas used by small boats, at the more contaminated sites concentrations are well above the environmental quality standard. An important factor in understanding the fate and bio-availability of TBT in this system is the role of sediments, as TBT is removed from the water column by particles which then settle and become components of the sediments. Whilst TBT concentrations in the sediments are gradually decreasing in most areas, the rate is slower than that observed in the water column with a half-time of removal of about 7 years. Thus, sediments are an important reservoir for TBT; an important consequence is that this is potentially an important route, for making TBT bio-available to organisms in sediments. Whilst some sediment organisms (such as nereid worms) do not greatly bio-accumulate TBT, many bivalves do; this is particularly so for *Mya arenaria* (Langston *et al.*, 1994). The deposit feeding clam *Scrobicularia plana* shows a strong correlation between sediment TBT concentration and TBT content in the organism, supporting this view of uptake from sediment-associated TBT (Langston *et al.*, 1994). *Scrobicularia plana* was common in the 1970s, but declined considerably in the 1980s.

The International Maritime Organization universal ban on TBT in antifouling paints for all vessels including large merchant ships, which is intended to come into operation in 2006, should help to reduce inputs of TBT to the estuary and the Test in particular. However, there still remains the problem of disposal of old antifouling materials removed during the refit of vessels. Consequently, it will take many years before the concentrations of tributyltin in sediments fall to negligible levels.

SUMMARY AND IDENTIFIED FUTURE RESEARCH NEEDS

The Solent and Southampton Water region appear to be relatively uncontaminated with dissolved trace metals, in relation to current EQS values, with the exception of tributyltin levels in the Test estuary. Metal concentrations offshore rapidly decline to values typical of uncontaminated coastal waters. There is clearly a need for the long-term monitoring of tributyl tin in the estuary, together with its impact on biota. The change in ship and boat antifouling materials, which will occur as TBT paints are phased out, may lead to replacement by other potentially harmful materials which are likely to be released into solution. From the point of view of metal-based antifouling materials, copper is of particular importance; concentrations of this in waters and biota will require careful monitoring, particularly close to marinas and other areas with a high vessel density. Speciation of dissolved metals is also an important issue, as regards the biological impact of metals.

The sediments in Southampton Water, particularly on the western shore, show the legacy of contamination from the early days of industrialisation in the area. Thus copper and, to lesser extent zinc and lead, concentrations are elevated in the organic rich muds of this zone. The reduction in industrial inputs of metals in recent years is demonstrated elegantly in sedimentary cores obtained from saltmarshes in the Hythe area. The impact of metal and hydrocarbon contamination on, in particular, molluscs in the sediments is reflected in the abundance and growth of these organisms relative to less contaminated areas. There have been a limited number of studies on metals in sediments and organisms in Poole Bay, Portsmouth Harbour and adjacent zones. Future work on these areas would be useful, in terms of assessing the overall environmental quality of the region. An important issue is whether or not the metals in sediments are effectively locked away on long time-scales, or if they may be released back into the water column in dissolved, potentially biologically available, forms; this would occur through slow diffusive mechanisms or through more episodic activities such as dredging. The implications of bottom sediment disturbance through fishing activities, which have similar implications to dredging, have recently been considered by Pilskaln *et al.* (1998). New data collection tools, such as *in situ* monitoring buoys for trace metals, are required to provide information on estuarine waters; this will allow the short time-and space-scale changes in metals to be followed and more fully understood. A long-term objective for the Solent region is to develop predictive models that will effectively describe the behaviour of metals (and other chemical species) in this system, and the understanding of processes such as sediment exchange will play an important role in developing such tools.

ACKNOWLEDGEMENTS

Information and advice provided by the EA (David Lowthion and Paul Salmon) were very helpful in the compilation of parts of this paper.

REFERENCES

Algan, A.O. 1993. Sedimentology and geochemistry of fine-grained sediments in the Solent Estuary. Ph.D. Thesis, University of Southampton, Southampton, 269 pp.

Armannsson, H., Burton, J.D., Jones, G.B. & Knap, A.H. 1985. Trace metals and hydrocarbons in sediments from the Southampton Water region, with particular reference to the influence of oil refinery effluent. *Marine Environmental Research*, **15**: 31-44.

Bryan, G.W. & Gibbs, P.E. 1991. Impact of low concentrations of tributyltin (TBT) on marine organisms: a review. In: *Metal ecotoxicology: concepts and applications*. Newman, M.C. & McIntosh, A.W. (Eds.), Lewis Publishers, Ann Arbor, 323-361.

Bryan, G.W., Gibbs, P.E., Hummerstone, L.G. & Burt, G.R. 1986. The decline of the gastropod *Nucella lapillus* around south-west England: evidence for the effects of tributyltin from antifouling paints. *Journal of the Marine Biological Association of the United Kingdom*, **66**: 611-640.

Bryan, G.W., Gibbs, P.E., Hummerstone, L.G. & Burt, G.R. 1987. The effects of tributyltin (TBT) accumulation on adult dog-whelks, *Nucella lapillus*: long-term field and laboratory experiments. *Journal of the Marine Biological Association of the United Kingdom*, **67**: 525-544.

Cundy, A.E., Croudace, I.W., Thomson, J. & Lewis, J.T. 1997. Reliability of salt marshes as "geochemical recorders" of pollution input: a case study from contrasting estuaries in southern England. *Environmental Science and Technology*, **31**: 1093-1101.

Dicks, B. & Levell, D. 1989. Refinery-effluent discharges into Milford Haven and Southampton Water. In: *Ecological impacts of the oil industry*. Dicks, B. (Ed.), Institute of Petroleum, John Wiley and Sons Ltd., 287-316.

Dolamore-Frank, J.A. 1984. The analysis, occurrence and chemical speciation of zinc and chromium in natural waters. Ph.D. Thesis, University of Southampton, Southampton, 305 pp.

Fang, T.-H. 1995. Studies on the behaviour of trace metals during mixing in some estuaries of the Solent region. PhD Thesis, University of Southampton, Southampton, 358 pp.

Hart, J.T. 1997. The record of deposition and the migration of elements in salt marshes. Ph.D. Thesis, Southampton University, Southampton, 225 pp.

Head, P.C. & Burton, J.D. 1970. Molybdenum in some ocean and estuarine waters. *Journal of the Marine Biological Association of the United Kingdom*, **50**: 439-448.

Holliday, L.M. & Liss, P.S. 1976. The behaviour of dissolved iron, manganese and zinc in the Beaulieu estuary, S. England. *Estuarine, Coastal and Shelf Science*, **4**: 349-353.

Houston, M., Lowthion, D. & Soulsby, P.G. 1983. The identification and evaluation of benthic macroinvertibrate assemblages in an industrial estuary- Southampton Water U.K. using a long term, low-level sampling strategy. *Marine Environmental Research*, **10**: 189-207.

Howard, A.G., Arbab-Zavar, M.H. & Apte, S. 1984. The behaviour of dissolved arsenic in the estuary of the River Beaulieu. *Estuarine, Coastal and Shelf Science*, **19**: 493-504.

Langston, W.J., Bryan, G.W., Burt, G.R. & Pope, N.D. 1994. *Effects of sediment metals on estuarine benthic organisms*. **203**, National Rivers Authority R & D Note, Plymouth, 141 pp.

Luoma, S.N. 1995. Prediction of metal toxicity in nature from bioassays: limitations and research needs. In: *Metal speciation and bioavailability in aquatic systems*. Tessier, A. & Turner, D.R. (Eds.), John Wiley and Sons, Chichester, 609-659.

Measures, C.I. & Burton, J.D. 1978. Behaviour and speciation of dissolved selenium in estuarine waters. *Nature*, **273**: 293-295.

Millward, G.E. & Turner, A. 1995. Trace metals in estuaries. In: *Trace elements in natural waters*. Salbu, S. & Steinnes, E. (Eds.), CRC Press, Boca Raton, 223-245.

Moore, R.M., Burton, J.D., Williams, P.J. le B. & Young, M.L. 1979. The behaviour of dissolved organic material, iron and manganese in estuarine mixing. *Geochimica et Cosmochimica Acta*, **43**: 919-926.

Mostafa, H.M. & Collins, K.J. 1995. Heavy metal concentrations in sea urchin tissue from Egypt, Ireland and United Kingdom. *Chemistry and Ecology*, **10**: 181-190.

National Rivers Authority, 1994. Implementation of EC Shellfish Waters Directive. *NRA Water Quality Series* No **16**, 72 pp.

Oyenekan, J.A. 1980. Community structure and production of the benthic macro infauna of Southampton Water. Ph.D. Thesis, University of Southampton, Southampton, 351 pp.

Phillips, A.J. 1980. *Distribution of chemical species. The Solent estuarine system: An assessment of present knowledge*, The Natural Environment Research Council Publications Series C No 22. NERC, 44-61.

Pilskaln, C.H., Churchill, J.H. & Mayers, L.M. 1998. Resuspension of sediment by bottom trawling in the Gulf of Maine and potential geochemical consequences. *Conservation Biology*, 12(6): 1-8.

Savari, A. 1988. Ecophysiology of the common cockle *(Cerastoderma edule L.)* in Southampton Water, with particular reference to pollution. PhD Thesis, University of Southampton, Southampton, 340 pp.

Savari, A., Lockwood, A.P.M. & Sheader, M. 1991. Effects of season and size (age) on heavy metal concentrations of the common cockle (*Cerastoderma edule* L.) from Southampton Water. *Journal of Molluscan Studies*, 57: 45-57.

Soulsby, P.G., Lowthion, D. & Houston, M. 1978. Observations on the effects of sewage discharged into a tidal harbour. *Marine Pollution Bulletin*, 9: 242-245.

Tankere, S.P.C. 1992. Studies of trace metals in an urbanized estuary (Southampton Water). M.Sc. Thesis, Université de Toulon et Var, Toulon, France, 50 pp.

Tappin, A.D., Hydes, D.J., Burton, J.D. & Statham, P.J. 1993. Concentrations, distributions and seasonal variability of dissolved Cd, Co, Cu, Mn, Ni, Pb, and Zn in the English Channel. *Continental Shelf Research*, 13: 941-969.

Waldock, M.J., Waite, M.E. & Thain, J.E. 1993. *The effect of the use of tributyltin (TBT) antifoulings on aquatic ecosystems in the UK*. Contract PECD 7/8/74, Department of the Environment, 234 pp.

Weir, J.L. 1995. A study on heavy metals (Cu, Cd, Zn, and Mn) in Southampton Water using two indicator species, a mollusc *Crepidula* fornicata (Linn.) (American slipper limpet) and an Echinoderm, *Psammechinus miliaris* (Gmelin.) (Green Sea Urchin), Department of Oceanography, University of Southampton, 79 pp.

Microbiological Quality of the Solent

David Lowthion

Environment Agency, 4 The Meadows, Waterberry Drive, Waterlooville, Hampshire, PO7 7XX, U.K.

INTRODUCTION

Public interest in the microbiology of estuarine and coastal waters is focused on those bacteria associated with sewage pollution primarily because of: (a) the public health implications associated with bathing, or other water contact sports, in water contaminated by sewage; and (b) the health implications related to the consumption of contaminated food products such as shellfish. Therefore, this paper is restricted to a discussion on the levels of bacteria that are most commonly regarded as indicators of sewage pollution i.e. total coliforms, faecal coliforms and faecal streptococci (enterococci). In the Solent area, monitoring activity for these bacteria is targeted at bathing waters, other recreational waters and shellfish waters.

There is a widespread belief that sea bathing may lead to illness, but evidence for such an association has proved elusive. No consistent correlation between any single microbiological indicator of water quality and disease has emerged from the many studies that have been undertaken world-wide (Kay *et al.*, 1994). Despite this limitation, microbiological standards and recommendations have been drawn up for bathing waters in Europe and north America. To qualify this statement, there is no doubt that bathing in seawater contaminated with sewage causes illness (particularly, gastrointestinal symptoms). The doubts relate to the parameters and the standards that we should seek to achieve, to minimise such health risks.

EC BATHING WATER DIRECTIVE

The two main objectives of the EC Bathing Water Directive are to improve, or maintain, the quality of bathing waters for amenity reasons, and to protect public health (EEC, 1976). The directive's microbiological standards provide benchmarks, against which to assess progress in achieving improvements in water quality. The public health relevance of these standards has been questioned by many authors. However, UK government research into the health effects of sea bathing have concluded that the imperative standards of the directive for total coliforms and enteroviruses (by implication, for faecal coliform bacteria), provide adequate protection to health and do not support the introduction of more stringent standards (Pike, 1994).

The Environment Agency has statutory responsibilities to sample and analyse bathing waters in accordance with the requirements of the Directive. The Agency also has statutory duties and powers to control discharges to waters, with respect to relevant water quality objectives. For bathing waters, water quality objectives were set out in the Bathing Water (Classification) Regulations, which came into force in August 1991.

Concentrations of faecal bacteria can be very variable, but generally reflect: (i) proximity to sources of human and other faecal matter; and (ii) the dispersion and dilution capabilities of the receiving waters. Levels in rivers typically exceed those in estuaries, which exceed those in coastal waters (Pike & Gale, 1993); this is a pattern that is followed in the Solent.

The continuous sewage discharges into the Solent and its adjacent estuaries (Figure 1) are the major source of faecal bacteria, although storm water discharges are significant during periods of wet weather (especially in waters close to the shore). The numbers of faecal coliform and faecal streptococci decline as the degree of conventional treatment is increased (Table 1). However, there are still large numbers of faecal bacteria present in full, secondary treated, sewage. To place this into context, a small untreated sewage discharge such as Cowes long sea outfall (LSO) will discharge a similar faecal coliform load to the much larger secondary treated effluent discharge to Stokes Bay. Stormwater, as discharged down combined sewer overflows during wet weather, has bacteriological levels about an order of magnitude less than secondary treated effluent.

Table 1 Typical bacterial concentrations in raw and treated sewage. Source: Department of the Environment "Revision of the Bathing Water Directive: Study of the cost implications to the UK sewerage undertakers and regulators" Final Report by Sir William Halcrow and Partners Ltd, February 1995.

Treatment Stage	Faecal coliforms (cfu/100ml)	Faecal Streptococci (cfu/100ml)
Raw Sewage	1×10^7	1×10^6
Stormwater	1×10^5	1×10^4
Primary Treated Sewage	5×10^6	5×10^5
Secondary Treated Sewage	1×10^6	1×10^5
Disinfected Final Effluent	5×10^2	5×10^1

The Directive specifies mandatory and imperative standards, as well as guideline standards for a number of microbiological and physico-chemical parameters. The microbiological standards are the most important (Table 2), but the standards are not absolute; compliance is determined on a percentage achievement basis. For the imperative standards, the Directive requires that 95% of samples must comply i.e. 19 samples out of the 20 taken. For the guideline standards, which are 20 times stricter than the imperative standards, the percentage compliance requirement is 80% for the coliforms and 90% for the streptococci.

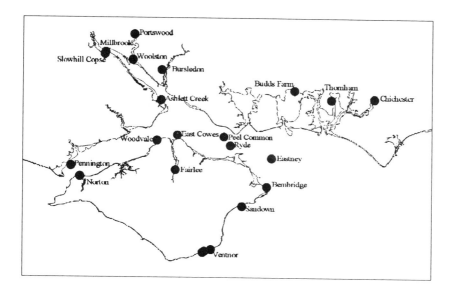

Figure 1 Location of major sewage discharges to the Solent.

Table 2 EC bathing water microbiological standards (EEC, 1976).

	Imperative	Guideline
Total coliforms (cfu/100ml)	10,000	500
Faecal coliforms (cfu/100ml)	2,000	100
Faecal streptococci (cfu/100ml)	-	100
Salmonella in 1 litre	0	-
Enterovirus (pfu/10litres)	0	-

The imperative coliform standards alone are generally used to assess compliance. The directive only requires *Salmonella* and Enterovirus to be monitored when an inspection of the bathing area shows that the substance may be present or that the quality of the water has deteriorated. The Department of Environment, Transport and the Regions (DETR) has indicated that a minimum of two samples should be analysed, at those bathing waters which failed the imperative coliform standards the previous year.

BATHING WATER QUALITY

The results obtained for bathing waters in the Solent, for the 1997 bathing season, are shown in Figure 2. Squares mark those bathing waters that failed to meet the imperative standards, - Southsea, Seagrove, Gurnard and Totland. Circles mark bathing waters that met the imperative standards but failed the guideline standards. Compliance with the guideline standards was achieved at all the bathing waters marked with diamonds. In 1998, all of these bathing waters complied with the imperative standards.

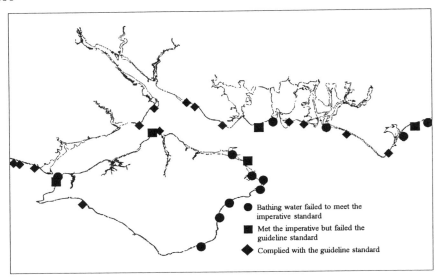

Figure 2 Compliance of bathing waters in the Solent during 1997.

Table 3 summarises compliance with the imperative and guideline standards, over the last decade, for Hampshire and Isle of Wight Bathing waters. In 1988, compliance with imperative standards was much lower (42%) than in subsequent years. However, this figure is misleading because of the way compliance was judged during that particular year. Prior to 1988, the bathing waters were only sampled 12 times per season; one failure was allowed, this being the UK Government's interpretation of '95% compliance'. In 1988, a stricter interpretation of percentage compliance was introduced and the sampling frequency was increased to 20 times per season. In 1988, only a small number of the bathing waters were sampled at the higher frequency and 36% of the local bathing waters had a single failure, in 12 samples. Many of these may have complied, if a full 20 samples had been completed.

In other years, compliance with imperative standards has been variable, peaking at 92% in 1993 but falling to 84% in 1997, and then rising to 100% in 1998. This 100% compliance rate is welcomed, but must be qualified by recognising that a number of bathing water improvement schemes are still to be completed; these include those at four of the bathing waters which failed in the previous two years.

Compliance with guideline standards cannot be accurately assessed prior to 1991, because the detection level of the faecal coliform analysis for Isle of Wight samples was too high (in relation to the guideline standard). In subsequent years compliance has also been variable and it is difficult to be certain if there is any evidence of improvement, although 1996 and 1998 were especially high.

Variation in compliance from year to year can be caused by a number of factors as outlined in Table 3.

Table 3 Bathing water compliance for Hampshire and Isle of Wight bathing waters (for definitions see text).

Year	Number of Bathing Waters	Comply with Imperative Standards	Comply with Guideline Standards
1988	24	42%	-
1989	24	67%	-
1990	25	88%	-
1991	25	80%	40%
1992	25	76%	44%
1993	25	92%	36%
1994	25	92%	24%
1995	25	88%	36%
1996	25	88%	72%
1997	25	84%	48%
1998	26	100%	58%

Completion of a remedial sewage scheme

Most of the major schemes dealing with continuous sewage discharges in the Solent were completed in the late 1980's, with the exception of the commissioning of: the Eastney LSO in May 1991; minor works associated with the Cowes scheme completed in 1992; and the recent diversion of discharges from Barton-on-Sea and Totland. Remedial schemes for Bembridge and Ventnor are still outstanding.

Weather conditions

Wet weather increases the likelihood of combined sewer overflows, into the bathing waters. Locally, it is these intermittent discharges that cause most of the bathing water failures. Southern Water Services Ltd are required to complete a large number of improvement schemes, by 2000. These schemes include improvements in screening to meet aesthetic standards; improvements to reduce the frequency of discharges; and relocation of outfalls, to below low water in some locations. Sunshine kills bacteria in seawater, so compliance may vary depending on the weather on a year-to-year basis.

Statistical variability

Some bathing waters have borderline quality and they pass the Directive's standards one year, but fail the next, or *vice-versa*. This variability is caused by a combination of the inherent variability of the data collected over the bathing season and the way compliance is assessed on a pass/fail basis. Changes in compliance may be caused simply, for no other reason, than by statistical chance.

A more consistent measure of long term bathing water improvements can be obtained by examining bathing water compliance over a longer term. Table 4 shows the proportion of beaches that passed imperative standards consistently for three consecutive years, failed once in three years, failed twice in three years, or failed consistently for all three years. The table shows a pattern of decrease in the numbers of bathing waters that consistently, or very regularly, fail standards.

The EU has announced recently their intention to revise thoroughly the Bathing Water Directive. Opinion polls suggest the public believe standards are not high enough, with a willingness to pay increased charges to fund further improvements (particularly, disinfection of effluents). Options to deliver higher standards may be included in Southern Water Services' asset management plans for the period 2000-2005, but final requirements will be determined by the Government.

Table 4 Consistency of compliance, over three year periods, for Hampshire and Isle of Wight bathing waters.

Years	No Failures	Failed 1 Year	Failed 2 Years	Failed 3 Years
1988-1990	9	7	7	2
1989-1991	16	4	3	2
1990-1992	19	1	2	3
1991-1993	18	2	4	1
1992-1994	18	4	3	0
1993-1995	21	1	2	1
1994-1996	20	3	1	1
1995-1997	18	5	1	1
1996-1998	20	5	1	0

RECREATIONAL WATER QUALITY

Bathing water improvements may do little to improve bacteriological quality in non-bathing areas, where other water contact sports are important. Many of our estuaries are used for dinghy sailing, canoeing and other activities; likewise, a number of activity training centres are located very close to sewage discharges to confined waters. Examples are centres at Woodmill (near Portswood Sewage Treatment Works (STW)) on the Itchen estuary), at the Itchen Bridge (near to Woolston STW), and at Dodnor (near the Fairlee STW, on the

Medina). Although these discharges are secondary treated, the bacteriological load is significant. These areas have been intensively monitored by Local Authorities and the Agency, typical results are summarised in Table 5.

The Solent itself is also used for water contact sports and these waters will have elevated levels of faecal bacteria in areas down-tide of the sewage outfalls. These sewage plumes can affect large areas, because of the strong tides within the Solent. However, the sewage is diluted rapidly as the discharge jets to the surface above a well-situated outfall. It is dispersed then by the tidal currents.

These waters are not monitored regularly for faecal bacteria. Limited data collected in winter, from 13 shellfish beds in the east and west Solent, indicate that geometric mean faecal coliform concentrations range between 85 and 209 cfu/100ml.

EC SHELLFISH DIRECTIVES AND QUALITY

The Solent has an expansive oyster fishery, with numerous harvesting areas. Standards for the quality required of shellfish waters are established in the EC Shellfish Waters Directive (EEC, 1979) whereas quality criteria for the shellfish itself (required to protect consumers of shellfish) are established in the EC Shellfish Hygiene Directive.

The Shellfish Waters Directive includes a numerical value for faecal coliforms; 300 per 100 ml of shellfish flesh and intervalvular fluid. This level is provided as a guideline, not an imperative standard; compliance is judged on a 75 percentile basis with quarterly sampling. Only 17 shellfish waters have been identified in England and Wales; five of these from within the Solent - Stanswood Bay, Lepe Middle Bank, Sowley Ground, Yarmouth Road and Newtown Bank. Twice per annum monitoring, over the period 1992 to 1998, indicates that the 75 percentile faecal coliform concentrations vary between 120 at Stanswood up to 430 at Sowley. These waters are now sampled monthly, by the Environment Agency.

The Shellfish Hygiene Directive does not impose any obligation to achieve or maintain a particular bacteriological quality, but classifies shellfish harvesting areas according to the bacteriological quality of the shellfish populations. The classification determines the conditions under which shellfish harvested from those waters can be offered for sale, as summarised below.

1. Category A waters - shellfish can be sold direct with no treatment. Shellfish must have less than 300 faecal coliforms/100g flesh

2. Category B waters - shellfish must be cleansed or relayed until they meet Category A standards. Shellfish must have less than 6000 faecal coliforms/100g flesh - 90 percentile compliance

Table 5 Microbiological quality of Solent estuaries and Harbours (based on faecal coliform bacteria). For locations, see maps in *Preface*.

Estuary	Location	Period	% Compliance with Imperative Standard	Geometric Mean Faecal Coliforms
Itchen Estuary	Woodmill	1992-1993	0%	9854
	Itchen Bridge	1992-1996	82%	1175
Hamble Estuary	Fairthorne Manor	1992-1996	70%	1098
	Upper Hamble CP	1992-1996	94%	207
Medina Estuary	Dodnor	1992-1996	47%	2069
	Folly Inn	1992-1996	81%	504
Portsmouth Harbour	Hardway	1991-1993	100%	29
	Stoke Lake	1991-1993	100%	27
Langstone Harbour	Langstone Bridge	1992-1996	88%	73
	Portsmouth Schools Sailing Centre	1992-1996	95%	57
	Entrance Channel	1992-1996	100%	36
Chichester Harbour	Fishery Buoy	1992-1998	100%	13
	Mill Rythe	1992-1998	100%	14
	Emsworth Channel	1992-1998	100%	17
	Camber Buoy	1992-1998	100%	14
	Thornham Channel	1992-1998	91%	72
	Cobner Buoy	1992-1998	100%	43
	Birdham Buoy	1992-1998	79%	330
	Dell Quay	1992-1998	63%	1057

3   Category C waters - shellfish must be relayed for at least two months, followed by sufficient treatment to achieve Category A standard, or heat-treated by an approved method. Shellfish must have less than 60,000 faecal coliforms/100g flesh

4   Shellfish harvesting is prohibited from areas which have a poorer quality than Category C

Locally, the Hygiene Directive applies to 11 production areas; these include a large number of shellfish beds, including oysters, hard clams and cockles. Each shellfish harvesting bed is sampled monthly, by either, the relevant Local Authority, or the Public Health Laboratory Service (on their behalf). These data are used by the Centre for Environment, Fisheries and Aquaculture Science (CEFAS) to classify the waters. The 1998 classification is summarised in Table 6.

Southampton Water, Chichester, Langstone and Portsmouth Harbours are classified Category B, except for areas close to the sewage discharges in Langstone and Chichester. Shellfish beds in the Solent are classified as Category B or C depending upon their location. Harvesting areas at Pennington and Lymington were prohibited in the 1997 classification, but have been upgraded to B Category, in 1998. The improvements are as a consequence of the application of secondary treatment to the sewage discharge at Pennington. The harvesting area at Yarmouth remains prohibited. Other harvesting areas in the Solent have generally been classified as Category C.

There is an obvious need for rationalisation of designations under the EC Shellfish Waters Directive and shellfish hygiene regulations. In the summer of 1998, Government consulted on proposals to designate additional waters, in which shellfish populations are currently harvested (or harvestable), under the Shellfish Waters Directive. It is expected that approximately 22 shellfish waters will be identified in the Solent and its adjacent estuaries and harbours, including the existing five designations. These shellfish waters will be areas with similar water quality, or similar water quality influences; hence, they may encompass one or more harvesting areas.

The Government has also consulted on standards and considers that an operational standard for faecal coliforms is essential in the drive towards meeting the Shellfish Waters Directive's secondary purpose, of contributing to public health. Government proposes that, in setting operational standards for faecal coliforms, the Environment Agency should seek to achieve a water quality standard for shellfish waters no less stringent than 300 /100 ml in 75% of samples. This is the water standard that has been calculated to be equivalent to Category B, within the Shellfish Hygiene Directive. This uniform approach takes no account of the different filtration and bacteriological accumulation rates, of shellfish species.

If progressed, the above proposals will result in substantial increases in the bacteriological monitoring of local estuarine waters and will establish a need for water quality improvement plans, for a number of fisheries. These improvements might involve UV disinfection of continuous sewage discharges, improvements to intermittent discharges (CSO's), and some attention to riverine inputs. The majority of the costs will be funded by the Water Industry, with most of the investment in the period 2000-2004.

URBAN WASTE WATER TREATMENT DIRECTIVE

The Urban Waste Water Treatment Directive established minimal levels of treatment for sewage discharges, based upon their size and the nature of the waters receiving the discharge (EEC, 1991). Major schemes for Portsmouth and the Isle of Wight will result in the cessation of five untreated discharges into the Solent (Bembridge, Ryde, East Cowes, Woodvale and Norton - for general location and main sites referenced refer to *Map A* and *Map B* in the *Preface*); those remaining will receive secondary treatment by 2000. The incidental improvements to offshore bacteriological quality within the Solent will be significant, benefiting recreational users and shell fisheries.

MONITORING AND RESEARCH NEEDS

Existing monitoring programmes and computer predictions provide adequate information on levels of faecal bacteria, in relation to the bathing and recreational uses of our waters. There will be little benefit in additional monitoring, except to investigate specific problem areas. Future research should focus on identifying sources of enteroviruses and other pathogens, their fate in the environment and their public health implications.

The Government proposes to include large numbers of shellfish waters within the scope of the EC shellfish waters directive and to ensure that all such waters meet operational faecal coliform standards necessary to ensure shellfish harvesting areas achieve Category B class. This will generate a need for substantive additional monitoring, as well as research to establish the significance of different sources of faecal pollution in some local shell fisheries.

Table 6 The 1998 classification of shellfish harvesting areas in the Solent – Designated bivalve mollusc production areas in England and Wales, September 1998. CEFAS, Weymouth Laboratory (Prov.=provisional)

Production Area	Bed Name	Species	Class
Chichester Harbour	Birdham Spit All other beds	O. edulis	C
Langstone Harbour	Broom Channel All other beds	O. edulis	B
Portsmouth Harbour	All beds	O. edulis	C
Southampton Water	All beds		B
Solent	Ryde Middle Bank, Kings Quay, Osbourne Bay, The Butts, Lee-on-Solent, Hill Head, Bramble Bank, Stanswood Bay, Lepe, East Sowley, Peel Bank, Mother Bank.	O. edulis	B
	Lymington Bank, Sowley and Pennington Bank.	O. edulis	B (Prov.)
	Spit Bank, Stokes Bay, Newtown Bank, Saltmead Ledge, Thorness, E and NE Ryde Middle, Ryde and SW of Gilkicker.	O. edulis	C
	Saltings (Lymington).	Cockles	C (Prov.)
Medina	Wharf	O. edulis	C
Newtown Harbour	Clamerkin Creek. Western Haven and Rivermouth inner.	O. edulis	B C
Totland Bay		O. edulis	C
Beaulieu	Needs Ore and Bucklers Hard.	O. edulis	C
Lymington River	All beds below Railway Bridge.	O. edulis	C
Keyhaven	Keyhaven River	O. edulis	C (Prov.)

REFERENCES

European Economic Community 1976. Council Directive of 8 December 1975 concerning the quality of bathing water (76/160/EEC). *Official Journal of the European Community*, L31, p1.

European Economic Community 1979. Council Directive of 30 October 1979 on the quality required of shellfish waters (79/923/EEC). Official Journal of the European Community, L281, p47.

European Economic Community 1991. Council Directive of 21 May 1991 concerning urban waste water treatment (91/271/EEC). *Official Journal of the European Community*, L135, p40.

European Economic Community 1991. Council Directive of 15 July 1991 laying down health conditions for the production and placing on the market of live bivalve molluscs (91/492/EEC). *Official Journal of the European Community*, L268, p1.

Kay, D., Fleisher, J.M., Salmon, R.L., Jones, F., Wyer, M.D., Godfree, A.F., Zelenauch-Jacquotte, Z., & Shore, R. 1994. Predicting likelihood of gastroenteritis from sea bathing: results from randomised exposure. *The Lancet*, **344**:905-909.

Pike, E.B. & Gale, P, 1993. *Development of Microbiological Standards*. WRc R&D Note 165, Water Research Centre plc, Medmenham, 103 pp.

Pike, E.B., 1994. *Health effects of sea bathing* (WMI 9021)-Phase III. WRc Report DoE 3412/2, Water Research Centre plc, Medmenham, 138 pp.

Behaviour of Organic Carbon in Southampton Water

Mark Varney

School of Ocean and Earth Sciences, Southampton Oceanography Centre, University of Southampton, European Way, Southampton, SO14 3ZH. U.K.

INTRODUCTION

Organic carbon in the waters and sediments of Southampton Water exhibits contrasting behavioural differences as illustrated in the following three examples presented here: plant litter, hydrocarbons, and volatile organic carbon compounds. Although much is known about the occurrence of these classes of compounds within Southampton Water together with their chemistry, more research is required to identify the mechanisms which determine their distributions.

PLANT LITTER

Plant litter is derived from terrestrial plants and trees; it is a surfactant that is most often noticed as foam generated by ledges and races, etc. Early analytical work discovered the presence of this material as interference when extracting plant pigments, using a solvent such as pentane. The plant material is neither soluble in the aqueous or organic layers, preferring to form a milky emulsion. On isolation, the plant litter exhibits a high molecular weight (typically 10^3 to 10^5 Daltons), and complex macromolecular polyelectrolyte chemistry. The foam is stable for many days and the solvent extract method now serves as the basis for its collection. Enrichment factors of organic carbon in the foam are between 200 and 5000 times greater than normal water column values.

The Function of Plant Litter

Principal functions of organic compounds in plants are as food reserves, structural materials, and antidessicants. Structural polysaccharides include: cellulose (the main component of most plants; pectin (present in woody tissues and fruit); and chitin (main component of fungi, some algae, as well as in arthropods and molluscs). These compounds are distinguished by the degree of chain branching and the variety of monosaccharide units they contain.

The Chemistry of Plant Litter

Infrared spectroscopic analysis of the plant extracts (Figure 1) has shown high degrees of hydroxylation and carboxylation (-OH and -COOH groups). The chemical composition is typically 600 µg C/l protein and 4500 µg C/l carbohydrate - principally saccharide in nature. There is a small lipid content (e.g. <0.1% fatty acids). Characterisation by Raman spectroscopy reveals that the origin is principally terrestrial. It is evident that the macromolecular material has undergone considerable oxidation: condensation and transformation changes by the time it impacts in the estuarine environment. As such, these changes explain the limited aqueous solubility and considerable resistance to further chemical change. Potentially, polysaccharides from plant litter constitute a very large proportion (up to 90%) of the total organic carbon within the estuarine environment.

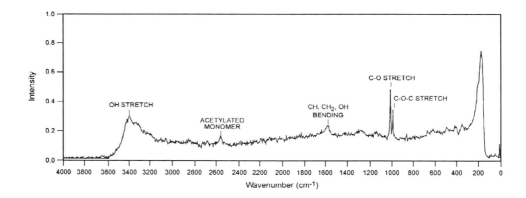

Figure 1 A typical infrared spectrum of the organic material extracted from the 'foam' found within the Eling Harbour area. The peaks can be assigned to stretching and vibrational frequencies of various carbon-carbon and carbon-hydrogen bonds, and groups containing oxygen. In this instance, it shows that the material has a high proportion of oxygen and is aliphatic in nature.

The Distribution of Plant Litter within Southampton Water

A high proportion of polysaccharide material within Southampton Water is of terrestrial origin with an obvious source from the New Forest river catchment area and intensive agricultural activities in the area. Despite this considerable material input, plant litter is not ubiquitous throughout Southampton Water. The highest concentrations (up to 90% of polysaccharide) are found adjacent to Marchwood (see Figure 3). The sediment surface has a distinct 'slimy' appearance; likewise, sediment cores often reveal a 'marbled' effect. This region is where freshwater first meets with the seawater (salinities are typically less than 25). Upstream, the freshwater is prevented from mixing by the tidal lagoon at Eling and by a weir at Redbridge (Figure 2). These physical barriers, restricting the freshwater, permit undiluted seawater to intrude long distances up Southampton Water. The polysaccharides are thought to be instantly removed from solution (polysaccharides are not water soluble) by a 'salting out' effect.

HYDROCARBONS

Hydrocarbons have been, and continue to be, discharged principally into Southampton Water from two major sources - industrial and domestic. There are several local, well-defined sources in and around Southampton Water, situated along its entire course (Figure 3). Domestic sources are principally internal combustion engine-derived, and include the supposedly inadvertent discharge of waste oil into sewage treatment systems. Industrial

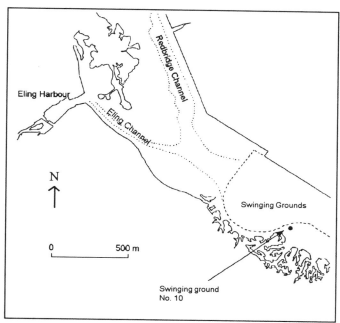

Figure 2 Schematic map of the top of Southampton Water showing the confluence of the River Test and Eling Channel. The Swinging Grounds are where large container ships and other cargo carriers are allowed to turn prior to departure. The dashed line indicates the 10 m (maintained) depth contour, while the dotted line is the approximate position of the 5 m depth contour, defining the two main channels.

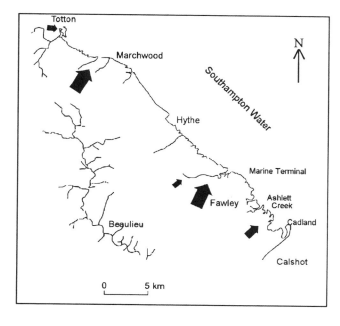

Figure 3 Schematic map of the lower part of Southampton Water, showing the major inputs of organic matter (sewage treatment works, surface run-offs, and discharge outfalls). The sizes of the arrows are an indication of the relative magnitudes of the inputs.

souces of hydrocarbons are well documented; however, because of the legislative restrictions on discharges into the aquatic environment, the inputs have declined in recent years. Environment Agency "Consent to Discharge" limits have been reduced from 10/20 ppm to under 5 ppm in just 5 years; this reduction is reflected in the sediment records.

Incorporation into sediments

Dissolved seawater concentrations of hydrocarbons display a clear downward trend (Figure 4). Higher concentrations of hydrocarbons (often parts per thousand) are found in the sediment; concentrations that have remained virtually constant over the past two decades (Figure 5). Hydrocarbons are not generally soluble in water and are preferentially adsorbed onto particles; this is followed by the slow accumulation into sediments. The partition coefficient (K_D) defines the concentration ratio of the chemical in water, relative to the sediment. Most hydrocarbons have small, but measurable solubilities and K_D's. Time is an important factor in the ultimate concentrations of hydrocarbons in sediments. Calculations have suggested that >99 % of hydrocarbons associated with suspended particulates will be removed from Southampton Water into the Solent system (and beyond) if the residence time of water within Southampton Water is less than 5 days. In other words, the concentration of hydrocarbons observed in the sediments is only a very small fraction of the huge quantities discharged into the water.

POLYNULEAR AROMATIC HYDROCARBON'S (PAH) IN SEDIMENTS

Polynuclear aromatic hydrocarbons (PAHs) are probably the most relevant of the compounds identified within the more general 'hydrocarbon' class of compounds. PAHs are often highly toxic, having only anthropogenic sources. Surface sediment concentrations at various locations around Southampton Water (Table 1) reveal that the proportion of PAHs to total particulate organic carbon (POC) can be alarmingly high. The quantity of PAHs and all other organic compounds is proportional to the total amount of particulate matter deposited

Table 1 Concentrations of selected PAHs in sediments (units µg/kg), and particulate organic carbon POC (expressed as a percentage of the total quantity of sediment). Location of sites is shown on Figure 3. * No data available

Compound	SG No 10	Hythe	Marine Terminal	Cadland
Fluoranthene	23	3	5	
Pyrene	18	3	4	74
Chrysene	9	*	*	47
Benz(a)pyrene	27	*	*	27
Total PAH	143	6	19	203
POC	12.6%	4.8%	5.8%	10.9%

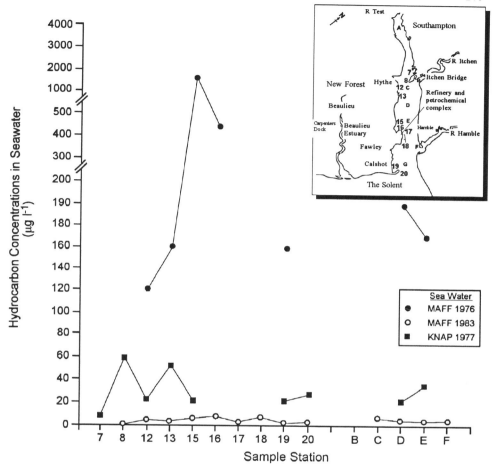

Figure 4 Temporal variation in seawater concentrations of hydrocarbons within Southampton Water. Closed and open circles are historical data from past monitoring exercises undertaken by MAFF (1976 and 1983), the closed squares are those abstracted from the PhD thesis of Knap (1977). Locations are indicated on the map (insert).

The quantity of PAHs and all other organic compounds is proportional to the total amount of particulate matter deposited over the years, as revealed by ultraviolet fluorescence of sediment core extracts (Figure 6). Clearly, the patterns of deposition are different between locations: those in the vicinity of sources continue to accumulate hydrocarbons; those farther away appear to contain less hydrocarbon within the surface sediments.

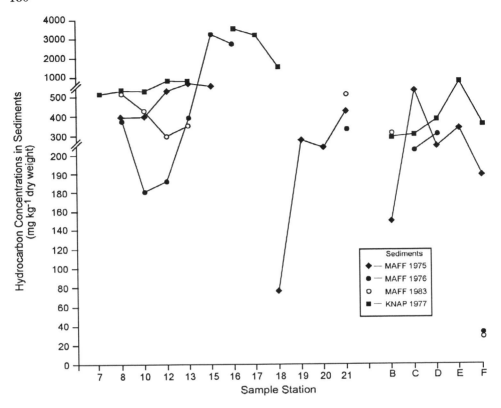

Figure 5 The historical concentrations of hydrocarbons found in surface sediments within Southampton Water have remained virtually constant over the past two decades.

Sediment cores were dated using the Pb^{210} method to enable identification of the historical inputs of hydrocarbons over the past 50 years (Figure 7). Since, Pb^{210} arises from Ra^{226} decay within the sediment (termed "supported" decay) and from the atmosphere (termed "unsupported" or "excess" radon). Assuming decay and sedimentation are constant, age-depth curves can be constructed (Figure 8). Interestingly, there have been different depositional rates over the years. The most recent surface layer between 0 and 5 cm was found to have a depositional rate of 1.2 mm/year. Slightly deeper, a greater amount of sediment between 5 and 20 cm was deposited at a slower rate of 5 mm/year. The point at 5 cm corresponds to an 'erosion' event (circa 1955-58), which makes sediment dating somewhat difficult. However, the levels of total hydrocarbon (THC) are directly proportional to inputs (which have decreased over recent years) and to sedimentation rates. Increased degradation within the top few cm's is unlikely; this is because of the varied, complex nature of the hydrocarbons, which do not break down easily. The date of the subsurface THC maximum in sediments (Figure 8) taken at Calshot corresponds to the post-war commencement of full-scale operations in the petrochemical refinery complex at Fawley.

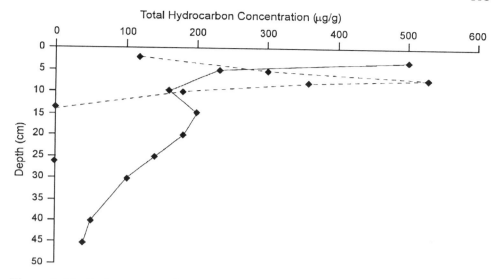

Figure 6 Vertical concentrations of total hydrocarbon concentrations (units µg/g) as revealed by ultraviolet analysis of solvent extracts taken from 5 cm intervals in a sediment core from Hythe (dashed line) and Ashlett Creek (solid line). For location of sites, refer to Figure 3.

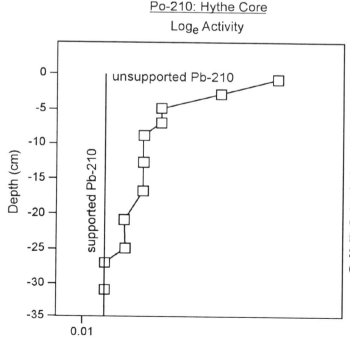

Figure 7 Sediments are dated using Pb^{210} analysis (actually measuring P_0). Natural amounts of Pb^{210} come from atmospheric inputs: when the amounts rise above that level they are termed 'unsupported' and indicate additional (or anthropogenic) injection (see text). Source: K. O'Sullivan (1988).

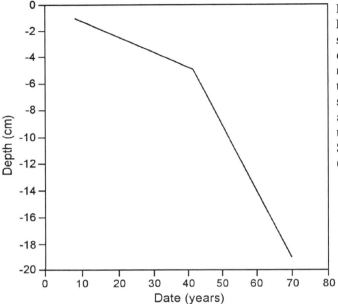

Figure 8 The quantity of Pb^{210} allows the age of the sediment layers to be dated; this reveals in most instances that the deposition of sediments has occurred at different rates over the past 50 years. Source: K. O'Sullivan (1988).

VOLATILE ORGANIC CARBON (VOC)

These compounds are of low molecular weight and exhibit low to medium polarity; they are typically in the range C_3 to C_{18}. There is an extremely wide range of compounds found, of varying concentrations (Figure 9). They are typically extracted by sparging with an inactive gas (helium or nitrogen) and cryogenic adsorption onto an active surface (such as Carbowax or Tenax GC). Gas chromatographic analysis is difficult, because of the coelution of similar compounds at low temperatures; it requires sophisticated automatic thermal desorption equipment. VOC compounds have high volatilities and evaporation is the major route for the loss of these types of compounds. Theoretical models estimate that half-lives within the water column are of the order of hours to days, although there are extremely high concentrations of VOC in Southampton Water. Past projects have routinely investigated 'suites' of VOC compounds over an 8 year period (Table 2). In contrast to other classes of compounds, the historical occurrence of VOC does not reflect any legislative decrease in emission (Figure 10). The behaviour of VOC reflects anthropogenic versus natural sources, and seasonal influences. The quantities of biogenic VOC are significantly seasonal.

Table 2 VOC estimated volatilisation rates (kg/day) for data collected over 3 month periods during the summer and winter period. Source: Bianchi & Varney, 1998b.

Compound	Summer	Winter
Benzene	2	100
Toluene	18	36
Hexane	2	75
Decane	1	10
Trimethylpentane	3	131
Chloroform	5	117
Trichloroethane	1	2
Tetrachloroethylene	3	35
Total VOC	320	2020

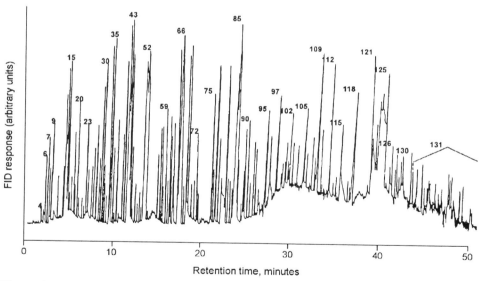

Figure 9 A typical chromatogram of a volatile organic carbon extract taken from a sediment sample in Southampton Water, adjacent to Hythe. Over 200 separate compounds (peaks) may be identified, which possess a mixture of boiling points and chemical functions (alkanes, branched, cyclic, oxygenates, sulphur-containing, etc.). Peaks are numbered sequentially. Source: Bianchi & Varney, 1998a.

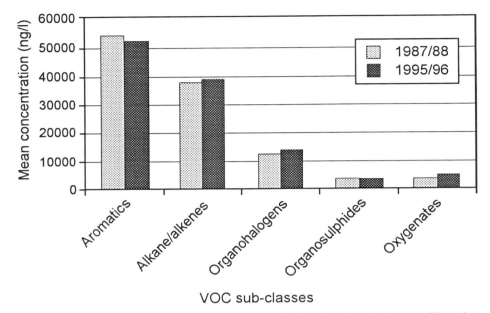

Figure 10 Individual compounds grouped into similar functional classes. The plot verifies that historical inputs of VOCs have not declined significantly, in extent. The data result from a multitude of surveys taken throughout the seasons, within the years 1987-88 and 1995-96. Source: Bianchi, 1990.

The short-term removal of VOC is significant. An important finding from this research is that the evaporative rate of loss is different between the summer and winter months (Table 2). Interestingly, rates are higher during the winter, which may be contrary to expectation, but is ultimately due to the increased turbulence and mixing processes in winter (stronger winds and higher waves). Overall, VOC removal processes are faster than simple dilution (freshwater flow). Removal processes act together, giving rise to an extremely dynamic system; especially in relation to interpreting the occurrences and distributions of VOC. Less than 5% of VOC is removed through adsorption onto particles; the majority is lost to the atmosphere, by evaporation or through direct solubilisation.

SUMMARY

Organic compounds have an extremely wide variety of sources and behaviour in Southampton Water. The sediments act as a large (but inefficient) collection device; the top few centimetres represent historical inputs from the last few decades. Because sediments are mobile and can be re-worked, they can act also as a conveyor belt. It is convenient to sample the sediments and extend the analogy to define the rates of input and mechanisms of transport, for a multitude of different compounds.

The largest input is entirely natural, but one which is least known and least studied. The impact of plant litter is greatest at the head of an estuarine system because of a 'salting out' effect; however, this is strongly seasonal. Our present knowledge of the association of other compounds, with plant litter, is inadequate but the best projections are that there is a very strong interaction. High sedimentary concentrations of plant litter (POC) may enhance the concentrations of other compounds, both natural and anthropogenic. Discharges or releases of other organic compounds may, therefore, be significantly affected by location.

Considerations to the future planned discharges or releases of compounds should include; residence times of water in Southampton Water (related to rainfall); suspended sediment concentrations (related to flow rates and meteorological conditions); and season (mixing processes).

ACKNOWLEDGEMENTS

I am indebted to Alex Bianchi for much of the data shown herein; similarly, to various undergraduate students who willingly and happily carried out the lengthy and laborious extraction procedures (as part of their final year practical projects).

REFERENCES

Bianchi, A.P. & Varney, M. 1989. A modified technique for the determination of trace organics in water using dynamic headspace and GCMS. *Journal of Chromatography,* **467**: 111-128.

Bianchi, A.P. & Varney, M. 1998a. Volatile organic compounds in the surface waters of a British estuary. Part 1 Occurrence, distribution and variation. *Water Research,* **32**(2): 352-370.

Bianchi, A.P. & Varney, M. 1998b. Volatile organic compounds in the surface waters of a British estuary. Part 2 Fate processes. *Water Research,* **32**(2): 371-379.

Bianchi, A.P. 1990. *Volatile and non-volatile organic compounds in the water and sediments of Southampton Water and the Beaulieu estuaries.* Unpublished Ph.D thesis, Open University.

Knap, A, 1977. *The fate of non volatile petroleum hydrocarbons in refinery effluent entering Southampton Water.* Unpublished Ph.D. Thesis, University of Southampton.

O'Sullivan, K. 1988. *Analysis of hydrocarbons in the sediments of Southampton Water using ultraviolet fluorescence and gas chromatography, and a study of their distribution using ^{210}Pb dating.* Unpublished BSc undergraduate project report, University of Southampton.

Sewage Contamination of Bathing Waters: Health Effects

Gareth Rees

Robens Centre for Public and Environmental Health, Farnborough College of Technology, Boundary Road, Farnborough, Hampshire, GU14 6SB, U.K.

INTRODUCTION

The issues surrounding recreational waters and potential adverse health outcomes are extremely complex. The economic benefits of recreational waters to coastal communities can be considerable. However, it also follows that any factor that intervenes to disrupt the level of recreational use of an area can have considerable adverse effects on those communities. The Solent region has 34 identified bathing waters, which attract a great deal of revenue to the area. Of those 34 bathing waters, 33 comply with the EU Bathing water directive, 24 of which attain the Guideline levels of the Directive, equivalent to the highest quality standard. Not withstanding the uncertainty regarding the acute and chronic health risks from exposure to recreational waters, the issue of water quality occupies an incredibly high position on the public agenda (Rees, 1993). The unexplained failures of Southsea beach to meet the EU guidelines and the associated media interest amply illustrate the point. The various types of recreational waters are not mutually exclusive, but different factors may gain prevalence in different types of waters – whether marine, fresh, or swimming pool waters. This paper will concern itself almost exclusively with coastal recreational waters and will discuss, in general terms, the items that need to be considered in assessing the likelihood of adverse health effects.

PUBLIC HEALTH ISSUES

Public health issues associated with recreational water use include drowning, trauma, sunburn and bathing associated illness. Drowning and physical injuries are frequently associated with water-based recreation; in the latter case these injuries include those sustained from diving and collision with submerged objects. Associated with bathing may be over-exposure to the sun, leading to sunburn and potentially more serious conditions including melanoma. Finally, bathing-associated illness is of particular importance where sewage is known to be discharged in the vicinity of a recreational water area. Such illnesses are generally relatively minor in severity and duration, although anecdotal evidence is regularly provided to support claims that illnesses that are more serious can be contracted via exposure to sewage-contaminated recreational waters.

SEWAGE DISCHARGES

Sewage discharges direct to the coast pose a range of potential risks to human health. Apart from bathing-associated illness, there may be the potential for sanitary items such as medical waste, carried by insufficiently treated sewage discharges, to cause illness, trauma or infection (see below). Certainly significant levels of needlestick injury have been attributed to coastal recreational activities, but not necessarily swimming (SWRHA 1991). Apart from adverse health issues, it is interesting to note that sewage pollution (as indicated by items of sanitary and medical waste) can have a drastic effect on the economies of coastal tourist locations. This pattern was noted in the 1980s in the bathing resorts of New Jersey, USA. An estimated $2 billion was lost due to sewage pollution: evidenced by a relatively small number of sanitary items and medical waste washing up on the beaches (Valle-Levinson & Swanson, 1991).

INFECTION

A pathogen can only exert an influence if a chain of infection can be demonstrated. There must be a *source* of the likely infected organism, a *depot* or *reservoir* for that likely infected organism to survive, a *route of spread* and finally a *susceptible host* (Cartwright, 1991). These four stages of the chain of infection can all be shown to occur in recreational waters. Sources include natural sources, agricultural waste, sewage and the bathers themselves. Of these, the major point source of likely pathogenic organisms is undoubtedly any sewage outfall discharging near to recreational water. The depot or reservoir is the capacity of the water body itself to support any microorganisms. Route of spread is primarily via ingestion, which will occur when bathers enter the water; it has been estimated that every time a bather immerses their head, whilst bathing, they ingest approximately 15 ml of water (Cartwright, 1991). The final link in the chain, a susceptible host, depends on the health of the individuals comprising the bathing population.

PATHOGENS – WATERBORNE DISEASE

A wide range of pathogens is found in the marine environment (Table 1). A more comprehensive list of pathogens, isolated from UK marine waters, is found elsewhere (Fewtrell & Jones, 1992). The prime sources of pathogens, believed to cause waterborne disease, are derived from the faeces and urine of warm-blooded animals (Knudson & Hartman, 1992). Enteric viruses have been proved to be a significant aetiological agent (Berg & Metcalf, 1978), but are difficult to detect in water (Marzouk et al., 1984). It has been estimated that more than 100 types of pathogenic viruses which may cause disease, occur in faecal-polluted water (Melnick, 1984; Havelaar, 1993). Melnick (1984) noted the Norwalk RNA viruses, in particular, to be linked with waterborne disease, especially diarrhoea and gastroenteritis. Cabelli (1983) and Havelaar (1993) supported this view by claiming the Norwalk virus, human rotavirus, hepatitis A virus, adenovirus and other enteric viruses are likely to be the main aetiological agents of waterborne diseases. As yet, no perfect indicator has yet been identified to accurately assess the presence of pathogens, although attempts have been made to model bacterial concentrations against illness rates using dose-response curves (Cabelli et al., 1982; Kay et al., 1994). For a detailed summary of epidemiological studies together with resultant models describing health risk from bacterial indicator levels, see Pruss (1996).

Table 1 Pathogens found in marine recreational waters and associated diseases.

	Pathogen	Diseases
Viruses	Poliovirus	Paralysis, meningitis, fever
	Echovirus	Meningitis, respiratory disease, rash, fever, gastroenteritis
	Coxsackievirus A	Respiratory disease, meningitis, fever, hand, foot and mouth disease
	Coxsackievirus B	Myocarditis, congenital heart irregularities, rash, fever, meningitis, respiratory disease
	Hepatitis A	Infectious hepatitis
	Rotavirus	Gastroenteritis, diarrhoea
	Reovirus	Unknown
	Parvovirus	Unknown
	Adenovirus	Respiratory Disease, conjunctivitis, gastroenteritis
	Cytomegalovirus	Hepatitis, infectious mononucleosis, immunological deficiency syndrome
	Papovirus	Associated with progressive multi-focal leukencephalopathy and immunosuppression
Bacteria	*Aeromonas* spp.	Wound infection, gastroenteritis
	Campylobacter spp.	Enteritis
	Clostridium spp.	Botulism, tetanus, gastroenteritis
	Escherichia coli	Gastroenteritis
	Plesiomonas spp.	Gastroenteritis, meningitis, cellulitis
	Pseudomonas spp.	Follicular dermatitis, ear, nose and throat infections
	Salmonella spp.	Enteric fever, gastroenteritis
	Shigella spp.	Bacillery dysentery
	Staphylococcus spp.	Soft tissue infections
	Vibrio spp.	Cholera, wound infections
	Yersinia spp.	Gastroenteritis
Protozoan	*Cryptosporidium* spp.	Gastroenteritis
	Giardia spp.	Gastroenteritis
Fungi	*Candida albicans*	Dermatitis, thrush

Sources: Philipp (1991); Bitton (1994)

EPIDEMIOLOGICAL INVESTIGATIONS

Outcomes of epidemiological studies, undertaken in a wide range of countries over a long period, show a certain degree of consistency in their results (Cabelli, *et al.*, 1982; Cabelli 1983; Cheung *et al.*, 1990; Fattal *et al.*, 1987; Pike, 1994; Seyfried *et al.*, 1985 a, b). Bathers always show an increased incidence of symptoms over non-bather control groups. The commonest symptoms recorded are those related to the eye, ear, upper respiratory tract and gastrointestinal tract. The only consistent relationship with water quality, as evidenced by sewage bacteria indicators, is that between gastrointestinal illness and water quality. All the diseases recorded in such studies are described as 'minor self-limiting symptomatic illnesses'. Such generic conclusions are generally accepted by all those involved in researching, or regulating, the issue. The nature and extent of the illnesses, the relationship between those symptoms and water quality and the indicators of water quality are all rather more contentious. Similarly, anecdotal evidence linking the onset of more serious infections, such as hepatitis A or aseptic meningitis, is not universally accepted.

It is also apparent that the outcomes of epidemiological studies can only be applied on a local basis. When further analyses of the results of these epidemiological studies are undertaken, it is then possible to identify equivocal situations, where the results can be interpreted in a number of ways. Thresholds that result in the onset of illness, derived from these epidemiological studies, vary enormously. For instance, in marine waters the thresholds for *E. coli* vary between 40-414 per 100 ml, whereas those for faecal streptococci vary from 6 up to 9,000 per 100 ml (Ferley *et al.*, 1989; Cheung *et al.*, 1990; Kay *et al.*, 1994; Pruss, 1996).

WATER QUALITY DETERMINATIONS

Once it is shown that there is a potential health risk attributable to water of poor quality, then it is necessary to examine the way in which the quality of a particular water body is determined, with reference to its microbiological components. Such an examination must also consider the factors which may affect that water quality; these include sampling procedures, location of the sampling sites and the techniques used to collect, process and store samples. Analytical conditions (such as incubation, resuscitation, and the composition of culture medium) have a marked effect on the ultimate determination of microbiological quality of water.

The capacity for water quality determinations to vary was the focus of a European Union sponsored report, investigating the performance of methods for microbiological examination of bathing water (Anon., 1995). Several recommendations came out of this international study, which are outlined below:

- Standards should be made available for the validation of all methods used to determine recreational water quality.

- Guarantees should be included on the composition of the media, used to make those determinations.

- Quality assurance schemes should be introduced, to guarantee the quality of all the measurements.

- A system should be defined that could extrapolate sampling data, to inform decision-makers about bathing water quality.
- Specifications for sampling, transporting and storage conditions are essential.

The Report went on to conclude that when a common method was used by skilled analysts, consistent results were produced (Anon., 1995). However, when the skilled analyst used their own individual methods and the results were compared, there was a degree of inconsistency - such findings make the comparison of bathing water quality data between and within different countries rather meaningless. The Report defined some of the analytical conditions, to ensure that in future there would be consistency in comparative water quality determinations (Anon., 1995).

There appears to be limited faith in the numbers generated, as a result of the variety of methods used by a variety of analysts, in the different countries (Fleisher, 1990). Confidence is further undermined when one considers that samples may be taken once every ten or more days: the results from which are then pooled, to provide a retrospective picture of bathing water quality. The numbers collected in this way are used as the basis for the standards applied to regulate bathing waters, posing serious questions as to the worth of those standards. The investment of huge sums of money based on compliance with such flawed microbiological standards is entirely inappropriate. There is also little real-time evidence of the likely risks that a bather may encounter in recreational water, if one considers the data so unpredictable.

PRESENT STANDARDS

The European Union, in introducing a Bathing Water Quality Directive (in 1976) went a long way towards protecting and improving the quality of coastal waters around the European Union (Anon., 1976). In 1994, the EU set about amending this Directive (Anon., 1994) and the process of amendment is still continuing amidst a scientific debate. The original idea of these amendments was to protect the bathing areas, primarily by preventing the discharge of sewage into designated bathing waters and by instituting adequate management policies. The proposed amendments suggested that the sewage discharge installations, together with the finance to achieve those goals, were linked with a microbiological monitoring programme. Although laudable, the fact that the outcome of microbiological monitoring programmes are, at best, very variable, suggests that major capital expenditure based around these numerical values is a 'high risk' strategy.

A number of dilemmas now beset those responsible for producing policy and maintaining standards. Firstly, all responsible authorities agree that the existing standards do not protect against the minor symptomatic illnesses associated with recreational waters influenced by sewage discharges. Secondly, tightening existing standards can only be achieved by a range of sewage treatment improvements; this will be extremely expensive. Thirdly, the health gain due to such tighter standards is unknown, since the relationship between health and current standards is not properly known. However, the general public expects standards to be set at levels that will protect them from illness. It is also important that science should drive the policy agenda that any revision of standards and any proposed management interventions should be based on sound scientific consensus.

Emphasising numerical standards can lead to a great deal of misinterpretation. There must be a uniform approach to arrive at a number that standards can be based; this is patently not the case in and around the member states of the European Union. There are definite uncertainties in the outcomes of sampling and analytical procedures, due to the different processes undertaken by the different authorities. Indicators of sewage contamination are rarely related consistently to levels of likely pathogens: studies variously indicate a strong relationship, or no relationship, between the particular indicator and the particular pathogen. Equally, the relationship between an indicator number and a measurable health effect shows virtually no consistency. One then has to step back and question whether a constant universally applicable number exists that can be applied in the context of recreational waters and health. The more that one bases standards on numbers with an inherent degree of uncertainty, the more one becomes a hostage to those numbers and their subsequent statistical analysis.

THE WAY FORWARD

It is becoming increasingly clear that a wide range of factors contribute to the likely health risk associated with recreational water exposure. Microbiology, although extremely important, is only one of those components. Instead of relying upon a system generated by numbers of dubious provenance, it would seem far more appropriate to:

- Undertake a full audit of the likely influences on a recreational water.
- Categorise those influences.
- Estimate a measure of the likely health risk.

The beach registration system trial, undertaken in the Black Sea Environment Programme (supported by the World Health Organisation (WHO), unpublished) set about the systematic registration of beaches, selecting the various sites, gathering all the data through both desk and field studies and determining what parameters need to be assessed at what sampling stations. This pilot programme went on further to specify sampling and analytical procedures and to define how the subsequent data should be processed. Beach management tools and a hierarchical beach classification system can emerge from such a process, leading to a full and real-time public information strategy.

In the practical assessment, leading to successful beach registration, a number of factors have to be considered. These factors include charting both point and non-point sources of pollution and characterising the nature of treatment in any sewage outfall, which may comprise the major point source of pollution. The effects of climate and oceanographic conditions on a particular water body must also be considered, as should the degree of development of the adjoining hinterland. The nature of the population load (total and seasonal) and their health characteristics is equally important. It is the population serviced by a sewage discharge that produces the sewage and therefore their number and health status will define the health risk posed to bathers – if they are excreting pathogens then the likelihood of infecting bathers is increased. Where possible and appropriate, the microbiological quality of the bathing water should also be assessed; however, it should not be treated as the sole arbiter of compliance with the standard. Visual indicators of pollution should also be an important component of any pollution assessment, particularly in the instance of sewage debris. Indications are that the EU, the WHO and other regulatory and policy organisations with an interest in protection of human health in recreational waters, are gradually shifting their perspective to adopt such a holistic beach registration scheme.

CONCLUSIONS

A move away from strict adherence to a numerical standard that is of dubious worth, to a much more holistic approach is the way forward in safeguarding recreational water users. Instead of a preoccupation with collecting numbers, processing those numbers and arriving at an answer based on those numbers, any such numbers generated should merely comprise part of the process. Where it is known that there is a sewage discharge of a particular volume and a particular level of treatment, a value judgement on how that is likely to impact on the recreational water should be made. However, such an assessment should not be made in isolation of other factors predisposing towards pollution. In a situation where there is no infrastructure appropriate to undertake microbiological determinations, the very presence of a sewage discharge can be taken as posing a particular level of health risk. There may well be other risk factors, of varying importance, to add to the total assessment.

It has been shown that where tertiary sewage treatment has been undertaken to ensure that the effluent is of high quality, run-off from surrounding land can still cause failures in relation to guideline standards (Wyer *et al.*, 1995). This factor is potentially very damaging to the economy supported, to whatever extent, by the bathing water concerned. Further, it indicates just how unpredictable microbiological data contributing to standards can be and how undue prominence should not be placed on the outcomes of the analysis of such inherently variable microbiological data. It is more appropriate to categorise the likely risk, based upon all the available information; this may mean, in some cases, ignoring microbiological data where it is considered to be spurious or derived from a source that is unlikely to pose a risk to human health.

In many countries where there is neither the infrastructure nor the resources available to undertake costly and extensive sampling programmes, it may yet be better to advocate a holistic approach (such as beach registration) rather than limited microbiological determinations. One would hope that this approach would give a truer picture of the quality of the Solent's resort beaches, allowing the public to make informed decisions. It may be that such an approach could help to put in context the inexplicable failures such as those at Southsea and provide everyone – regulators and public alike – with the facts.

REFERENCES

Anon. 1976. Council Directive of 8 December 1975 concerning the quality of bathing waters. *Official Journal of the European Communities,* No L31 5.2.76: 1-7.

Anon. 1994. Proposal for a Council Directive concerning the quality of bathing water. *Official Journal of the European Communities* No C112, 22.4.94: 3-10.

Anon. 1995. *Seawater Microbiology. Performance of Methods for the Microbiological Examination of Bathing Water.* Part I. EUR 16601.EN. ISBN 1018-5593, 62 pp.

Berg, G. & Metcalf, T.G. 1978. Indicators of viruses in waters. In: *Indicators of viruses in water and food.* E. Berg (Ed.), Michigan: *Ann Arbor Science:* 267-298.

Bitton, G. 1994. *Wastewater Microbiology.* Wiley Series in Ecological & Applied Microbiology. Wiley-Liss. ISBN 0-471-30985-0, 478 pp.

Cabelli, V. J. 1983. *Health Effects Criteria for Marine Recreational Waters,* EPA-600/1- 80-031, US Environmental Protection Agency, Health Effects Research laboratory, Triangle Park, North Carolina 27711, 98 pp.

Cabelli, V.J., Dufour, A.P., McCabe, L.J. & Levin, M.A. 1982. Swimming associated gastroenteritis and water quality, *American Journal of Epidemiology,* **115**: 606-16.

Cartwright, R.Y. 1991. Recreational waters: a health risk for the nineties? In: *Health-Related Water Microbiology.* Morris, R., Alexander, L.M., Wyn-Jones, P. & Sellwood, J. (Eds.), International Association of Water Pollution Control and Research, Glasgow, 3-5 September 1991: 1-10.

Cheung, W.H.S., Chang, K.C.K., Hung, R.P.S. & Keevens, W.W.L. 1990. Health effects of beach water pollution in Hong Kong. *Epidemiology and Infection,* **105:** 139-162.

Fattal, B., Peleg-Olevsky, E., Agurski, T. & Shuval, H.I. 1987. The association between seawater pollution as measured by bacterial indicators and morbidity among bathers at Mediterranean bathing beaches of Israel. *Chemopshere,* **16**, 565-570.

Ferley, J.P., Zmirou, D., Balducci, F., Baleux, B., Fera, P., Larbaight, G., Jacq, E., Moissonnier, B., Blineau, A. & Boudot, J. 1989. Epidemiological significance of microbiological criteria for river recreational waters. *International Journal of Epidemiology,* **18**: 198-205.

Fewtrell, I. & Jones, F. 1992. Microbiological aspects and possible health risks of recreational water. In: *Recreational Water Quality Management volume 1 coastal waters.* Kay (Ed.), Ellis Horwood. p220.

Fleisher, J.M. 1990. Conducting recreational water quality surveys: some problems and suggested remedies. *Marine Pollution Bulletin,* **21** (12): 562-567.

Havelaar, A. 1993. Bacteriophages as models of human enteric viruses in the Environment. *ASM News,* **59**: 614-619.

Kay, D., Fleishrer, J.M., Salmon, R., Jones, F., Wyer, M.D., Godfree, A., Zelanauch-Jaquotte, Z. & Shore, R. 1994. Predicting the likelihood of gastro-enteritis from sea bathing: results from randomised exposure. *The Lancet,* **34**: 905-909.

Kay, D., Fleishrer, J.M., Salmon, R., Jones, F., Wyer, M.D., Godfree, A., Zelanauch-Jaquotte, Z. & Shore, R. 1994. Bathing: water quality. *The Lancet,* **344**: 905-909.

Knudson, L.M. & Hartman, P.A. 1992. Routine procedure for isolation and identification of enterococci and faecal streptococci. *Applied Environmental Microbiology,* **58**: 3027-3031.

Marzouk, Y., Manor, Y. & Halmut, T. 1984. Problems encountered in the detection of viruses in water. *Monographs in Virology,* **15**: 131-133.

Melnick, J.L. 1984. Etiological agents and their potential for causing waterborne virus disease. *Monographs in Virology,* **15**: 1-16

Philipp, R. 1991. Risk assessment and microbial hazards associated with recreational water sports. *Reviews in Medical Microbiology,* **2**: 208-214.

Philipp, R. 1993. Community needlestick accident data and trends in environmental quality. *Public Health,* **107**: 363-369

Pike, E.B. 1994. Health effects of sea bathing (WM1 9021) Phase III. WRc Report. Medmenham. 138 pp.

Pruss, A. 1996. *Background paper on health effects of exposure to recreational water microbiological aspects of uncontrolled waters.* WHO European Centre for Environment and Health. ICP EUD 022 DL96/1.

Rees, G. 1993. Health implications of sewage in coastal waters - the British case, *Marine Pollution Bulletin,* **26** (1): 14-19.

Seyfried, P.L., Tobin, R.S., Brown, N.E & Ness, P.F. 1985a. A prospective study of swimming-related illness II. Morbidity and the microbiological quality of water. *American Journal of Public Health,* **75**: 1071-1075.

Seyfried, P.L., Tobin, R.S., Brown, N.E & Ness, P.F. 1985b. A prospective study of swimming-related illness II. Morbidity and the microbiological quality of water. *American Journal of Public Health,* **75**: 1071-1075.

SWRHA, 1991. Health hazards associated with the recreational use of water. South Western Regional Health Authority, 57 pp.

Valle-Levinson, A. & Swanson, R.L. 1991. Wind induced scattering of medically-related and sewage-related floatables. *Marine Technology Society Journal,* **25**: 49-56.

Wyer, M.D., Kay, D., Jackson, G.F., Dawson, H.M., Yeo, J. & Tanguy, L. 1995. Indicator organism sources and coastal water quality: a catchment study on the Island of Jersey. *Journal of Applied Bacteriology,* **78**: 290-296.

SECTION 3

Short Contributions

M₂ Tidally-Induced Water Mass Transport and Water Exchange in Southampton Water and the Solent

Lei Shi and D. A. Purdie

School of Ocean and Earth Science, Southampton Oceanography Centre, University of Southampton, European Way, Southampton, SO14 3ZH, U.K.

INTRODUCTION

Water mass transport is the main physical factor determining the fate of pollutants (e.g. heavy metals, nutrients, Biological Oxygen Demand etc.) in the tidally dominated Solent estuarine system. The estuarine bathymetry and tidal oscillation make the prediction of pollutant dispersal in the estuary complex, even after a single tidal cycle. The conventional time-averaged Eulerian residual current may not exactly represent the water mass transport in the Solent. In fact, residual currents for water mass transport is better represented by the Lagrangian residual. Longuet-Higgins (1969) pointed out that residual currents (Lagrangian residual), of time-varying ocean currents, is the sum of the Eulerian residual and Stokes drift. Much research concerning residual currents has been conducted since then and Cheng *et al*. (1986) gave a more accurate formula where:

Lagrangian residual = Eulerian residual + Stokes drift + Lagrangian drift

One approach to estimate the Lagrangian residual and water mass transport is to use a series of drogue buoy deployments; however, this approach is time-consuming and limited by the number of drogues available. An alternative approach (Awaji *et al*., 1980) is to use a hydrodynamic model and to apply a particle tracking or tracer approach. A 2-D and 3-D numerical model has been developed to examine the water mass transport and water exchange in Southampton Water and the Solent. Lagrangian residual currents have been calculated by using the particle tracking method, and a water exchange ratio between Southampton Water and the Solent has been estimated. A brief discussion of the model results is presented here.

HYDRODYNAMIC MODEL

The hydrodynamic model is a 3-D hybrid finite element baroclinic model. The model uses the mode split method, with the external 2-D model solving the long wave tidal propagation equation. Solution for the vertical profile of velocity is obtained from a velocity-based model formulation. Vertical diffusion coefficients are parameterized using a modified Mellor-Yamada (Mellor & Yamada, 1982) two equation (q^2 - q^2l) turbulence closure model. The model addresses the intertidal flat i.e. the water-land boundary will change with the water level.

The model area and grid, which cover about 3150 km², is shown on Figure 1. The flexible triangular grid is well suited to the complex boundary and bathymetry of Southampton Water and the Solent. Figure 2a and b show the tidal flux through sections across the west Solent (at Hurst Castle) and east Solent (at Spithead) at hourly intervals over the M_2 tidal cycle. From east Solent to west Solent, the net tidal flux over one tidal cycle is about 38 km^3. The tidal prism across Hurst Castle is about 600 km^3 and the tidal flux across Spithead is about 480 km^3. For general location and main sites referenced, refer to *Map A* and *Map B* in the *Preface*.

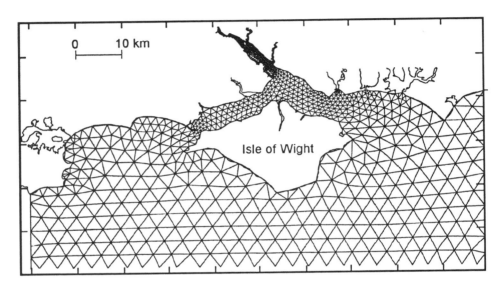

Figure 1 Model area and finite element grid of the Solent estuarine system.

LAGRANGIAN RESIDUAL CURRENT

As a straightforward approach, a particle tracking method was used to estimate the Lagrangian residual. Tracers are released at different phases of the M_2 tide, and their displacement, averaged over one tidal cycle, gives the average displacement. For the average distance moved by a particle in one tidal cycle, the Lagrangian residual current (Figure 3) can be calculated.

There is no significant residual current in Southampton Water. However, the residual current in the Solent is dominated by a westward flowing current. In some areas of the Solent there are regional features, e.g. an eastward residual current along the north part of the west Solent especially at Hurst Castle and in Stanswood Bay (from Stone Point to Calshot Castle), and a clockwise residual gyre at the mouth of Southampton Water. The existence of this residual gyre means that at the section across from Calshot Castle, residual current enters Southampton Water from the west and leaves Southampton Water from the east.

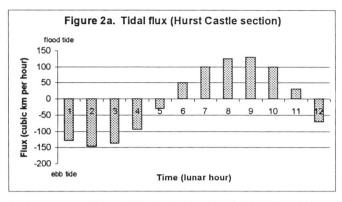

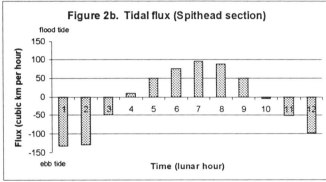

Figure 2 Tidal flux measured across (a) Hurst Castle section (west Solent); and (b) Spithead Section (east Solent).

WATER EXCHANGE BETWEEN SOUTHAMPTON AND THE SOLENT

Figure 4 (a-c) show the water exchange between Southampton Water and the Solent, with tracers released at mid-flood tide. Figure 4a shows the current advection of particles to Southampton Water, during the flood flow. Particles are advected back by ebb flow (Figure 4b), then, after the current turns, particles are moved into Southampton Water. Following one tidal cycle (Figure 4c), some of the water mass is advected out of Southampton Water by the Lagrangian residual current, along the north coast; some intrudes into Southampton Water, along Calshot Spit.

As each labelled particle represents a water column, we can obtain the volume of water exchanged between Southampton Water and the Solent.

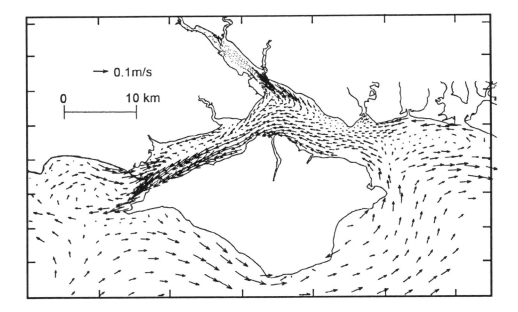

Figure 3 Mean M2 tidally induced Lagrangian residual current in the Solent estuarine system.

The exchange ratio, indicating the magnitude of tidal exchange is defined as:

$$\text{Exchange ratio} = V_{res}/V_{max}$$

Where V_{max} denotes the maximum volume of water flowing into Southampton Water over a tidal cycle, and V_{res} the exchanged volume of water into the estuary after one tidal cycle. For one tidal cycle the exchange ratio between Southampton Water and the Solent is about 20-25%.

DISCUSSION AND CONCLUSIONS

As part of an effort to examine the effects of natural processes and human activity on estuarine water quality, a 2-D and 3-D numerical model has been used to illustrate the water mass transport and water exchange in Southampton Water and the Solent. Model results reveal that there is a net tidal flux from the east Solent to the west Solent. The quantity of the net tidal flux in the Solent is about 38 km³ per M_2 tidal cycle. Using the particle tracking method, it was confirmed that a strong westward M_2 tidally induced residual current exist in the Solent and a residual gyre occurs near the mouth of Southampton Water. This gyre plays a significant role in the water mass exchange between Southampton Water and the Solent.

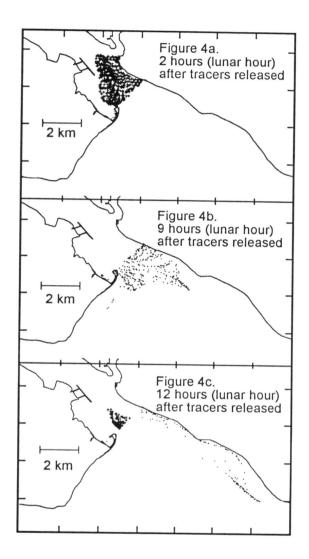

Figure 4 Water exchange between Southampton Water and the Solent, through a section across the estuary at Calshot Castle. All tracers are released at mid-flood tide. The positions of the tracers are shown at certain times (lunar hour) after release: (a) two hours; (b) nine hours; and (c) 12 hours. + represents the water mass from the Solent, entering Southampton Waters; • represents the water mass originating from Southampton Water, entering the Solent.

From the calculated tidally-induced residual current and water mass exchange in Southampton Water and the Solent, the whole Solent estuarine system can be divided into three parts, as outlined below.

1. Weak Water Exchange: Southampton Water, where the tidally-induced Lagrangian residual, as expected, is very small.

2. Medium Water Exchange: the connection between Southampton Water and the Solent, where no net tidal flux occurs, but there exists a significant water exchange ratio of about 20-25%.

3. Strong Water Exchange: the Solent, where the net east-westward tidal flux is 38 km^3 per M_2 tidal cycle (compared to the mean freshwater discharge from the river Test, and river Itchen, of 0.38 and 0.14 km^3 per M_2 tidal cycle, respectively).

Since the tidally-induced Lagrangian residual circulation is small in Southampton Water, it means that the water exchange depends on the gravity estuarine circulation and wind-driven circulation. Previous surveys have suggested that near-bottom hypoxia can occur in the upper most stratified part of the estuary, during the summer, when the gravity estuarine circulation is weak. Further research to define gravity estuarine circulation and the wind-induced circulation and their quantification are needed.

REFERENCES

Awaji, T., Imasato, N. & Kunishi, H. 1980. Tidal exchange through a strait: a numerical experiment using a simple model basin. *Journal of Physical Oceanography*, **10**: 1499-1508

Cheng, R.T., Feng, S. & Xi, P. 1986. On Lagrangian residual ellipse. In: *Physics of Shallow Estuaries and Bays*. Van de Kreeke, J. (Ed.), Springer-Verlag, 102-113.

Longuet-Higgins, M.S. 1969. On the transportation of mass by time-varying ocean currents. *Deep-Sea Research*, **16**: 31-47.

Mellor, G.L. & Yamada, T. 1982. Development of a turbulence closure model for geophysical fluid problems. *Review of Geophysics and Space Physics*, **20**: 851-875.

Fluxes of Dissolved Inorganic Phosphorus to the Solent from the River Itchen during 1995, 1996 and 1998

P. N. Wright[1], Jian Xiong, and D. J. Hydes.

Southampton Oceanography Centre, European Way, Southampton, SO14 3ZH, U.K.

[1] *Now at: Maritime Faculty, Southampton Institute, East Park Terrace, Southampton, SO14 DYN, U.K.*

INTRODUCTION

In many aquatic systems, the rate of primary productivity and the accumulation of phytoplankton biomass are influenced by the concentration and flux of phosphorus (Lebo, 1990). The biological removal of phosphate (here referred to as Dissolved Inorganic Phosphorus, or DIP) in estuaries includes uptake by both phytoplankton and bacteria, whilst man impacts upon DIP concentrations by enrichment through inputs of wastewater (Lebo & Sharp, 1992).

Recent studies have suggested that nutrient levels have a minor role to play in bloom events in Southampton Water, and that tidal processes are more important (Wright *et al.*, 1997; Hydes, *this volume*). These events appear to be short-lived and so have only a limited impact on modifying the flux of nutrients to the Solent. At the same time, it appears that significant quantities of DIP enter the estuary from riverine sources, which are further enhanced by point source inputs from waste water treatment works.

METHOD

Thirty two sampling surveys were undertaken between Calshot Buoy and Woodmill Lane, at the tidal limit of the Itchen Estuary (for general location and main sites referenced refer to *Maps A* and *B* in the *Preface*). Water samples were collected by hand, through the deckwash unit of the launch (R.V. Bill Conway), situated about 1 m below the water line. Salinity and location were recorded. A small sub-sample was taken and filtered through Whatman GF/F filters, into virgin 30 ml 'diluvials'.

On return to the laboratory, the samples were stored in the dark at 4°C, until analysis, which always occurred within 48 hours. DIP was analysed by the molybdate blue method, on a Berkard Scientific SFA-2 Autoanalyser (Wright & Hydes, 1997).

A theoretical riverine end member was calculated for each survey, using regression and extrapolation of the DIP and salinity data, as per Officer (1979). This theoretical end member was then combined with a monthly flow, measured at Riverside Park by the Environment Agency (EA). The Harmonised Monitoring station at Gater's Mill was interrogated for DIP levels in the river, just above the tidal limit. When combined with the flow data, this gave an actual riverine flux entering the estuary which could be compared to that calculated from the SONUS (Southern Nutrients Study) data set.

RESULTS

Figure 1 shows fluxes of DIP calculated from EA data, the SONUS surveys, and the differences between the two data sets. A clear seasonal pattern can be observed within the EA data, showing peaks through the months of high flow (e.g. December 1995 at 6500 kg), and low fluxes during low flow months (e.g. August 1995, at 2000 kg). This seasonality is not as apparent in the SONUS data set, although the magnitudes of the fluxes are much greater. Both data sets suggest that DIP fluxes have increased between 1995 and 1998, from an average monthly flux of 3751 kg in 1995, to 4670 kg in 1995, and 5412 kg in 1998.

The percentage difference between the two data sets suggest that the flux calculated from the SONUS surveys is much greater than that entering the river during periods of low flow. This implies that the relative difference between the phosphate levels in the river and in the estuary is much greater during the summer months. Similarly, during the winter, the levels of phosphate found in the river are closer to those estimated from the calculations on the SONUS data.

DISCUSSION

Long-term data from the EA suggest that, after a rise in phosphate levels during the late seventies, concentrations peaked and fell from 1985 onwards. However, these data also imply that since the early nineties phosphate levels have started to rise again; this explains why fluxes of DIP, in both data sets, have risen from 1995 to 1998. It is unclear as to whether this represents enrichment in the type of waste being added to the upper reaches of the Itchen, or is just a reflection of increased output from wastewater works in the upper catchment.

It is clear that the seasonality of the DIP flux in the EA data is related to flow conditions; and that although levels in the river do fluctuate, the dominant control on the flux is discharge of freshwater into the estuary. However, this seasonality is masked in the SONUS data, since low flow and low freshwater DIP fluxes are enhanced through inputs of wastewater near the tidal limit, at Portswood. During periods of low river flow, whilst wastewater discharge decrease, these point sources become relatively more important, since they are less diluted by the low freshwater discharge. Thus, whilst riverine DIP flux varies as flow varies, DIP fluxes from the estuary are maintained, by enhanced inputs of phosphate rich waste water. Since it appears that biological uptake processes, such as algal growth, only occur over small time-scales, it would appear that much of this flux reaches the Solent unmodified.

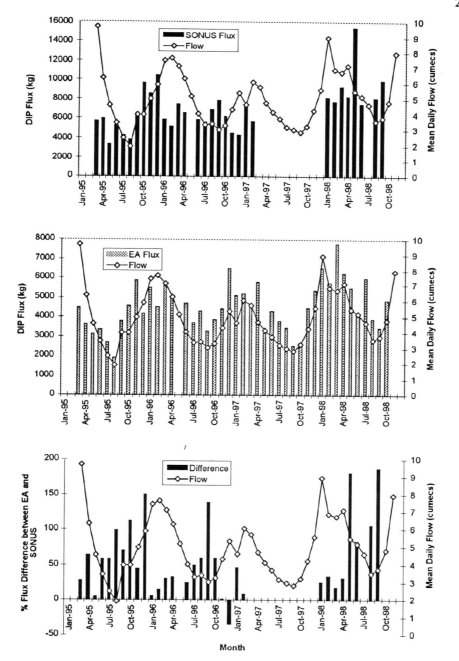

Figure 1 DIP fluxes calculated from SONUS survey data, EA fresh water data and the differences between the two calculations.

REFERENCES

Hyde, D.J. (this volume). Nutrients in the Solent. In: *Solent Science – A Review*, Collins, M.B. & Ansell, K (Eds.), Proceedings in Marine Science Series, Elsevier, Amsterdam.

Lebo, M.E. 1990. Phosphate uptake along a coastal - estuary. *Limnology and Oceanography*, **35**: 1279-1289.

Lebo, M.E & Sham, J.H. 1992. Modelling phosphorus cycling in a well mixed coastal plain estuary. *Estuarine, Coastal and Shelf Science*, **35**: 235-252.

Officer, C.B. 1979. Discussion on the behaviour of non conservative dissolved constituents in estuaries. *Estuarine, Coastal and Marine Science*, **9**: 91-94.

Wright, P.N. & Hydes, D.J. 1997. *Report on the methods used over the duration of the Southern Nutrients (SONUS) Project*. 1995-1997. Unpublished Report. Southampton Oceanography Centre Research and Consultancy Report No 8. 130 pp.

Wright, P.N., Hydes, D.J., Lauria, M-L., Sharples, J. & Purdie, D.A. 1997. Results from data buoy measurements of processes related to phytoplankton production in a temperate latitude estuary with high nutrient inputs: Southampton Water, UK. *Deutsche Hydrographische Zeitschrift*, **49**: 201-210.

Evaluation of the Environmental Risk of the Use of Preservative-Treated Wood in Langstone Harbour

S.M. Cragg, C.J. Brown, A. Praël and R.A. Eaton

Institute of Marine Sciences, University of Portsmouth, Ferry Road, Portsmouth, PO4 9LY, U.K.

INTRODUCTION

Timbers of wharves, piers, groynes, navigation posts and marinas are immersed in Solent waters. Boring invertebrates can severely damage submerged wood. Environmental concerns may limit the use of durable tropical timbers, but pressure treatment with a combination of water-soluble compounds of Copper (Cu), Chromium (Cr) and Arsenic (As), referred to as CCA, provides long-term leaching-resistant protection. Studies undertaken in the USA have demonstrated the environmental impacts of treated wood (Weis & Weis, 1996), but the methodology examined worst-case scenarios (Albuquerque & Cragg, 1995). The European Union PINTO, standing for the Preservative Impact on Non-Target Organisms Project, aimed to provide a realistic assessment of environmental risks from treated wood in coastal installations (Cragg, 1995), together with a suite of bioassays for testing the environmental risks of wood treatments (Brown *et al.*, 1998).

METHODS

Settlement and growth of fouling organisms on treated wood panels was determined at Langstone Harbour (for general location refer to *Map A* and *Map B* in the *Preface*) and other sites in European coastal waters. The leaching of Cu, Cr and As from treated wood was measured. The activity of *Crassostrea gigas* larvae, exposed to leachate (in spectrometer cuvettes), was filmed.

RESULTS

The rate of leaching of Cu, Cr and As declined rapidly after exposure to seawater (Albuquerque *et al.*, 1996). No negative effects on the fouling community could be detected on wood panels treated with CCA preservative and exposed in Langstone Harbour; similarly found at other sites located between Sweden and Greece. No evidence of inhibition of settlement by fouling organisms on CCA-treated panels was detected, or was there evidence of lesser diversity or biomass on treated panels. Indeed, the treatment appeared to enhance settlement of barnacles at a number of sites (see, for example, Figure 1). Three and seven day-old oyster larvae, exposed to CCA-leachates, swam on average 2 to 3 times faster than those in normal seawater. They also showed an increased level of up and down movement (Figure 2).

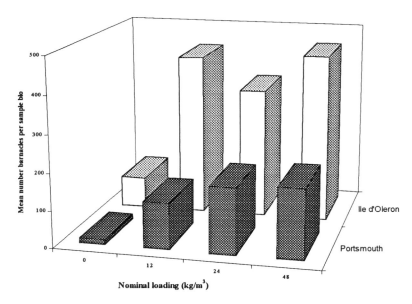

Figure 1 Effect of CCA loading on settlement of *Elminius modestus* at Langstone Harbour and in western France (Ile d'Oleron).

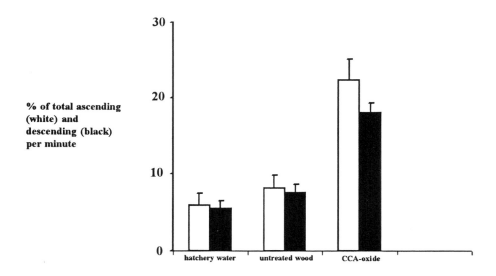

Figure 2 Effect of leachate from CCA-treated and untreated wood on the activity of oyster larvae.

DISCUSSION AND CONCLUSIONS

The absence of inhibition of fouling organisms by CCA runs counter to the observations of Weis & Weis (1996). CCA-treated wood in well-flushed conditions does not appear to have the same adverse effects as those reported from stagnant conditions. The difference between the two sets of observations may be due to the more open water circulation at the PINTO sites. It is likely that new constructions, incorporating wooden components, will continue to be installed along the shoreline of the Solent. Even components above the waterline are subject to leaching. To ensure a suitable service life, and to promote the proper use of timber resources, such construction should be protected from biodeterioration by suitable leach-resistant treatment. The PINTO project has developed components of an environmental impact testing protocol, for evaluating such treatments. In addition to the oyster larval swimming behaviour study, described above, an algal bioassay and an assay of effects on thraustrochytrid motility was included (see Brown *et al.*, 1998). *Artemia* was found insufficiently sensitive to metals in leachate for use in such a protocol, while the oyster D-larva embryological test offers a more sensitive bioassay.

Further research on the aquacultural use of timber is required. As aquaculture activities are likely to provide a growing proportion of protein derived from coastal areas and as the Solent has areas suitable for aquaculture, now that more stringent water quality requirements are being imposed, such research would be particularly timely.

REFERENCES

Albuquerque, R.M. & Cragg, S.M. 1995. Evaluation of impact of CCA-treated wood on the marine environment. In: *3rd International Symposium Wood Preservation: The Challenge, Safety-Environment*, Cannes-Mandelieu, France, 224-236.

Albuquerque, R.M., Cragg, S.M. & Icely, J.D. 1996. Leaching from CCA-treated wood submerged in seawater: effects of high loadings, and a comparison between laboratory and marine conditions. *International Research Group Wood Preservation* IRG/WP/96-50023, 17 pp.

Brown, C.J., Fletcher, R.L. and Eaton, R.A. 1998. Bioassays for rapid assessment of heavy metal toxicity in seawater. *International Research Group Wood Preservation* IRG/WP/96-50076, 15 pp.

Cragg, S.M. 1995. Impact on non-target organisms of anti-marine wood-borer treatments. In: *Marine Sciences and Technologies: 2nd MAST days and Euromar market, Vol. II*, Luxembourg, Office for Official Publications of the European Communities, 1227-1236.

Weis, J.S. & Weis, P. 1996. The effects of using wood treated with chromated copper arsenate in shallow -water environments; a review. *Estuaries* **19**: 306-310.

The Prevention of Biofilm Formation and Marine Settlement on Protective Coatings Prepared from Low-Surface-Energy Materials

P. Graham,[1] M. Stone,[1] A. Thorpe,[1] I. Joint,[2] T. Nevell,[1] J. Tsibouklis[1]

[1]*School of Pharmacy and Biomedical Sciences, University of Portsmouth, Portsmouth, PO1 2DT, U.K.*

[2]*Plymouth Marine Laboratory, Prospect Place, The Hoe, Plymouth, PL1 3DH, U.K.*

INTRODUCTION

The commercial implications of low-surface-energy polymeric materials, with good film-forming characteristics, become apparent when the lack of universally applicable and environmentally friendly protection against marine biological fouling is considered.

It has been suggested that low-surface-energy polymers must posses a flexible linear backbone, onto which side-chains with low intermolecular interactions is attached *via* suitable linking groups (Kobayashi & Owen, 1995). In this work, we consider two compounds that conform to these molecular design requirements, namely: poly(1H,1H,2H,2H-perfluorododecyl acrylate), *PFA9* and poly(methyl propenoxy-1H,1H,2H,2H-perfluorododecyl siloxane), *PFE9*.

EXPERIMENTAL METHODS

Both compounds are readily accessible. *PFA9* is prepared by the bulk polymerisation of acrylate esters (Stone *et al.*, 1998); whereas the synthesis of *PFE9* (Thorpe *et al.*, 1998) involves a hydrosilylation reaction.

Films of these materials were deposited on glass and/or poly(methylmethacrylate)-supporting substrates (10 x 10 x 1 mm). The acrylates were deposited from the melt or from $CF_2ClCFCl_2$ (0.1% w/w) solution, whereas (the liquid) silicone samples formed into glassy structures after thermal crosslinking at 105°C for 16 hours. The surface-free energies of polymer samples were determined by contact-angle goniometry, using a Kruss G10 goniometer interfaced to image capture software; the instrument was fitted with an enclosed thermostated cell. Bacterial cells for colonisation experiments on film structures were grown to stationary phase in nutrient broth: marine *Pseudomonas* were injected (1:100 v/v) into an autoclaved MG medium, whereas sulphate-reducing bacteria were injected (1:10 v/v) into a sealed vessel containing autoclaved Postgate C medium (pH 7.5).

RESULTS AND DISCUSSION

Surface energies were evaluated using the surface-tension-component theory (Good & van Oss, 1991) (Table 1).

Table 1 Advancing contact angles for water, diiodomethane (DIM) and ethylene glycol (EG) on silicone (PFE9) and acrylate (PFA9) film structures, together with corresponding surface energies, γ_s, for these materials: $\gamma_s = \gamma_s^{LW} + (\gamma_s^+ \gamma_s^-)^{1/2}$; γ^{LW} Liftshitz/van der Waals component; γ^+ Lewis-acid component; γ^- Lewis-base component.

Sample	n	Contact angle, θ (°)			Surface energy (mJ/m²)			
		H_2O	DIM	EG	γ_s^{LW}	γ_s^+	γ_s^-	γ_s
PFE9	9	109	95	94	10.6	1.3	0.5	12.2
PFA9	9	125	112	120	5.0	0.1	1.6	5.6

Even after prolonged exposure (two weeks) to cultures of sulphate-reducing bacteria and mixed marine *Pseudomonas,* no evidence of any permanent bacterial attachment to *PFA9* and *PFE9* surfaces could be identified by scanning-electron- and optical- microscopes; in each case, the controls were extensively colonised. Atomic Force Microscopy (AFM) investigations revealed that this effect was due to the inability of the bacterial exopolymer to wet or become attached to the low-energy surface. This effect is particularly evident with *Pseudomonas*-derived exopolymers that are observed in the tapping mode AFM image as non-wetting droplets on the surface of *PFA9* films.

ACKNOWLEDGEMENTS

The financial support from EPSRC/MTD Ltd (research studentships to MS and AT) and NERC (research studentship to PG) is gratefully acknowledged.

REFERENCES

Good, R.G. & van Oss, C.J. 1991. The modern theory of contact angles and the hydrogen bond components of surface energies. In: *Modern Approaches to Wettability: Theory and Applications.* Schrader, M.E. & Loeb, G. (Eds.), Plenum Press, NY, 1-27.

Kobayashi, H. & Owen, M.J. 1995. Surface properties of fluorosilicones. *Trends in Polymer Science,* **3**: 330-335.

Stone, M., Nevell T.G. & Tsibouklis, J. 1998. Surface energy characteristics of poly(perfluoroacrylate) film structures. *Materials Letters,* **37** (1-2): 102-105.

Thorpe, A.A., Nevell, T.G. & Tsibouklis, J. 1998. Surface energy characteristics of poly(methylpropenoxyfluoro alkylsiloxane) film structures. *Applied Surface Science,* **136**: 99-104.

SECTION 3

Workshop Findings

Findings of the Water Quality and Chemistry Workshops

Session Leader: David Hydes *(Southampton Oceanography Centre)*

INTRODUCTION

Four workshops were held to consider water quality issues, as outlined below:

1. Eutrophication
2. Water Quality and Shell Fisheries
3. Anti-fouling measures - TBT and Alternatives
4. Bathing Water Quality and Public Health

The most obvious indication within the region of the effects of eutrophication is the extensive weed growth on mud flats in the harbours. There is specific concern about the link between shellfish quality and potential changes, with planned changes in the sewage discharges into the Solent system. There is a general concern about the effects of anti-fouling agents, on the biodiversity of the Solent ecosystems, but there is a lack of specific knowledge for the different areas of the Solent. The maintenance of good bathing water quality is a general cause for concern. Some of the potential problems in the Solent region are being addressed, by changes in the location of the various sewage discharges. However, the present practices of bathing water quality monitoring are suspect, with regard to their ability to provide the best indications of the likely risk to both bathers and other water users.

A feeling expressed in all the Workshops was a need for better linkage of the work being undertaken by the different groups (higher education institutes, Environment Agency, local, port, harbour and regional authorities); a suitable mechanism needs to be found to allow this to happen. Initially, there is a need to increase the awareness of the research activities of the various groups; this could be undertaken through a Solent Research Meeting, which could be held in conjunction with the annual Solent Water Quality Conference. Better awareness could be fostered also by the establishment of a Discussion Group/Bulletin board web page.

EUTROPHICATION

Rapporteur: David Hydes *(Southampton Oceanography Centre)*

Main Issues

Within the terms of the Urban Waste Water Treatment Directive, Southampton Water is not an eutrophic environment. Levels of nutrients in the rivers are moderate, for those draining populated areas with intensive agriculture. Sewage discharges of phosphate and ammonia are greater than the river water inputs to the estuary. Dilution appears to be too rapid for significant de-oxygenation to occur, due to the nitrification of this ammonia. Clear symptoms of eutrophication are present in the harbours (Langstone, Chichester and Portsmouth), in of eutrophication are present in the harbours (Langstone, Chichester and Portsmouth), in the form of the extensive beds of macro-algae.

The Solent is an ideal location to undertake basic research, to assist in the understanding of processes in eutrophic waters, because of: the range of environments in the area; the studies which have already been undertaken and the potential to improve on the links that already exist between the various academic, agency and governments bodies working in the area.

Specific scientific areas of concern are outlined below.

- The lack of understanding and, therefore, inability to predict, the occurrence of algal blooms. A particular example are the red/brown tides that were common in the estuaries in the 1980s; these were absent in recent years, but returned this year (1998).
- The extent to which the low-frequency monitoring that has been carried out, so far, defines correctly the magnitude and duration of blooms in Southampton Water.
- The effects of the Southampton Water discharge on the nutrient supply to the harbours has not been investigated.
- As far as it is known, there is a connection between macro-algal weed growth and nutrient supply. However, the details of the main supply routes of nutrients supporting growth are, at different times of year, not well defined; nor are the criteria for the establishment of algal colonies, on particular substrates.
- The important processes which are known to be enhanced in saltmarsh environments, such as denitrification, have not be quantified for the Solent area.

Research Priorities

- All of the above topics outlined above require investigation. The critical areas of research are those which are relevant to the two current EU directives on "Nitrates" and "Urban Waste Water Treatment". Consideration must be given also to the requirements of the OSPARCOM (Convention for the Protection of the Marine Environment of the North-East Atlantic) strategy, on the reduction of eutrophication.
- Research undertaken in this particular area could be assisted by improved integration of studies being carried out by the Environment Agency, University of Portsmouth, Southampton Oceanography Centre and Southampton Institute, etc.
- As in many areas of environmental research, the system could be better understood if consistent long-term records were available. There is a need for both the establishment of a readily-accessible review of what is available, which could then be collated into a central database. At the same time, to establish monitoring which can provide pertinent and consistently-collected information, over periods of time, allowing the identification of trends in eutrophication processes.

Management Priorities

The essential priority is the management of the system, in line with the current EU directives. Studies to be undertaken are summarised below.

Blooms

- Short-term studies: continuous chlorophyll/nutrient monitoring at key sites, to increase the understanding of bloom occurrence - in relation to information that has previously been derived from low-frequency wide area monitoring.
- Long-term studies: monitoring, to understand factors controlling irregular blooms.

Nutrients

- Short-term studies: to improve the understanding of nutrient fluxes between various estuaries in the Solent, especially the flow from Southampton Water into the harbours.
- Medium-term: to assess the impact of the relocation of sewage discharges into the Solent - particularly the likely increase in nitrogen discharges into Southampton Water, following the ending of sewage sludge dumping in the Solent.

Other Considerations

A problem which is being investigated world-wide is the degree to which the intensification of total biomass, in blooms, is associated with an increased risk of the enhancement of the abundance of toxic algal cells. Such events are considered to be related, in some areas, to changes in the ratios of the available nutrients (nitrogen, phosphorus and silicon), as well as a simple increase in concentrations.

WATER QUALITY AND SHELLFISHERIES

Rapporteur: Ron Lee *(Ministry of Agriculture, Food and Fisheries)*

Main Issues

- There are difficulties in both the identification of sources of shellfish (for their assessment against hygiene criteria) and the accurate gauging of productivity, which is estimated from records of landings. This problem arises because the required information has to be recorded by the harvester and small-scale operations, whose catch accounts for a large proportion of the total catch; these are not required to keep and return such records.
- There is confusion regarding the relative roles of the Shellfish Waters Directive and the Shellfish Hygiene Directive. Consequently, it might be an improvement if the two directives where combined.

- Proposed new regulations, that form part of the EU Shell Fish Waters Directive, may constitute a 'driver' for investment in the upgrading of the level of sewage treatment that has to be applied, before discharge or the relocation of discharges. A point that should be considered is the 'cost-benefit analysis' of the relative costs of improved sewage treatment, compared to any increase in the value of the shell-fishery that would result from the work. Such discussion requires the collaboration of the interested parties: Environment Agency, Southern Water, Local Authorities, CEFAS and the shellfish industry.
- There is a need to increase public knowledge and awareness, as only a limited amount of general information is available presently on water quality in shellfish areas.
- There is generally a poor understanding of the impacts of fishing practices, such as trawling and oyster dredging, on benthic ecosystems.

Further research requirements

- At present, two almost directly opposed arguments can be advanced about the likely effects of the proximity of a shellfish bed to a sewage discharge. One is that the high nutrient and organic content of the discharge will provide an enhanced food source and improve the yield; alternatively, such a benefit may be outweighed by the increased risk of bacterial and viral contamination. Research is required to assess the validity of these competing claims.
- Specifically, there is a need to investigate the likely effects of the planned new combined sewage outflow systems in the Solent, on water quality around the shellfish beds of the region.
- To monitor compliance with regulations, it would useful if methods were available for tracing the origin of shellfish; this may be possible using molecular techniques or chemical finger-printing.

Management priorities

The monitoring of existing sewage improvements, in order to provide additional information to enable the improved targeting of future investment.

ANTI-FOULING MEASURES (TBT AND ALTERNATIVES)

Rapporteur: Sue Lewey *(Southampton Institute of Higher Education)*

Main Issues

Direct studies of the impact of pollution from anti-fouling material on the Solent ecosystem have been limited in their extent. There is some evidence originating from studies undertaken, on Southampton Water bottom sediments that concentrations of trace elements such as copper have decreased, in recent years. Such trace elements have been used as anti-foulants.

Changes are likely in the near future, as the use of TBT on all vessels will cease in 2006, in response to a ban introduced by the International Maritime Organisation. At present, different regulations apply to leisure and commercial craft.

Desirable alternatives to the use of anti-fouling measures are regular defouling, the development of non-toxic fouling release surfaces and, possibly natural product-based anti-fouling coating materials. Studies are required of the likely impact of interim alternative anti-fouling materials (e.g. Cu-based, Irgarol). Research in these areas is being undertaken locally at the University of Portsmouth.

Research Priorities

- Within the Solent area, there is a need to assess the impact that anti-fouling treatments may have had (and may be having) on marine bio-diversity. Sub-regions need to be considered separately. The effects are likely to be small in the open waters of the Solent, but may be greater in the container ports, naval harbours and enclosed marinas. The short-term monitoring of environmental impact (including sub-lethal and synergistic effects) of existing and future anti-fouling materials needs to be considered; similarly, the long-term fate of existing and future use of anti-fouling agents in the differing biogeochemical environments of the Solent.

- Adopting a wider perspective, developments are required of: (i) physical de-fouling methods (for both small and large vessels); (ii) fouling release surfaces, both natural and synthetic; and (iii) of electrical and other novel anti-fouling systems.

- Best practices need to be developed, for the disposal of de-fouling waste.

- Procedures need to be developed, for the identifiction of toxic components in visiting ship's anti-foulants.

BATHING WATER QUALITY AND PUBLIC HEALTH

Rapporteur: Gareth Rees *(Robens Institute, Farnborough College of Technology)*

Main Issues

- Flaws in the present monitoring systems.
- The need for set standards for the collection, storage and analysis of water quality information.
- The need for emphasis on recreational waters, not just bathing waters.

Research Requirements

- Identification of the most significant indicators of risk i.e. not just micro-biological data.
- Consideration of the standards most appropriate for coastal waters, if bathing water standards are not really appropriate.

Management priorities

- The need to monitor a wider range of waters.
- Improvement in public awareness.
- The identification of all recreational waters within the Solent system.

SECTION 4

Biodiversity and Conservation

Viewpoint: Conservation, Policy and Management of Maritime Biodiversity

Dan Laffoley

Head of Marine Conservation, English Nature, Northminster House, Peterborough, PE1 1UA, U.K.

INTRODUCTION

This paper provides a viewpoint from the Head of Marine Conservation, to explain the broader context of the work of English Nature in managing maritime biodiversity. In order to undertake this objective, some of the driving forces behind the conservation and management of biodiversity in the U.K. are examined. The reasons for undertaking certain studies and deciding how various initiatives have arisen are also outlined.

BACKGROUND

At the broadest level, English Nature's work on maritime conservation is guided by the concept of ecologically sustainable development. A key principle here is the maintenance of biological diversity i.e. the variety of life we see all around us. The seas around the United Kingdom contain a tremendous variety of marine habitats; it is the greatest of any European country with an Atlantic coastline. The habitats range from highly exposed locations, such as Rockall (the remotest part and famed as a name in shipping forecasts), through to more sheltered environments such as the comparatively calmer waters of the Isles of Scilly. In the latter location, conditions allow dense beds of the sea grass *Zostera* to flourish and particular communities of bivalve molluscs, worms and echinoderms to thrive in the shallow water sandflats and on the shore. Species vary from those that have a widespread distribution around the whole coast, such as kelp forests, through to colourful species that have a restricted distribution; this is due to their specific requirements, such as the solitary corals which flourish in warm waters at certain places in southwest Britain.

The maintenance of this biodiversity is a 'core test' as to whether ecologically sustainable development is being achieved. In contrast to land, the highly linked nature of the sea results in a widespread distribution of most marine plants and animals. Few marine species are restricted to the coastal waters of the U.K. At a broad level, this means that a single development or operation in the nearshore zone may not, on its own, compromise the overall maritime biodiversity of this Country. The cumulative effects of these activities are, however, of major concern.

Another principle that shapes and guides English Nature's maritime work is the maintenance of ecological integrity i.e. the ability of our coasts and seas to support and maintain the full range of habitats and species that occur in a natural state. This approach is both in terms of ecosystem structure (such as the variety of species) and ecosystem function (such as primary production). The maintenance of the ecological integrity of our coastal waters is a fundamental priority. Practically all other environmental values and societies uses of our coastal waters are dependent, in some way upon maintaining the ecological integrity of the environment.

HUMAN ACTIVITIES

The level of threat posed by human activities, to the ecological integrity of maritime ecosystems, can be assessed broadly by their magnitude and the direct or indirect impacts they have on the particular aspect of the under consideration; for example, the discharges of non-toxic substances from land, such as run-off and sewage. Around the world, the discharge of nutrients into coastal waters is considered one of the principle threats to such systems, making up 70% of marine pollution; this is particularly important in areas where restricted exchanges of water occur. Recently, in terms of the Solent System, there was a Government announcement that Chichester and Langstone Harbours have now been defined as 'eutrophic sensitive areas'.

Tackling water quality remains a very large and difficult problem. Much action and media attention focuses on the quality of bathing waters, or litter on beaches. Considerable amounts of money are being spent on cleaning up discharges; however, consideration should be given also to the more 'invisible', yet more crucial, effects from chemical groups such as endocrine disrupters. What of water quality standards for coastal waters and how do they relate to the water quality needed to maintain healthy plant and animal populations on the coast and in the seas?

Coastal development is another particular issue. Over half of the world's population lives within the coastal zone. Similarly, ever-larger numbers of people want to spend holidays by the sea; the U.K. is no exception. With such pressures come the associated infrastructure developments, ranging from coastal defences to land reclamation, to provide further capacity for industry, ports, housing, recreation and leisure. Unless carefully planned, all can have negative impacts on wildlife and the very beauty of the coast that many value.

Large-scale physical disturbance to the marine environment is another area of concern; frequently, this manifests itself in terms of fishing, or to be more precise, overfishing. Considerable concern for this issue is regularly expressed, whether from the conservation side or from commercial interests. Common ground exists between many different groups. Many calls are being made for changes to enable fish stocks and the environment to recover. Exactly the best way to decide this has yet to be agreed: agreement simply seems so hard to achieve.

Another threat is frequently seen to be the shipping and transport of hazardous substances. The wreck of the *Sea Empress*, in southwest Wales, acts as a continued reminder of the impact that shipping can have on the environment and commercial interests. Recent challenges include how to tackle 'flags of convenience', the ignorance of the rules and pollution of the seas, together with the risks posed by the introduction of non-native species in ballast waters. Consideration should be given to the commercial and conservation implications when things go wrong.

Finally, there are real global issues of concern, such as climate change. At present, in terms of marine conservation, this is a rather neglected area. Attention has been focussed upon the effects of rising sea levels on coastal habitats, but what of the effects on marine life? It may be that we are some way away from understanding the impact of climate change on ocean currents. The conservation perspective is important though, not the least because many of our key marine wildlife sites have strong links to biogeography.

The management responses to all these sorts of threats have arisen from a variety of sources, not only as a result of the scale of the problems but also as a result of the scale of solutions needed. Hence, initiatives have developed at international, European and U.K. levels. Some are 'area specific', whilst others deal with wider issues.

MANAGEMENT RESPONSES

Global Approach

At a truly global level, in 1992, at the Earth Summit held in Rio, John Major and over 150 other Heads of State or Government signed the Convention on Biological Diversity; this is a good example of a response to deal with wider issues. At the heart of this Convention is Article 6A; this requires the preparation of national strategies, plans, or programmes for the conservation and sustainable use of biological diversity (FCO, 1995). The Government's response to the Convention on Biological Diversity has been the development of the U.K. Biodiversity Action Plan (DoE, 1994). This plan includes some 59 steps to be taken, to maintain, restore and manage the biodiversity of our Country. Since then, to support its implementation, just under 400 individual plans have been, or are in the process of being, prepared for the conservation of particular habitats and species; of these, nearly 80 are designated for maritime habitats and species.

European Approach

A more regional and area-specific management response is seen from the European Commission; it has acted to maintain biodiversity across Member States by introducing the Habitats Directive (CEC, 1992). This initiative, forming just part of the actions set out in the U.K. Biodiversity Action Plan, came into force in 1992. The Directive requires Member States to identify, designate and manage Special Areas of Conservation (SAC) for particular habitats and species. As such, this Directive should be viewed as a significant site-based strategy for achieving some of the aims of the Convention on Biological Diversity, at a European level. Implementing this Directive has resulted in the U.K. and the other Members States putting forward sites to the European Commission, to form an international network of protected areas. Any Member State with coastline has had to consider proposing sites for certain marine habitats and species. Some 40 marine areas are being put forward by the U.K.

U.K. Approach

Providing support for the Habitats Directive, at the U.K. level, is area-based conservation initiatives, these pre-date the Directive, such as Site of Special Scientific Interests (SSSI) and Marine Nature Reserves (MNR). A SSSI is the term used to denote an area of land notified under the Wildlife and Countryside Act, because of its special nature conservation interest. Presently, there are over 3,900 designated SSSIs in England, covering nearly a million hectares. Based upon planning legislation, they are limited in the protection they can give to maritime biodiversity; usually, this is restricted to above Mean Low Water in England. Outside this area, until the implementation of the Habitats Directive, declaring a 'statutory marine nature reserve' was the only mechanism available to protect and manage important marine sites. The legislation proved difficult to apply, progress has been slow and, so far, only three have been declared. The implications of the Habitats Directive, against this background, are considerable. An almost overnight change from these few nationally important sites, to a potential series of nearly 40 internationally important Marine Protected Areas.

Progress on marine conservation, however, does not appear to be restricted to the actions outlined above. As pointers to the future, other initiatives may be, or are, on their way; these will further impact upon management responses to particular issues. More sites may be identified through the continued drive from some part of the community, for proper recognition of the many sites that are of national importance for their marine wildlife. This will only be achieved, however, through new domestic legislation. Government is in the process of setting up a Working Group to look into this and related issues. What the group will recommend is at present unclear. Farther offshore, the Convention for the Protection of the Marine Environment of the North East Atlantic (OSPAR) is a significant initiative, operating outside Territorial Waters. This initiative involves obligations to: protect and conserve ecosystems and biological diversity; where practical, restore maritime areas that have been adversely affected by human activities; and control human activities of particular concern.

European Fisheries Policy

Lastly, the European Communities Common Fisheries Policy is up for review and, although it is unclear what the outcome will be, what is clear is that improvements in how we manage fisheries must be sought. Environmental considerations should be at the core of this thinking. Any resulting actions should also have positive benefits for marine conservation and biodiversity. Distinct interest is being shown across Europe, in closing nearshore areas of the sea, to assist with broader management of marine areas and to benefit fishing stocks.

Other initiatives

Other initiatives are being developed which, when implemented, will have further far-reaching effects on how we manage our resources. A key initiative from Europe is the proposed Water Framework Directive. This Directive will have a major impact on how we mange water resources across the Community. Sustainable use will be at its heart, achieved by a combination of management plans for river basins, monitoring and actions to achieve good ecological status for all surface and ground waters. Moreover, the latest news is that it may have a particular focus on SACs.

CONCLUSIONS

Three conclusions can be outlined; they all relate to how we view maritime biodiversity, together with changes that are needed if we are going to do things better and become more successful in our endeavours.

Firstly, there are particular issues surrounding the management of dynamic systems: the coasts and seas are under continual change. The extent and distribution of habitats may vary, even between each tide, or may take decades or longer to alter, as landmasses sink and sea levels rise. Such dynamics present particular management problems. Some habitats are naturally temporary, such as saline lagoons. Other habitats may move, over time, to cover other areas of interest. Such habitats include shingle banks, moving landward by the action of the sea, to cover freshwater habitats or saltmarshes that are important in their own right. The conflict is between allowing change to occur and the legal obligations to maintain at least the existing area and quality of each of the habitats involved. All these considerations are very difficult to reconcile, with site maps having solid boundaries drawn around the features of interest. Indeed, hard work and a degree of innovation will be needed, to ensure that management approaches keep pace with these changes. Such approaches are required if we are to maintain, let alone restore, such maritime biodiversity as is found in the Solent region, and other key areas in the U.K., into the next millennium.

Secondly, there are information requirements. With any initiatives charged with seeking sustainable management solutions, comes the need for basic data and information to underpin the decision-making process. Such an approach often involves baseline information on the extent and distribution of marine plants and animals. However, it also needs to include an understanding of key ecological processes, such as environmental change, and the links between threatening processes and key conservation features. Clearly, we are not starting with a 'clean slate'. Much valuable research has been undertaken and published. What needs to be done though, is to understand the management needs behind these major initiatives and to ensure that research and development across Government departments, agencies and institutions is better co-ordinated to provide the answers that everyone needs.

Finally, there is the matter of the role of informed judgement. In an ideal world, it should be possible to produce the necessary data and information, to answer all such questions. However, such conditions do not exist, even with better co-ordinated research activities. Nonetheless, there is a strong desire to have things quantified. Industry wants the comfort of hard facts and figures, on which to make environmental decisions. Nevertheless, in some, perhaps many, instances management processes will need to run on informed opinion supported by the very best available scientific information. Both the cost and time-scales involved will prohibit any other response. To use the words of the current Secretary for the Environment (Michael Meacher) at a recent European conference, "informed, subjective judgement must play a part". Even with facts and figures, uncertainty will still be present;

it is something everyone will have to learn to live with, whilst efforts are made to improve the science base. Such an improvement will be a difficult process. Often long-time series of data are going to be necessary, to understand how the maritime environment functions and responds to particular pressures. Some of these processes are known, but many await discovery. Years may be required to establish basic confidence in some areas, which will also bring inevitable pressures over sustaining funding for such initiatives. Often, there may not be 'overnight fixes', but events such as the Solent Science Conference help to move us forward with maximum speed to achieve the end results we all desire.

REFERENCES

DoE et al. 1994 Biodiversity: the UK action plan [and] summary report. Department of the Environment, Command paper Cm. 2428

CEC. 1992. Council directive 92/43/EEC: on the conservation of natural habitats and of wild fauna and flora. Council of the European Communities. *Official Journal of the European Communities*, L206/7.

FCO 1995. Convention on biological diversity open for signature at Rio de Janeiro, June 1992. Foreign and Commonwealth Office, Command paper, Cm.2915. London, HMSO. Treaty Series No. 51

Coastal Habitats of the Solent

Sarah L. Fowler

The Nature Conservation Bureau Ltd, 36 Kingfisher Court, Hambridge Road, Newbury, Berkshire, RG14 5SJ, U.K.

INTRODUCTION

The coastal habitats of the Solent area, together with its estuaries, intertidal and subtidal habitats, make up a complex and inter-linked system of international nature conservation importance. Despite the Solent's heavy industrial and recreational use, it includes some very important natural and undisturbed lengths of coast, with unusual examples of natural gradations from maritime to coastal and marine habitats; these have been lost from most other sheltered coastal areas in southern Britain.

The national nature conservation importance of the Solent coast's rare habitats, animals and plants is recognised by the designation of most of the high quality undeveloped and semi-natural coastlines as Sites of Special Scientific Interest (SSSIs). Many of these SSSIs have also long been known as Special Protection Areas (SPAs) and/or Ramsar sites, in recognition of their international importance. Large areas with high quality lagoonal, estuarine and marine habitats are now proposed as Special Areas of Conservation (SACs); these will contribute towards the establishment and protection of the important habitats across Europe (Natura 2000 network). Other areas of coastal habitat are of local conservation importance and have been identified as Sites of Interest for Nature Conservation (SINC). Finally, many coastal areas receive further protection, through their designation and management as National or Local Nature Reserves (NNRs or LNRs).

Most of the features of nature conservation importance are known to be interdependent; as such, they cannot be managed in isolation, however rigorous the protection of individual sites. For example, bird populations move between the harbours to feed and may roost on adjacent landward areas (otherwise of no apparent ecological value and unprotected by conservation designations). Lagoonal species diversity at individual sites appears to be dependent upon the numbers of lagoons nearby. Sublittoral, littoral and coastal habitats are linked closely through the migration of marine organisms, coastal sediment processes and water quality. Even those coastal habitats or areas of intrinsically low interest, when viewed in isolation, are usually of importance as an integral component of the whole dynamic ecosystem.

This contribution briefly describes the major coastal habitat types and complexes in the Solent area; it summarises their nature conservation importance in a regional, national and international context. The major threats to the future of these habitats are highlighted, drawing attention to the danger of considering each habitat in isolation when the future of all is so closely interdependent. In conclusion, taking into account the special nature of its coastal habitats and the threats they face, priorities are suggested for a research and management agenda for the Solent. Without a co-ordinated approach to future research and conservation management, it will be difficult to ensure that future generations of wildlife and local people will continue to benefit from the rich matrix of coastal habitats that still survive, at the end of the 21st century.

PHYSICAL AND ANTHROPOGENIC FACTORS

The great diversity of the Solent's coastal habitats and associated communities is influenced heavily by the diverse physical environment of the area, the wide range of exposure to wave action and tidal currents present, its varied geology and geomorphology, and its relatively warm climate. An understanding of these physical factors (described in other papers in *this volume*) is essential in understanding why the Solent's wildlife is so special in a regional, national and international context. For example, the Needles, in the west, exposed to prevailing winds from the Atlantic approaches to the English Channel, are not only more exposed to wave action than any site farther east along the Channel, but also more exposed than any North Sea coast, south of the Isles of Shetland. From this extreme, a declining range of exposure to wave action is experienced on other Solent cliffs and headlands, with extremely sheltered conditions occurring within some of the harbours. The Solent is also notable for its range of geology and geological landforms. These, combined with a maritime influence and warm south coast climate, support the great variety of coastal habitats, species and communities. For general location map and main sites referenced, refer to *Maps A* and *B* in the *Preface*.

In addition to physical factors, the Solent's history of coastal land management and creation of coastal habitats has been of considerable importance in influencing the range of coastal habitats present today. Former agricultural practices i.e. claiming large areas of upper saltmarsh behind embankments to yield high quality grazing land, have produced some of the highest diversity grasslands present on the south coast. The excavation and management of former salterns, mariculture ponds, and the construction of seawalls have, together with natural coastal processes, produced among the greatest density of coastal lagoonal features in Britain; this has resulted in the diversity of lagoonal species in the Solent. The long history of shipping activity in the area has resulted, in turn, in the introduction of large numbers of non-native species, sometimes with detrimental effect, and certainly with consequences that are of scientific importance. Unusually, for such a heavily populated and developed area, man's influence has been relatively small in many of the Solent's inlets. A remarkably large number of examples of natural transitions between marine, coastal and terrestrial habitats (for example, intertidal habitats grading into saltmarsh, and saltmarsh to ancient semi-natural woodland or grassland) still remain here, but are absent from most other coastal systems in England.

The major challenge to the future extent and biodiversity of the Solent coastal habitats is a combination of both natural and anthropogenic factors: relative sea level rise in the Solent is beginning to take effect (Tubbs, 1995; see also Bray *et al., this volume*).

COASTAL HABITATS

The following sections (derived mainly from information collated by Fowler (1995) and presented in Barne *et al.* (1996)) describe briefly the main coastal habitats of the Solent, although it can be difficult to determine precisely how far inland these habitats extend. Maritime grassland, affected by salt spray and influenced by the coastal climate on top of exposed western coasts and cliffs, may extend large distances inland. Similarly, seawater influence penetrates a long way inland in some sheltered low-lying areas; here rivers, ditches and grassland still experience occasional seawater incursions. These trends may increase, as a result of rising sea level rise and increased storminess.

Estuaries

Estuaries are formed where rivers widen as they meet the sea, and a mixing of freshwater and salt water occurs. Typically, they contain a gradation of salinity from freshwater within the inflowing rivers to entirely saline waters near the mouth, and a range of habitats and associated species characteristic of this complex environment. The coastal and marine habitats of the Solent estuaries are described individually below and by Collins & Mallinson (*this volume*). Identifying and considering the component habitats of estuaries in isolation, however, runs the risk of overlooking their importance as complex functioning ecosystems; they are comprised of a large number of interdependent components, marine, coastal, geomorphological and physical processes.

The Solent is one of the most important areas in the UK for estuarine habitats. It is of international conservation significance, notably for:

- the variation of estuarine types and large number of estuaries present;
- the presence of important marine communities (Collins & Mallinson, *this volume*);
- the historical record of saltmarsh development; and
- the large areas of saltmarsh and coastal grazing marsh, shingle spits and large numbers of lagoonal systems.

This mosaic of estuarine habitats supports internationally important populations of breeding, migrating and wintering birds (Burges, *this volume*) and many other rare species of animal and plant.

All the Solent estuaries are classified, in geomorphological terms, as either coastal plain or bar-built systems, closed by sand or shingle spits. They are mostly shallow and predominantly sediment-filled, with a relatively small freshwater inflow and strong marine influence. Most are relatively small, in comparison with others in the UK, with a total estuarine area of just under 14,000 ha, or 2.65% of the total British estuary resource (Davidson *et al.*, 1991). Several are individually of international importance. The largest estuarine area is made up of the three large inter-connecting harbours of Chichester, Langstone and Portsmouth. This complex is of international importance for its saltmarsh, extensive sediment communities and the huge populations of birds that it supports (Burges, *this volume*).

Langstone and Chichester harbours are also unusually natural and undisturbed, apart from a trend towards eutrophication (Hydes, *this volume*), as is the remarkably pristine Newtown estuary, where eutrophication is not a problem; very few estuaries in southern Britain are as unpolluted as this site. Beaulieu River, partly closed by a vegetated shingle spit, is of particular nature conservation significance for the way in which its very natural and dynamic estuarine character grades into semi-natural terrestrial habitats. It is also noted for its bird populations. This site is substantially unaltered by man, which is now rare, particularly on the south coast (Davidson, 1996).

Most of the Solent estuaries have large human populations nearby. Substantial parts of their coastlines are urban and artificially-defended. There is a long history of enclosure of saltmarshes and other tidal land in most estuaries, particularly in their upper reaches. Some areas of saltmarsh reclaimed from Langstone Harbour, Beaulieu River and Lymington River are now among the largest areas of coastal grazing marshes on the south coast. Land-claim, often for port, industrial and waste disposal, has affected coastal and marine habitats in Langstone Harbour, Portsmouth Harbour, Southampton Water (where development has greatly reduced the size of the upper estuary) and Bembridge Harbour. Industry is concentrated around the naval docks and installations of Portsmouth Harbour and the extensive port and dock complexes, oil industries and power station developments of Southampton Water. Only Chichester Harbour, the Beaulieu River and the Newtown Estuary have retained a predominantly rural hinterland with natural shoreline transitions, although some very important areas of natural transition do occur elsewhere (including within Southampton Water). Almost all estuaries in the region, with the exception of the Newtown Estuary, are used for a wide variety of land and water-based leisure and recreational activities (Barne *et al.*, 1996).

Saltmarshes

Saltmarsh vegetation develops between the mean high water of spring and neap tides, where sediment settlement occurs in sheltered conditions. As the sediment accumulates, the level of the marsh rises and different successional stages of vegetation develop. Because saltmarsh species vary in their tolerance of inundation by salt water, species-poor communities develop at levels flooded by most tides with the most diverse vegetation at the highest levels, where tidal inundation occurs only infrequently. The most botanically diverse saltmarshes are absent on shores where the construction of coastal defences has removed the highest vegetation zones.

The Solent contains about 6% of the total area of saltmarsh in Great Britain (Burd, 1989). The region is one of the best areas in the United Kingdom for this habitat, containing many important examples of natural transitions between saltmarsh and landward habitats, rare elsewhere, and other unique features. Chichester Harbour contains one of the ten largest saltmarshes in Britain and the largest on the south coast of England. The intertidal area between Keyhaven and Lymington supports the third largest saltmarsh on the south coast. More than half of the Solent's saltmarshes is dominated by *Spartina anglica* (cord grass), but there are also some examples that support more typical and diverse communities. For example, the River Medina contains some of the best examples of mature mixed saltmarsh in southern Britain. Important transitional marshes, grading away from the coast into shingle, freshwater, swamp, heathland, grassland, scrub and oak woodland, also occur in the Solent; these are sometimes grazed, and often of high invertebrate biodiversity. Such transitions are absent from most other regions because of heavy grazing (in the north) or enclosure (in most other regions of England and Wales).

Most of the Solent's saltmarshes are relatively young. They were formed as a result of the introduction of cord grass (*Spartina alterniflora*) by ship from America, hybridisation with native *Spartina maritima* and, in due course, the rapid expansion of the aggressive polyploid hybrid *Spartina anglica* (Doody, 1996; Raybould *et al.*, *this volume*). Rapid expansion of *S. anglica* led to it becoming dominant throughout the Solent; this has been followed by die-back over large areas. Hythe and Calshot marshes are of international importance, as the site where hybridisation of introduced and native species of cord grass *Spartina* eventually gave rise to the fertile polyploid *S. anglica*; this is now introduced to promote the growth of saltmarshes all over the world. The progenitors of *S. anglica* still occur in the area, with *S. alterniflora* at Marchwood and *S. maritima* now confined to Hayling Island and Newport Harbour. The Solent area has been the subject of extensive scientific study (Doody, 1984; Gray & Benham, 1990), and is unmatched as a genetic resource for the genus.

Despite the long history of research into Solent saltmarshes, together with their recognised nature conservation and coastal protection importance, there is still no definitive baseline for the extent, diversity, function and quality of this habitat. Saltmarsh is considered to be of great importance as a source of organic material to adjacent marine and estuarine communities, likewise to provide a cleaning function by absorbing nutrients, heavy metals and oils from the estuarine system. However, the scale of their contribution in the Solent is unknown. Causes of *Spartina* die-back are also largely unknown. This information is essential for planning for sea level rise, the re-creation of lost habitat, and managed retreat. Sea level rise poses a particular threat to some of the important saltmarsh communities in the Solent. The numerous examples of natural transition between saltmarsh and other coastal habitats here present an important and unusual opportunity for monitoring the natural landward movement of coastal communities. Without managed retreat, however, the overall extent and biodiversity of this habitat will decline as sea level rise 'squeezes' upper saltmarsh communities against seawalls.

Grazing marshes

Grazing marshes are flat, low-lying areas of grassland drained by a network of ditches. Many grazing marshes were created originally, or expanded, by the enclosure of former estuarine saltmarshes. A few still grade into saltmarsh habitat to seaward and lowland wet grassland, reedbed, freshwater marshes, fen meadows, mires and ancient woodland to landward. However, most grazing marshes are now isolated by seawalls and agricultural, coastal, or urban development. Although the majority of this habitat has now been converted to intensive agriculture, or developed for housing or industry, the areas that remain are of national nature conservation importance. They support an important grazing marsh flora, rare invertebrates (particularly associated with drainage ditches), and breeding birds; at the same time, they are used by internationally important populations of wintering wildfowl (Burges, *this volume*).

Extensive areas of grazing marsh and brackish ditches occur on the Solent coast, particularly around Langstone Harbour, Beaulieu River, Brading Harbour and the Lymington River, which are the largest and most important grazing marshes remaining along the south coast. Grazing marsh habitat is particularly species-rich in the Solent, compared with other southern sites, with many unusual species present. Farlington Marshes is the richest grazing marsh in Hampshire, with a gradation of types of calcareous grassland, including over 50 grass species (Rose, 1996) and nationally-scarce plants and invertebrates. The drainage ditches often receive saltwater seepage through seawalls and embankments. The range of ditch habitats, from brackish to saltwater, may support brackish plants and large numbers of rare and endangered invertebrates; these include species characteristic of coastal lagoons and nationally-notable species of the beetle *Coleoptera* spp. and fly *Diptera* spp. (Parsons & Forster, 1996).

Although almost all of the remaining Solent grazing marsh is designated as SSSI, there is no overall inventory and baseline of the extent and quality of this habitat in the area from which to determine future research, monitoring and management requirements. Certainly, habitat quality is threatened by pollution of ditches from excess nutrients in agricultural and other land runoff, and changes in land and water management e.g. improved drainage and conversion to arable or intensive grassland management. Habitat loss could also arise from the upgrading of coastal defences. Managed retreat, if attempted in the Solent to overcome the problems of coastal erosion and saltmarsh retreat caused by sea level rise, could well result in the net loss of grazing marsh, unless equivalent areas of habitat are created elsewhere in suitable locations. Without reference to a detailed baseline, planners and managers run the risk of sacrificing higher quality grazing marsh, in order to recreate less biodiverse saltmarsh or lagoonal habitats.

Sea cliffs and maritime grassland

Hampshire and the Isle of Wight, including the Solent, provide one of the most important areas in the United Kingdom for vegetated sea cliffs. These cliffs, all formed through the action of the sea, are variable in height and range in slope from 15° to the vertical. The range and diversity of these sea cliffs, from hard chalk at the entrance to the Solent (the Needles, Foreland and Culver Cliff), to soft, eroding and slumping cliffs elsewhere, is of national importance. Rare species of plant and invertebrates, nationally-important geological exposures, and seabird colonies occur here. The cliffs often grade into rich maritime grassland, heathland and scrub at the cliff top. On the Isle of Wight, the combination of chalk exposures and coastal climate results in some of the best examples of maritime grassland, in the country. Extensive rich calcareous grasslands with unusual species, such as the endemic and internationally-protected early gentian *Gentianella anglica* (Chatters, 1994), occur on steep slopes above hard and soft cliffs. This grassland is a superb example of a vegetation type long-since destroyed elsewhere by agricultural improvement and industrial development. Such rich cliff-top grasslands and associated invertebrate communities are of international significance.

Unprotected examples of soft cliffs cover just 256 km of the English coast, with 49 km of this (nearly 20%) in Hampshire and the Isle of Wight (Moffat, 1994). Such cliffs are liable to falls and slumping; they are of particular interest, not only for their geology and geomorphology, but also for their vegetation and invertebrate communities (Hodgetts, 1996; Morgan, 1996). Different species assemblages occur on different cliff types, with some species restricted to limestone, chalk or crumbling cliffs. South-facing soft-rock and crumbling cliffs provide particularly important microhabitats for invertebrates, ensuring that pioneer food plants do not become overgrown by other vegetation, or by providing ideal nesting situations for burrowing bees and wasps and an abundant supply of nectar sources. Some invertebrates prefer dry substrates and sparsely-vegetated conditions, whilst others require wet flushes at the base of clay cliffs. Seepages and trickles can add diversity to the habitat for a number of invertebrates. The Solent's cliff invertebrate fauna includes many nationally-rare and notable species (Parsons & Foster, 1996).

Moving down the cliffs to the transition with the intertidal zone, the Isle of Wight contains 5% of the European extent of littoral chalk with the outcrops at Culver Cliff, and between the Needles and Freshwater Bay. These are two of the most important chalk cliff and upper littoral cave sites in Britain for the biodiversity of their unique chalk-boring algal communities (Fowler & Tittley, 1993).

Coastal protection works which isolate the cliff face from the intertidal zone, may obscure important geological structures, fossil beds, and upper shore and splash zone communities. Stabilisation of cliffs encourages overgrowth of the open habitats that are important for rare invertebrates and colonising plants by scrub; it starves coastal features downstream of the sediments released by cliff erosion. Conversely, where there is only a narrow fringe of maritime grassland along the cliff top, bordering arable land or other habitats of low conservation value, continued natural erosion of the cliff top will eventually results in the loss of this habitat.

Research is required to clarify the extent and relative quality of cliff and cliff-top habitats and the processes influencing them (including physical coastal processes and man-induced change). This information will help to determine the most important cliff sources of sediments to the Solent and adjacent areas; similarly, to develop a strategy for cliff habitat conservation. It will be particularly important to identify opportunities for improving cliff biodiversity, within shoreline management plan policies.

Coastal sand and shingle habitats

Much of the Solent coast is dominated by shingle and sand habitats, wherever physical conditions have allowed large quantities of these sediments to accumulate. Most harbours and estuaries are protected from the open sea by multiple ridge sand dune or shingle spit systems, shingle islets or breached barrier bars; there are numerous examples of transitions, from shingle and sand to saltmarsh and lagoons (Randall 1996). Shingle beaches and other coastal shingle structures develop in high-energy environments, but may persist as fossil formations long after the physical conditions and shingle deposits (which enabled them to form) exist.

Although many Solent shingle habitats are mobile and unvegetated, a wide representation of nationally rare shingle vegetation communities occurs. These range from classic pioneer species (e.g. sea pea *Lathyrus japonicus* and little robin *Geranium pupureum* subsp. *forsteri*), to well-established vegetation and invertebrate communities on stable shingle and sand further from the sea, including gravel terraces in the Test valley. Although coastal shingle is a relatively barren habitat, it supports a surprising range of invertebrates, with some species occurring only in this habitat. Coastal shingle at Browndown is of national importance for invertebrate conservation. Sand dune communities include mobile accreting wind-blown dunes, dune slacks, and dune heath (Dargie, 1996; Radley, 1994), with some of the vegetation and invertebrate communities of highest biodiversity being associated with a mixture of sand and shingle.

Because vegetated shingle is a nationally-rare habitat, whilst sand dunes are scarce along the south coast, all examples of these habitats in the Solent area are of regional importance for the characteristic vegetation and invertebrates that they support. Some sites, however, are of national importance. The multiple shingle ridge system of Browndown is one of the ten most important shingle sites in Britain and of national importance (Sneddon & Randall, 1994). The shingle spit and sand dunes at St Helen's Duver were assessed by Sneddon (1992) as the 13th most important shingle structure in Britain. St Helen's, Hayling Island and Pagham Harbour are of national importance for their biodiversity. These sites are thought to support the richest flora recorded on mosaics of sand dune and shingle habitats in the UK (Sanderson, 1997 and 1998), despite habitat damage at Hayling Island. Other sites of significance include the mobile shingle spit at Hurst Castle, which is slowly moving into the Solent over the saltmarsh that it protects; and several vegetated shingle beaches.

Most of the Solent's coastal shingle and sand dune sites are subject to high levels of recreational use, including some of the highest dune visitor numbers in Britain; this results in damage, from trampling and some vehicle access. Browndown is used for military training and has been affected by clearance of vegetation and some structural changes, although recreational access is limited. Many sites are affected by coastal engineering, whilst some shingle sites are designated for gravel extraction and waste disposal; sand extraction also occurs. Aggregate extraction and dredging operations offshore may affect also these structures, indirectly.

Despite a history of detailed study of these habitats, there is still no overall baseline for the extent and biodiversity importance of the full range of sand dune and shingle habitats present in the Solent area. The dynamic processes influencing these habitats and their vegetation communities are also not fully understood, although it is accepted that the free-functioning of coastal processes supplying sediments to these structures is vital. As a result, the potential impact of man's activities and the scope for retaining, restoring and improving the nature conservation status and functional role of sand and shingle structures are uncertain: an example being the allowance of the accretion of new dune systems and their movement inland. Future research is needed to address this particular problem.

Coastal lagoons

Coastal lagoons are water bodies that are either completely or partially separated from the nearby sea, but which receive some input of seawater providing saline or brackish conditions. Such areas are colonised by only a small number of species, many of them lagoonal specialists seldom found elsewhere (Bamber *et al.*, 1992). True lagoons are separated from the sea by a natural sedimentary barrier and are of high nature conservation importance; however, these are very rare in Britain and Europe. Artificial lagoons include modified natural sites or ponds wholly created by man, that provide a similar habitat and often contain a comparable range of species; they are also of high nature conservation importance (Sheader & Sheader, 1989). Brackish lagoon habitat is rare internationally and a priority for conservation under Annex 1 of the Habitats Directive (CEC, 1992).

The Solent coast is one of the most important areas for brackish lagoons in the United Kingdom, with 15% (over 100 ha) of the national resource (excluding the Dorset Fleet) and the highest density of lagoons in the country. Of these, nearly fifty lagoons, totalling 76 ha, are of some significance for nature conservation (Smith & Laffoley, 1992). Two of the UK's (forty-one) natural coastal lagoons occur here (Barnes, 1989). The Solent is of international importance because of the large number of lagoon sites and complexes it contains, the wide range of examples present, and the high proportion of these that are of international, national or regional nature conservation importance. A significant proportion of Solent lagoonal habitat is a candidate for a Special Area of Conservation.

Many lagoons in the Solent area contain nationally-rare and protected species, such as foxtail stonewort *Lamprothamnion papulosum*, the startlet sea anemone *Nematostella vectensis* (whose type locality is at Bembridge) and the lagoon shrimp *Gammarus insensibilis*. The diversity of lagoonal specialists, including rare species (protected under the Wildlife and Countryside Act, 1981), is correlated to the density of lagoons nearby (Bamber & Barnes, 1996; Barnes, 1988). A number of lagoons in the area have six or more lagoonal specialist species in their communities, Eight Acre Pond, Hampshire, includes ten such species, four of them protected (listed on Schedule 5 of the Wildlife and Countryside Act, 1981). Despite their apparent isolation, therefore, lagoon sites are interdependent. They have the potential to act as sources of recruitment of lagoonal species for neighbouring lagoon habitats, newly established or formerly of poor quality, through percolation, overtopping or sluice connection with the sea. Some lagoonal species may also occur in brackish drainage ditches (e.g. in grazing marshes); however, to date, these have not been surveyed adequately.

Lagoons are usually relatively short-lived, because of natural coastal dynamic changes and man's activities. Some lagoons will progress naturally from brackish conditions to become fully freshwater or completely overgrown by *Phragmites* marsh. Others may return to fully saline or estuarine conditions, as their barriers are eroded and lost. Consequently, many of the former lagoonal features bordering the Solent are now freshwater, estuarine or marine. In natural conditions, new lagoons continually become established: a range of lagoonal habitats is usually present, holding species characteristic of each successional stage in their development. The widespread influence of man on most parts of the Solent, particularly through the construction of coastal works, has prevented these natural processes from creating new lagoons; it has infilled or completely changed the form of most natural lagoons. As a result, natural lagoons are scarce; correspondingly, it is unlikely that many will arise through natural coastal processes. Conversely, though, the very high density of lagoonal habitats in the Solent is largely due to the activity of man in the coastal environment (whether constructing salterns or mariculture ponds, boating lakes, or excavating borrow pits when improving coastal defences). In order to retain and augment the coastal lagoon habitats of the Solent coast, management action will be needed to retain these ephemeral habitats; this will change existing low value coastal ponds into brackish sites, and create lagoon sites *de novo* during coastal engineering projects.

Lagoons in the Solent area have received a great deal of attention, including inventory and survey work both as a result of projects undertaken by local Universities, Societies and Laboratories, and the national survey commissioned by the former Nature Conservancy Council (e.g. Bamber & Barnes, 1996; Sheader & Sheader, 1989). Most of the work undertaken, however, has been related to survey rather than on the biology and habitat requirements of the rare and protected species that they support, or on the processes influencing the colonisation of new or existing habitats. Future research needs to identify the environmental factors that make some lagoons of particularly high nature conservation importance and encourage colonisation, in order to guide management and creation of this habitat.

Coastal species

Solent coastal habitats not only have a particularly high biodiversity, but also support maritime and coastal plants and animals that are individually of high nature conservation importance in Britain; some of them have been recorded rarely in the country. These habitats include species with a predominantly Mediterranean distribution (reaching their northern limit of distribution on the south coast of England) and nationally-rare species, characteristic of maritime habitats. The area is of international importance, for its many rare and scarce species of flowering plants; it is of national importance for its lower plants. Thirty-nine scarce plant species of coastal habitats ('scarce' plants are defined as those known from up to one hundred 10 km squares in Great Britain) are present here; this represents more than half of the total of sixty-one such scarce plants in the country (Stewart *et al.*, 1994). Many of these are found in transitions between different habitats and communities, for example, between saltmarsh and grazing marsh, or shingle, sand dune and heathland.

The Solent coast also supports a particularly high species diversity of maritime invertebrates, including numerous scarcer species; it is of national importance, for the conservation of a range of many invertebrates (Parsons & Foster, 1996). A number of species, whose distribution is correlated strongly with a southern climate and coastal habitats, are known to be restricted to this and neighbouring areas; they reach the eastern limits of their distribution here, along the English coast of the Channel. Other species have a very restricted distribution on the south coast, including the Solent area.

TRENDS AFFECTING COASTAL HABITATS

Man's activities on the coast (as noted above), including the construction of coastal defences, cliff stabilisation and land claim, have resulted in the interruption of many natural coastal processes; these have, in turn, resulted in coastal habitat loss and change. Key coastal habitats in the Solent area are still at risk. The partly man-made origin of some coastal habitats, together with the constraints on natural processes over much of the remainder of the coastline means that very few coastal habitats are now able to develop as a result of natural progression. Indeed, natural progression may even threaten important habitats; for example, the development of freshwater or reed habitat in saline lagoons, and the trend towards scrubbing over ungrazed grassland, or stabilised cliffs. Recreational activity continues to result in the deterioration of some coastal habitats, whilst a small amount of land claim for various forms of development continues. Low-lying coastal habitats are threatened increasingly by relative sea level rise, resulting in coastal squeeze and seaward erosion, and possibly increasing storminess (see Banyard & Fowler, Bray *et al.;* McCue; Price & Townend, *this volume*). Managed retreat and/or upgrading of coastal defences will be required to address this problem, with significant impacts on existing habitats likely, as well as potential opportunities for coastal habitat creation *de novo*.

The recommended, although largely untested, response to rising sea level and coastal squeeze is now, where possible, managed retreat combined with improved protection of the low-lying land that is of highest value (usually in economic terms – land of high biodiversity value, but low economic value, rarely receives costly protection). Managed retreat allows seawater to penetrate through abandoned seawalls to the replacement coastal defences, 'set back' further inland. This allows new intertidal and saltmarsh communities to develop in front of the 'set back' line of defence. This pattern of development ideally takes place using low-lying coastal areas of low economic and nature conservation value, for the creation of new saltmarsh and intertidal habitat. In the Solent area, target locations for managed retreat may have to include some areas that are already of high nature conservation importance (e.g. for grazing marsh or coastal lagoons). Some of these habitats may be created elsewhere, but this may not always be possible. The challenge will lie in defining priorities for habitat creation to maximise biodiversity and retain the special nature of the Solent coast (whilst contributing also to the regional and national coastal habitat priorities). The research agenda for the Solent must, as a matter of urgency, tackle the task of identifying the most valuable coastal habitats, and the favoured opportunities for managed retreat and the creation of new or replacement habitats in areas that are to be 'set back'.

A RESEARCH AGENDA FOR COASTAL HABITATS

The coastal habitats of the Solent area and its estuaries, intertidal and subtidal habitats make up a complex, interdependent, and inter-linked system of international nature conservation importance. Bird populations move between the Solent harbours to feed and may breed or roost on saltmarsh, grazing marsh, shingle and other coastal habitats. Coastal lagoon species diversity at individual sites appears to be dependent upon the numbers of nearby lagoons. The species biodiversity of mosaics of coastal habitats and transitions between habitats is often greater than the combined biodiversity of similar but separate areas of habitat; these marine and coastal habitats are linked closely through coastal sediment processes and water quality. Even those areas that are of intrinsically low interest, if viewed in isolation, are usually of importance as an integral component of the whole dynamic system. Because of the interdependent nature of the area and its habitats, it is important that the whole coastal system of the Solent area is managed as a single dynamic unit. However, this is extremely difficult to undertake without a greatly improved knowledge of the dynamics and processes that control the components of the system; and indeed, without a complete baseline of information on the precise distribution, composition and quality of existing habitats. Guidance is urgently needed on a strategy for addressing coastal habitat conservation priorities. In addition, the development of a research and management strategy is required for the retention and re-creation, where possible, of the full range of the Solent's coastal habitats and communities. This guidance must recognise and take into account the interdependence of many coastal habitats; further, considering the ways in which changes in one area, or habitat, may affect other natural features.

A considerable amount of survey work has been undertaken in the Solent, but (with a few exceptions) this has rarely been as part of an overall strategy aiming to provide a complete overview for any one habitat; it has even more rarely considered interactions between interdependent habitats. A research agenda for the Solent should aim to complete a full inventory and database for all major coastal habitats. Such an agenda should incorporate survey information on the distribution and quality of habitats, their vulnerability and sensitivity to coastal sediment processes, coastal change, management and other activities, and the existence of relevant research and monitoring data. Where information on any of these factors is incomplete, research or survey work should be targeted at filling these particular gaps in the information. Information on habitat distribution and quality will allow landowners, managers and planners to begin to identify areas that could be suitable for managed retreat; this would incorporate the set-back of coastal and maritime habitats and re-creation of important landward habitats. The aim should be, wherever possible, to maintain the total area of each habitat now present; similarly, to implement appropriate habitat re-creation against a background of sea level rise and unavoidable development of the coast.

Despite the history of research in the Solent area, understanding of the complex functioning of coastal habitats, species and communities and their interrelationships is still lacking. This subject is particularly important now that coastal habitat is increasingly being constrained between a narrowing intertidal area and development inland. A collaborative approach between planners, managers and the research community will be required to identify those areas requiring improved co-ordination of research efforts in order to achieve these objectives.

Finally, managed retreat, whilst still a largely untested and poorly-understood management option, is likely to be essential for the maintenance of coastal habitats and functions in the Solent. Lessons can be learnt from other areas where managed retreat is being undertaken, but it will be necessary to begin to apply this form of habitat creation in the Solent before it has been fully tested elsewhere, if coastal biodiversity is to be retained. A carefully targeted collaborative research programme should be established from the outset of any managed retreat programme, to monitor progress with habitat creation or re-establishment and to enable the lessons learnt here to be applied elsewhere.

ACKNOWLEDGEMENTS

This paper draws heavily on a review of the marine and coastal habitats of the Solent area (Fowler, 1995) funded by English Nature and the Solent Protection Society on behalf of the Solent Forum Nature Conservation Topic Group. I am grateful to the staff of English Nature and the Joint Nature Conservation Committee, who provided advice, unpublished material and generous assistance during that study. I would particularly like to acknowledge the help of R. Bamber, J. Barne, C. Chatters, P. Gilliland, A. Inder, D. Johnson, J. Maskrey, R. Page, I. Pearson, C. Robson, and C. Tubbs.

REFERENCES

Bamber, R.N. & Barnes, R.S.K. 1996. Coastal Lagoons. In: *British Coasts and Seas. Region 9: Southern England. Hayling Island to Lyme Regis.* Barne, J. *et al.* (Eds.), Joint Nature Conservation Committee, Peterborough, UK, 249 pp.

Bamber, R.N., Batten, S.D. & Bridgwater, N.D. 1992. On the ecology of brackish water lagoons in Great Britain. *Aquatic Conservation: Marine and Freshwater Ecosystems,* **2**: 65-94.

Banyard, L. & Fowler, R. (*this volume*). Lee-on-the-Solent Coast Protection Scheme. In: *Solent Science - A Review.* Collins, M.B. & Ansell, K. (Eds.), Proceedings in Marine Science Series, Elsevier, Amsterdam.

Barne, J.H., Robson, C.F., Kaznowska, S.S., Doody, J.P. & Davidson, N.C. (Eds.). 1996. *Coasts and Seas of the United Kingdom. Region 9: Southern England. Hayling Island to Lyme Regis.* Joint Nature Conservation Committee. Peterborough, UK, 249 pp.

Barnes, R.S.K. 1988. The faunas of land-locked lagoons: chance differences and the problems of dispersal. *Estuarine, Coastal and Shelf Science,* **26**, 309-318.

Barnes, R.S.K. 1989. The coastal lagoons of Britain: an overview and conservation appraisal. *Biological Conservation,* **49**: 295-313.

Bray, M.J., Hooke, J.M. & Carter, D.J. (*this volume*). Sea-level rise in the Solent region. In: *Solent Science - A Review.* Collins, M.B. & Ansell, K. (Eds.), Proceedings in Marine Science Series, Elsevier, Amsterdam.

Burd, F. 1989. The saltmarsh survey of Great Britain: an inventory of British saltmarshes. Nature Conservancy Council *Research and Survey in Nature Conservation* No. 17. Peterborough.

Burges, D. (*this volume*). Ornithology of the Solent. In: *Solent Science - A Review.* Collins, M.B. & Ansell, K. (Eds.), Proceedings in Marine Science Series, Elsevier, Amsterdam.

CEC 1992. Council directive 92/43/EEC: on the conservation of natural habitats and of wild fauna and flora. Council of the European Communities. *Official Journal of the European Communities,* L206/7.

Chatters, C. 1994. *Gentianella anglica* Early gentian. In: *Scarce Plants in Britain.* Stewart, A., Pearman, D.A. & Preston, C.D. (Eds.). Peterborough, Joint Nature Conservation Committee/Institute of Terrestrial Ecology/Botanical Society of the British Isles, 192-194.

Collins, K.J. & Mallinson, J.J. (*this volume*). Marine habitats and communities. In: *Solent Science - A Review.* Collins, M.B. & Ansell, K. (Eds.), Proceedings in Marine Science Series, Elsevier, Amsterdam.

Dargie, T.C.D. 1996. Sand dunes. In: *British Coasts and Seas. Region 9: Southern England. Hayling Island to Lyme Regis.* Barne, J. *et al.* (Eds.), Joint Nature Conservation Committee, Peterborough, UK, 249 pp.

Davidson, N.C. 1996. Estuaries. In: *British Coasts and Seas. Region 9: Southern England. Hayling Island to Lyme Regis.* Barne, J. *et al.* (Eds.), Joint Nature Conservation Committee, Peterborough, UK, 249 pp.

Davidson, N.C., Laffoley, D., Doody, J.P., Way, L.S., Gordon, J., Key, R., Drake, C.M., Pienkowski, M.W., Mitchell, R.M. & Duff, K.L. 1991. *Nature conservation and estuaries in Great Britain.* Nature Conservancy Council, Peterborough, UK.

Doody, J.P. (Ed.). 1984. *Spartina anglica* in Great Britain. *Focus on nature conservation*, **5**, Nature Conservancy Council, Attingham Park, 72 pp.

Doody, J.P. 1996. Introduction to the Region. In: *British Coasts and Seas. Region 9: Southern England. Hayling Island to Lyme Regis.* Barne, J. *et al.* (Eds.), Joint Nature Conservation Committee, Peterborough, UK, 249 pp.

Fowler, S.L. 1995. *Review of nature conservation features and information within the Solent and Isle of Wight Sensitive Marine Area.* Unpublished report of a desk study carried out under contract to English Nature and the Solent Protection Society, on behalf of the Solent Forum Nature Conservation Topic Group. Nature Conservation Bureau, Newbury, Berkshire, UK, 85 pp.

Fowler, S.L. & Tittley, I. 1993. The Marine nature conservation importance of British coastal chalk cliff habitats. English Nature Research Report No. 32. Peterborough.

Gray, A.J. & Benham, P.E.M. (Eds.). 1990. *Spartina anglica - a research review.* Institute of Terrestrial Ecology, HMSO, London, 79 pp.

Hodgetts, N.G. 1996. Terrestrial lower plants. In: *British Coasts and Seas. Region 9: Southern England. Hayling Island to Lyme Regis.* Barne, J. *et al.* (Eds.), Joint Nature Conservation Committee, Peterborough, UK, 249 pp.

Hydes, D. (*this volume*). Nutrients in the Solent. In: *Solent Science - A Review.* Collins, M.B. & Ansell, K. (Eds.), Proceedings in Marine Science Series, Elsevier, Amsterdam.

McCue, J. (*this volume*). Shoreline Management Plans – A Science or an Art? In: *Solent Science - A Review.* Collins, M.B. & Ansell, K. (Eds.), Proceedings in Marine Science Series, Elsevier, Amsterdam.

Moffat, A.M. (Ed.). 1994. *Priorities for Habitat conservation in England.* English Nature Research Report No. 97, Peterborough, 46 pp.

Morgan, V. 1996. Terrestrial higher plants. In: *British Coasts and Seas. Region 9: Southern England. Hayling Island to Lyme Regis.* Barne, J. *et al.* (Eds.), Joint Nature Conservation Committee, Peterborough, UK, 249 pp.

Parsons, M.S. & Foster, A.P. 1996. Coastal invertebrate fauna. In: *British Coasts and Seas. Region 9: Southern England. Hayling Island to Lyme Regis.* Barne, J. *et al.* (Eds.), Joint Nature Conservation Committee, Peterborough, UK, 249 pp.

Price, D. & Townend, I. (*this volume*) Hydrodynamic, sediment process and morphological modelling. In: *Solent Science - A Review.* Collins, M.B. & Ansell, K. (Eds.), Proceedings in Marine Science Series, Elsevier, Amsterdam.

Radley, G.P. 1994. *Sand dune vegetation survey of Great Britain: a National Survey. Part 1 - England*. Joint Nature Conservation Committee, Peterborough, UK, 126 pp.

Randall, R.E. 1996. Vegetated Shingle Structures and Shorelines. In: *British Coasts and Seas. Region 9: Southern England. Hayling Island to Lyme Regis*. Barne, J. et al. (Eds.), Joint Nature Conservation Committee, Peterborough, UK, 249 pp.

Raybould, A.F., Gray, A.J. & Hornby, D.D. (*this volume*). The evolution and current status of the saltmarsh grass, *Spartina anglica*, in the Solent. In: *Solent Science - A Review*. Collins, M.B. & Ansell, K. (Eds.), Proceedings in Marine Science Series, Elsevier, Amsterdam.

Rose, F. 1996. The habitats and vegetation of present-day Hampshire. In: *The Flora of Hampshire*. Brewis, A., Bowman, P. & Rose, F. (Eds.), Harley Books, Colchester, Essex, UK, 16-62.

Sanderson, N.A. 1997. *Vegetation survey of Sandy Point, Hampshire*. Report to Hampshire County Council from Botanical Survey and Assessment, 22 pp.

Sanderson, N.A. 1998. *A Review of the extent, conservation interest and management of lowland acid grassland. Volume 1: Overview*. Research Report No. 259. English Nature, Peterborough, 184 pp.

Sheader, M. & Sheader, A. 1989. *The coastal saline ponds of England and Wales: an overview*. Report to the Nature Conservancy Council. CSD Report No. 1009, 34 pp.

Smith, B.P. & Laffoley, D. 1992. *A directory of saline lagoons and lagoon like habitats in England*. Peterborough, English Nature.

Sneddon, P. 1992. Variations in Shingle vegetation around the British coastline. Ph.D. Thesis (unpublished) University of Cambridge.

Sneddon P & Randall R.E. 1994. *Coastal Vegetated shingle structures of Great Britain: Appendix 3: Shingle Sites in England*, JNCC, Peterborough. (Main report is dated 1993), 104 pp.

Stewart, A., Pearman, D.A. & Preston, C.D. (Eds.) 1994. *Scarce Plants in Britain*. Peterborough, Joint Nature Conservation Committee/Institute of Terrestrial Ecology/Botanical Society of the British Isles, 515 pp.

Tubbs, C.R. 1995. Sea level change and estuaries. *British Wildlife*, 6, 168-176.

Marine Habitats and Communities

Ken J. Collins and Jenny J. Mallinson

School of Ocean and Earth Science, Southampton Oceanography Centre, University of Southampton, European Way, Southampton, SO14 3ZH, U.K.

INTRODUCTION

The Solent region contains a wide diversity of marine habitats, from exposed rock to sheltered muds, reflecting the complexity of the local coast. Rocky headlands and associated reefs are a feature of the coast of the Isle of Wight while extensive rocky reefs occur to the south of Selsey Bill. The Solent is the only major sheltered channel that separates an island from a mainland, in European Waters. Whilst it is sheltered, fast tidal currents flow through it, particularly in the western arm. The extensive shallow harbours (Portsmouth, Langstone and Chichester) and estuaries of the region provide intertidal mudflats and salt marshes of national and international importance for the wildfowl and waders that they support. Lying in the central English Channel, the region is in a transition zone between Lusitanean (warm temperate) and Boreal (cold temperate) provinces, supporting examples of flora and fauna from both.

One of the first comprehensive overviews of the marine fauna of the Solent region was made by Thorpe (1980), who provides an extensive literature survey and description for the Solent, Southampton Water and the three harbours (Portsmouth, Langstone and Chichester). The Joint Nature Conservation Committee (JNCC) surveys of harbours, rias and estuaries in southern Britain included a sublittoral survey in the area and further reviewed the Solent area literature (Dixon & Moore, 1987). Covey (1988) builds on this earlier study, including a description of the shores of the Isle of Wight as part of a major JNCC Marine Nature Conservation Review of the UK coast.

English Nature's 'sensitive marine areas' initiative identified 27 sites that were considered to contain nationally important sub-tidal marine wildlife areas (English Nature, 1993). One of these sites was the Solent area, recognising its unique location and wide diversity of habitats. Fowler (1995) considered the marine nature conservation features of the region and identified a number of key intertidal and sub-tidal habitats, together with a provisional level of importance (international, national, and regional), reflecting the rarity of the biotopes. This listing is summarised in Table 1.

This contribution describes the range of marine habitats and communities within the region, drawing on personal experience of working and diving in the area over the past 30 years. For general location and main sites referenced, refer to *Map A* and *Map B* in the *Preface*.

Table 1 Suggested list of intertidal and sub-tidal habitats of marine conservation importance, adapted from Fowler (1995).

Site Name	Main Features	Importance
Bembridge Ledges	Limestone platforms, large rock pools & sands	International
Culver Cliff	Chalk cliffs and caves, specialised algal communities	International
Needles	Chalk cliffs and caves, specialised algal communities	International
Chichester Harbour	Un-dredged, stable & species rich mud & sand	National
Freshwater	Rich intertidal chalk with diverse flora	National
Langstone Harbour	Subtidal channels probably similar to Chichester Harbour	National
Newtown Harbour	Sheltered, undeveloped and unpolluted	National
Portsmouth Harbour	Affected by urban, ports and industrial activity	National
St Helens/ Ryde	Intertidal sediments with a diverse infauna and sea grass (*Zostera*) present	National
Stanswood Bay	Mixed sediments, high diversity. Native oyster bed	National
The Mixon Hole	Clay cliff topped by limestone slabs	National
Yar Estuary	Boulder and mixed sediment with a rich and diverse fauna	National
Bracklesham Bay	Sublittoral spherical boulders	Regional
Hurst Spit	Deep, scoured and current exposed 'hole'	Regional
NW coast, Isle of Wight	Limestone ledges and boulders, steep subtidal clay	Regional
Osborne Bay	Sediment shores, shallow subtidal with diverse infauna	Regional
Selsey reefs	Extensive subtidal limestone bedrock reefs and boulders	Regional
Solent Breezes	Sediment shores, muds and sand to muddy gravel	Regional
Solent Forts	Very diverse algal and epifaunal communities	Regional
South Wight	Subtidal bedrock and boulder reefs with rich algae	Regional
St Helen's Road	Very soft mud, *Maxmulleria lankestri* community	Regional

HABITAT TYPES

The substratum types (Figure 1) control the distribution of habitats and their associated communities. Greater detail of the sediment types is shown in the British Geological Survey seabed sediment map of the area (BGS, 1990). Apart from direct sampling of the seabed, remote methods such as side scan sonar and, more recently, digital echo sounding techniques (ROXANNE) in an area to the south of the Isle of Wight, have been used (Southeran & Foster-Smith, 1995). The characteristic sediments of the sheltered harbours and Solent are fine silts and muds. The tidally swept western Solent has deposits of gravel and small stones; this is in common with much of the Channel seabed beyond the Isle of Wight. To the east, the entrance to the eastern Solent and Bracklesham Bay have extensive areas of sand and gravel. Natural rock outcrops occur mainly on the exposed headlands of the Isle of Wight and to the south of Selsey Bill. Man-made structures, harbour walls, jetties and cables provide extensive additional hard substrate.

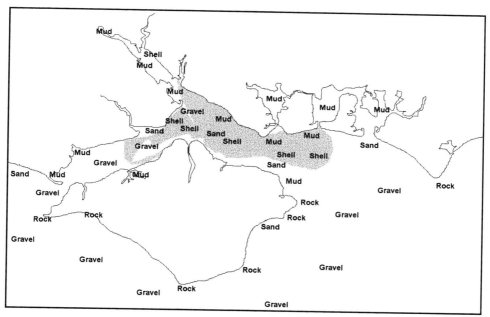

Figure 1 Distribution of substrate types plus the extent of the *Crepidula* association (shaded) reported by Barnes et al. (1973).

Chalk

Chalk bedrock and boulders occur to the west (Needles to Freshwater Bay) and east (Culver Cliff) of the Isle of Wight (Fowler & Tittley, 1993). Strong currents, which sweep around these headlands, prevent the accumulation of silts, and the intervening sediments are clean gravels. The shallow water chalk supports a heavy algal growth on upper surfaces, dominated by the brown algae *Laminaria* spp. and *Halidrys siliquosa* and a variety of smaller red algae including *Dilsea carnosa*, *Halurus flosculosus*, *Halurus equisetifolius* and

Phyllophora crispa. Much of the exposed chalk surface is covered with pink encrusting algae, probably *Lithothamnion* spp. Below 9 m depth, the hydroids *Tubularia indivisa* and *Kirchenpauria pinnata* become visually dominant over the few small species of red algae, such as *Phyllophora* spp. The stable current-swept surfaces provide ideal settlement sites for sponges, the most commonly found being *Dysidea fragilis, Halichondria panicea* and *Esperiopsis fucorum*. Another sponge, *Ciocalypta penicillus*, is found in its characteristic habitat, where the base of the rock meets mobile sediment. Anemones, *Anemonia viridis, Urticina felina* and *Cereus pedunculatus* are also prominent.

Limestone

Limestone occurs extensively off Bembridge and Selsey. The littoral Bembridge limestone pavement does not extend far into the sublittoral, but remnants of it, as broken slabs and isolated boulders, occur all along its edge. This is arguably the most scenic sublittoral area in the Solent region, supporting the widest diversity of species (Collins & Mallinson, 1988; Collins *et al.*, 1989). There is a characteristic depth-related gradation of the large brown algae on upper surfaces of rocks and boulders from the littoral to sub-littoral: *Fucus serratus* gives way to the kelps, first *Laminaria digitata* and then *Laminaria hyperborea*; the latter is found with *Halidrys siliquosa,* which generally extends somewhat deeper. Deeper still, red algae predominate, particularly *Heterosiphonia plumosa, Plocamium cartilagineum, Delessaria sanguinea* and *Cryptopleura ramosa*. Below this depth (10 m), animal species are dominant, with the occasional red algae such as *Phyllophora crispa*. The shallow broken pavement areas support a greater diversity of species than the isolated boulders; the unusual brown alga, *Zanardinia prototypus*, is found in such an area.

As the algal cover diminishes with depth, hydroids, bryozoans and sponges become more prominent; these include the hydroid, *Halecium halecinum,* and the bryozoans, *Flustra foliacea* and *Bugula plumosa*. The sponge *Halichondria panicea*, is the most massive, covering the upper surfaces of some small boulders. Another spreading form, *Esperiopsis fucorum*, although thinner, is found in similar situations. The sides of boulders, even in relatively shallow water, are fauna-dominated, with a continuous cover of bryozoans, hydroids, sponges, worms and ascidians. *Bugula turbinata* is the most prominent bryozoan here. Typical sponges found are *Dysidea fragilis, Hemimycale columella* and occasionally branching species, such as *Haliclona oculata* and *Raspailia hispida*. A number of ascidians share this habitat including *Clavelina lepadiformis* and *Morchelium argus*. The continuous cover on the rocks and boulders supports a wide range of mobile species, particularly molluscs and small crabs.

Deeper limestone outcrops at the Nab and Princessa Shoals, as well as components of the Wealden beds to the south of Culver Spit. These sites are generally below the depth for significant algal growth (10 m). In its place are erect hydroids *Nemertesia antennina* and *Tubularia* spp., and foliose byrozoa *Flustra foliacea* and *Bugula* spp. Sponges are a characteristic component of deep rock and boulder epifauna, usually in low density but showing a wide diversity. The more frequently encountered species are *Dysidea fragilis* and *Halichondria panicea*. Another prominent characteristic species of rock fauna is the soft coral, *Alcyonium digitatum*. Sloping rock surfaces in areas of strong currents sometimes support dense blankets of the small tunicate, *Dendrodoa grossularia* often containing specimens of *Polycarpa rustica*.

Clay

Clay bedrock outcrops on the northwestern and eastern coasts of the Isle of Wight and, most spectacularly, forms 10 m high cliffs in the Mixon Hole to the south of Selsey. These areas were generally found to be barren. The clay is extensively bored by the piddock *Pholas dactylus* and, generally, is too soft to support epifauna.

Artificial Substrates

Across the eastern Solent a string of forts were built on the sedimentary seabed providing islands of hard substrate with a notable species composition (Dixon & Moore, 1987). The excavation of the wooden warship "Mary Rose" provided a temporary habitat (Collins & Mallinson, 1984). Farther inshore, within the sheltered estuaries and harbours, port developments, piers and harbour walls have provided hard substrates in areas of otherwise soft mud. The fauna and flora in Southampton Docks have been described by Collins & Mallinson (1987). Here, there is clear zonation, particularly obvious on the vertical faces of walls, piles, and chains. From high to low water zones, the cover changes from predominantly barnacles and green (*Enteromorpha* spp. *Ulva* spp.) and brown *(Fucus* spp.) algae to calcareous tube worms such as *Hydroides enigmaticus*. Below this depth, sponges, (*Halichondria* spp., *Suberites masa*) and ascidians (*Styela clava, Ciona intestinalis*) take over with seasonal growth of hydroids (*Obelia longissima*) and byrozoa (*Bugula* spp.). In the upper reaches of the estuary, the plumose anemone, *Metridium senile* is also common in this zone. The walls of Southampton Docks have been classified as an example of the very sheltered circalittoral rock biotope more usually found in Scottish sea lochs (Connor *et al.*, 1997).

Gravels & Cobble

There is a variable mixture of sand, gravel and cobble, often mixed with shell, over much of the area. Stable gravel provides a firm substrate for settlement and consequently has a richer flora and fauna than the sand. There are extensive areas of pure cobble in the East Solent towards Selsey Bill (Collins & Mallinson 1983). The prominent anemones *Anemonia viridis* and *Urticina felina* are typical of the gravel areas. *Cereus pedunculatus* is also commonly found on this substrate. Where conditions are siltier, *Cerianthus lloydii* is more typical. Gravels containing a sand fraction support the sand mason worm, *Lanice conchilega*. Where finer sediments are associated with gravel, the peacock worm, *Sabella* spp., occurs. The slipper limpet, *Crepidula fornicata,* is a characteristic species of the Solent region. Its shells can form the main hard substrate in areas of soft sediments, is commonly associated with gravel and often achieves dominance.

Exposed gravel stones are often colonised by tubeworms, such as *Pomatoceros triqueter*, and by barnacles (*Balanus* spp.). On the deeper areas of stable gravel and cobbles, the hydroids *Halecium halecinum, Kirchenpaueria pinnata* and *Nemertesia antennina* are found. The foliose bryozoan *Flustra foliacea* occupies a similar niche.

Plate 1 Much of the Solent benthos has been transformed by the introduction of the slipper limpet, *Crepiduala fornicata*. Here, their shells cover a silty sand seabed in the East Solent, with an oyster (*Ostrea edulis*) in the foreground.

Plate 2 Limestone bedrock fauna, below 10 m depth off Bembridge, Isle of Wight, showing a dense cover of sponges, hydroids, ascidians and bryozoans. The most prominent species are the orange sponge (*Asperiopsis fucorum*) and the double white fans of the tube worm *Bispira voluticornis*.

In areas of strong tidal flow, algae are seldom found on the gravel; this is presumably because of current dragging. By contrast, in more sheltered areas, a substantial proportion of the algae found occurs on gravel and cobble substrates. The most striking area of weed growth on gravel is found to the north of St. Helens Fort where there is total cover by fronds of *Laminaria saccharina*, reaching several metres in length, during the summer.

Mobile species commonly found on the gravel are the gastropods *Buccinum undatum*, *Gibbula cineraria* and *Hinia reticulata*, and the crabs *Pagurus* spp. and *Liocarcinus* spp. Where foliose cover is available, there is an increased range of mobile fauna, including the spider crabs *Macropodia rostrata* and *Pisa tetraodon*. A species typical of cobble areas is *Galathea squamifera*, living on the underside of the larger rounded flint stones. At one site to the south of Culver Spit, the gravel contains live maerl (*Phymatolithon and Lithothamnion* spp.). On stony gravel seabeds, the bryozoan *Alcyonidium* spp. grows extensively, blanketing the seabed in late summer.

Sand

Areas of clean sand extend across Bracklesham Bay from Whitecliff Bay on the Isle of Wight. This sand is very mobile with ripples inshore and deep waves in the tidal mainstream. In the more stable areas the burrowing worms *Lanice conchilega* and *Arenicola marina* are found. Mobile species typically found in such areas include hermit crabs, *Pagurus* spp., and gastropods such as *Hinia reticulata* and *Buccinum undatum*. A number of species are sand tolerant but require some underlying stone or hard substrate for attachment; these include the anemones, *Urticina felina* and *Cereus pedunculatus*.

Mud & Silt

Generally, the muds and silts of the Solent region contain chains of slipper limpets (*Crepidula fornicata*) which provide attachment for other organisms. These include hydroids, *Kirchenpaueria pinnata* and *Hydrallmania falcata*, and sponges *Halichondria* spp. and *Suberites* spp. A number of small crab species, *Pisidia longicornis, Macropodia rostrata* and *Pagurus bernhardus*, are found in the cover provided by the slipper limpet shell epifauna. There are patches of oysters, *Ostrea edulis*, providing a resource for the local fishery. North of Bembridge, in St. Helen's Road, there is a unique area of fine, deep, soft mud without any hard substrate. This sediment type supports a large population of the echiurid worm *Maxmuelleria lankesteri*, in densities of several per square metre, with the amphipod *Ampelisca diadema* and the bivalve mollusc *Nucula nitidosa*. Muddy sediments usually support an infauna dominated by polychaetes, with oligochaetes and small molluscs present.

NOTABLE SPECIES

The Slipper limpet, *Crepidula fornicata*, is found throughout the area and, in some places, so numerous that its living chains and dead shells have modified the sediment type. Barnes *et al.* (1973) mapped a *Crepidula* dominated association, throughout most of the Solent (Figure 1).

An example of a less common species, which is of national conservation importance, is the sea grass, *Zostera marina*. This marine flowering plant is an important food for some seabirds and waders, and, in addition, supports a rich and diverse community. This sea grass occurs within the Bembridge Ledges lagoon system and patchily along the north coast of the Isle of Wight.

One of the notable findings of the Collins & Mallinson (1989) survey off Bembridge was the discovery of a previously unknown population of the echiurid worm, *Maxmuelleria lankesteri*, in the very soft muds of the St Helen's Road anchorage. The presence of both the soft muds and the echiurid worm is more typical of deep Scottish sea lochs (Hughes *et al.*, 1993).

The use of tributyltin (TBT) antifouling paints and the high density of ships in the area have had a marked effect on the population biology and physiology of the Dog whelk, *Nucella lapillus*. Herbert (1988) surveyed the dog whelk population densities around the Isle of Wight and found it absent from much of the north (Solent) coast: those populations close to the Solent showed a high degree of imposex. Since the ban on the use of TBT for pleasure craft, Bray & Herbert (1998) found some evidence for recovery. Future monitoring of this species will be invaluable.

ALIEN SPECIES

As a major international port, the Solent region has a long history of receiving alien species through fouling species on ships hulls; larvae in ballast waters; and introductions of organisms in association with mariculture. For example the slipper limpet, *Crepidula fornicata*, is considered to have been introduced with American oysters at the end of the last century. *Crepidula* and the ascidian, *Styela clava* (introduced in the 1950s) are now both so widespread as to be characteristic of the region. Recent notable seaweed introductions have been Japanese seaweed, *Sargassum muticum* (1971) and the kelp, *Undaria pinnatifida* (1994). The introduction of alien species is summarised in Table 2.

MARINE MAMMALS

Marine mammals are rarely seen in the Solent and Southampton Water, so any observations (Table 3) are notable. Dolphins are routinely observed off Durlston Head, Dorset where there is a monitoring programme[1]. Sightings in Poole Bay and off Brighton suggest there may be interchange between these locations. Observations in the Channel are presently being collated[2]. The juvenile seal, which appeared at Southampton Town Quay, in 1997, probably originated from one of the small groups resident in Chichester or Poole Harbours. Adult individuals of this species are occasionally seen at Calshot and Durlston Head. With widespread interest in dolphins and seals there is, potentially, a lot of information that can be gained from members of the public and people working on and around the estuary. What is required is a well-publicised contact point for reporting these observations. The Southampton Oceanography Centre has proved to be a natural point of contact for local sightings[3].

---

[1] Durlston Country Park (tel: 01929 424443).
[2] Andy Williams, British Marine Life Rescue (tel: 02392 552631)
[3] Jenny Mallinson, Southampton Oceanography Centre (tel: 02380 596299).

Table 2 List of alien species introduced into the Solent area, from Thorp (1980) and Eno et al. (1997)

Group	Species	Location & Date	Source
Annelida	*Ficopomatus enigmatica*	Chichester Harbour 1974	Thorp unpublished
	Pileolaria rosepigmentata	Portsmouth Harbour 1975	Knight-Jones et al. 1975
	Janua (Dexiospira) braziliensis	Portsmouth Harbour 1975	Knight-Jones et al. 1975
	Hydroides ezoensis Okuda	Southampton Water 1976	Thorpe et al. 1987
Crustacea	*Acartia tonsa*	Southampton Water 1954	Conover 1957
	A. grani	Southampton Water 1956	Lance & Raymont 1964
	Limnoria tripunctata	Southampton Water	Eltringham & Hockley 1958
	L. quadripunctata	Southampton Water	Eltringham & Hockley 1958
	Elminius modestus	Widespread 1945	Bishop 1947
Mollusca	*Mercenaria mercenaria*	Southampton Water 1925	Mitchell 1974
	Crepidula fornicata	Widespread 1880-90	Holme 1961
	Petricola phaladiformis	Lee-on-the-Solent	Duval 1963
Tunicata	*Styela clava*	Widespread 1960	Houghton & Millar 1960
Algae	*Sargassum muticum*	Bembridge 1971	Farnham et al. 1973
	Undaria pinnatifidia	Hamble 1994	Fletcher & Manfredi 1995
	Grateloupia doryphora	Southsea 1969	Farnham & Irvine 1973
	Grateloupia filicina var. *luxurians*	Solent before 1947	Farnham & Irvine 1968
	Asparagopsis armata	Solent 1973	Irvine et al. 1975
	Codium fragile subsp. *tomentosoides*	Devon 1939	Farnham 1980

Table 3 Records of marine mammals within Southampton Water and the Solent.

Date	Description of sightings
1975	2 bottlenose dolphins stayed in the Test arm of Southampton Water for 4 days.
1993-4	1 common dolphin resident at Calshot (particularly Calshot Spit buoy) for over six months before moving up the estuary to Marchwood where it sustained a minor propeller injury and departed soon after.
1997	A juvenile female common seal hauled out at Town Quay exhausted and dehydrated in late September.
1998	4 dolphins (probably bottlenose) were seen on one day in June off Netley.

FUTURE DIRECTIONS

The direction of future effort with regards to marine habitats and communities should address the following issues:

WHAT and WHERE?
- JNCC biotope classification
- Remote surveys
- Ground truth surveys
- Geographic Information System

WHEN?
- Monitoring - changes, aliens

There is a vast amount of information on the local marine habitats and biota held by local research bodies and authorities. Geographical Information Systems (GIS) provide a means to bring this data together in a readily accessible and understandable form. Once gaps have been identified in the data, additional surveys should be carried out. These surveys can be most efficiently achieved by employing remote methods such as ROXANNE or side scan sonar to identify the distribution of seabed types. However, such measurements must be verified by *in situ* sampling, such as grab, dredge, or diving. There is a national JNNC led initiative to classify marine habitats and communities into a set of well-defined biotopes (Connor *et al.*, 1997). This system was developed originally in Scottish and Northern Irish waters, and it currently shows a poor correlation with south coast habitats. Further work is needed to adapt this classification to the Solent communities.

The Solent system is dynamic and hence survey results may become less relevant to the current situation with the passage of time. The establishment of a Solent Special Area of Conservation (SAC) will require management of its key habitats to ensure that there is no degradation. This requires routine survey. Researchers at Portsmouth University have discovered numerous alien species arriving in the area, some of which have had profound effects on the local ecology. Constant monitoring is required to determine the effects of those that are here and those that may arrive in the future.

Since the Solent Science Conference, English Nature have commissioned a collaborative project between the Geodata Unit and School of Ocean and Earth Science (University of Southampton), School of Biological Sciences (University of Portsmouth), Environmental Monitoring Unit and Associated British Ports. This project will bring together existing data on seabed types and principal organisms into a Solent GIS.

REFERENCES

Barnes, R. S. K., Coughlan, J. & Holmes, N. J. 1973 A preliminary survey of the macroscopic bottom fauna of the Solent with particular reference to *Crepidula fornicata* and *Ostrea edulis*. *Proceedings of the malacological Society London*, **40**: 253–275.

Bishop, M.W.H. 1947. Establishment of an immigrant barnacle in British waters. *Nature*, **159**: 501.

Bray, S. & Herbert, R. J. H. 1998. A reassessment of the populations of the dog-whelk (*Nucella lapillus*) on the Isle of Wight following legislation restricting the use of TBT antifouling paints. *Proceedings of the Isle of Wight Natural History and Archaeolgy Society* **14**: 23–40.

BGS. 1990. Sea bed sediments and quaternary geology, Wight Sheet 50°N - 02°W, 1:250,000 series. British Geological Survey.

Collins, K.J. & Mallinson, J.J. 1984. Colonisation of the Mary Rose excavation. *Progress in Underwater Science*, **9**: 67-74.

Collins, K.J. & Mallinson, J.J. 1983. *Sublittoral Survey from Selsey Bill to the East Solent*. Report to the Nature Conservancy Council. Contract No: HF3-11-04, 28.

Collins, K.J. & Mallinson, J J. 1987. *Marine flora and fauna of Southampton Docks*. Report to the Nature Conservancy Council. Contract No: HF3-11-52 (7), 26.

Collins, K.J. & Mallinson, J.J. 1988. *Marine flora and fauna off Bembridge, Isle of Wight; sublittoral survey and review of existing knowledge*. Report to the Nature Conservancy Council. Contract No: HF3-11-52 (1), 42.

Collins, K.J. & Mallinson, J.J. 1989 *Marine Flora and Fauna off Bembridge, Isle of Wight; offshore sublittoral survey*. Report to the Nature Conservancy Council. Contract No: HF3-11-57 (3), 17.

Collins, K. J., Herbert R.J.H. & Mallinson, J. J. 1989 Marine flora and fauna of Bembridge and St Helens, Isle of Wight. *Proceedings of the Isle of Wight Natural History and Archaeological Society*, **9**: 41-85.

Conner, D.W., Dalkin, M.J., Hill, T.O., Holt, R.H.F. & Sanderson, W.G. 1997. *Marine biotope classification for Britain and Ireland Volume 2. Sublittoral biotopes*. JNCC Marine Nature Conservation Review Report No.230, 448.

Conover, R.J. 1957. Notes on the seasonal distribution of zooplankton in Southampton water with special reference to the genus *Acartia*. *Annals and Magazine of Natural History London*, Series 12, **10**: 63-67.

Covey, R. 1998. Eastern Channel (Folkestone to Durlston Head, MNCR Sector 7): The Solent System. In: *Marine Nature Conservation Review. Benthic marine ecosystems of Great Britain and the north-east Atlantic*, Hiscock, K. (Ed.), Joint Nature Conservation Committee, Peterborough,.204-207.

Dixon, I.M.T & Moore J. 1987. *Surveys of harbours, rias and estuaries in southern Britain: the Solent system*. Report to the Nature Conservancy Council by the Field Studies Council, Oil Pollution Research Unit, Pembroke, CSD Report No.723, 100.

Duval, D.M. 1963. The biology of *Petricola phaladiformis* Lamark (Lamellibranchiata: Petricolida). *Proceedings of the malacological Society London,* **35**: 89-100.

Eltringham, S.K. & Hockley, A.R. 1958. Coexistence of three species of the wood-boring isopod *Limnoria* in Southampton Water. *Nature*, **243**: 231-2.

English Nature 1993. *Important areas for marine wildlife around England. Draft August 1993*. English Nature, Peterbrough, 200.

Eno, N.C., Clark, R.A. & Sanderson, W.G. 1997. *Non-native marine species in British Waters: a review and directory*. Joint Nature Conservation Committee, Peterborough, 152 pp.

Farnham, W.F. 1980. Studies on aliens in the marine flora of southern England. In: *The shore environment, volume 2: ecosystems*, Price, J.H., Irvine, D.E.G. & Farnham, W.F. (Eds.). Systematics Association Special Volume, No. 17B. Academic Press, London, 875-914.

Farnham, W.F., Fletcher, R.L. & Irvine, I.M. 1973. Attached *Sargassum* found in Britain. *Nature*, **243**: 231-232.

Farnham, W.F. & Irvine, L.M. 1968. Occurrence of unusually large plants of *Grateloupia* in the vicinity of Portsmouth. *Nature*, **219**: 744-746.

Farnham, W.F. & Irvine, L.M. 1973. The addition of a foliose species of *Grateloupia* in the British Marine flora. *British Phycological Journal*, **8**: 208-209.

Fletcher, R.L & Manfredi, C. 1995. The occurrence of *Undaria pinnatifidia* (Phaeophyceae, Laminariales) on the south coast of England. *Botanica Marina*, **38**: 355-358.

Fowler, S.L. 1995. *Review of nature conservation features and information within the Solent and Isle of Wight Sensitive Marine Area*. Report to English Nature by Nature Conservation Bureau Ltd, 50 pp.

Fowler, S.L. & Tittley, L. 1993. *The marine nature conservation importance of British chalk cliffs habitats*. English Nature Research Report No. 32, 50.

Herbert, R. J. H. 1988. A survey of the dog whelk *Nucella lapillus* (L.) around the Isle of Wight. *Proceedings of the Isle of Wight Natural History and Archaeological Society* **11**: 40-42.

Herbert, R.J.H. 1998. Isle of Wight Marine Biological report 1996 and 1977. *Proceedings of the Isle of Wight Natural History and Archaeological Society,* **14**: 91-94.

Holme, N. A 1961. The bottom fauna of the English Channel. Journal of the Marine Biological Association, U. K. **41**: 397 – 461.

Houghton, D.R. & Millar, R.H. 1960. Spread of *Styella mammiculata* Carlisle. *Nature*, **185**: 862.

Hughes, D.J., Ansell, A.D., Atkinson, R.J.A. and Nickell, L.A. 1993. Underwater television observations of the surface activity of the echiurid worm *Maxmuelleria lankesteri* (Echiura: Bonelliidae). *Journal of Natural History.* 27, 219-248.

Knight-Jones, P., Knight-Jones, E.W., Thorp, C.H. & Gray, P.W.G. 1975. Immigrant spirorbids (Polycheata: Sedentaria) on the Japanese *Sargassum* at Portsmouth, England. *Zoological Scripta*, **4**: 145-149.

Lance, J. & Raymont, J.E.G. 1964. Occurrence of the copepod *Acartia grani* G.O.Sars in Southampton water. *Annals and Magazine of Natural History London*, Ser.13, **7**: 619-624.

Mitchell, R. 1974. Aspects of the ecology of the lamellibranch *Mercenaria mercenaria* (L.) in British waters. *Hydrobiological Bulletin*, **8**:124-138.

Southeran, I. & Foster-Smith, R. 1995. *Mapping the benthic biotopes of the Isle of Wight*. English Nature Research Report No.120.

Thorp, C.H. 1980. The benthos of the Solent. In: *The Solent Estuarine System, an assessment of present knowledge*. NERC publication series C No 22 November 1980, 76-85.

Thorpe, C.H., Pyne, S. & West, S.A. 1987. *Hydroides ezoensis* Okuda, a fouling serpulid new to British coastal waters. *Journal of Natural History*, **21**: 863-877.

Ornithology of the Solent

Dave Burges

Royal Society for the Protection of Birds, South East England Office, 2nd Floor, Frederick House, 42 Frederick Place, Brighton, BN1 4EA, U.K.

INTRODUCTION

The birds using the complex of coastal sites, from Pagham Harbour in West Sussex to Hurst Spit in Hampshire, have attracted for many years, the interest of amateur and professional ornithologists alike. Indeed, the concept of high and low tide counts was pioneered on the south coast harbours (Clark & Eyre, 1993). As a result of these initially *ad hoc*, and now more formalised bird-monitoring programmes, a reasonable understanding of overall bird numbers in the Solent has been developed. These data allow the monitoring of broad changes at the site level, harbour or estuary, through high tide counts (i.e. birds roosting); and highlight the use of individual mudflats through low tide counts (i.e. birds feeding).

Inevitably, this information provokes many more questions. In general terms, these questions fall into two categories: firstly, how important is a particular mudflat or harbour for a species or group of species; and secondly, how do the birds actually use the complex of habitats available to them? Common sense and casual observation suggest that some sites appear to be more important for some species than others. Additionally there is increasing evidence of significant movement within and between sites during tidal cycles.

The answers to these questions are of ecological interest in their own right. However, the same data are also required to assess, for example, the implications of sea-level rise on coastal habitats, the impact of development proposals on particular sites, and the need to manage the sometimes conflicting interests of birds and people at specific locations.

Several key themes emerge from the above summary. Initially, there is the desirability of considering the south coast harbours and estuaries as a single large site, the 'Greater Solent', which recognises the complex bird usage of these sites. Secondly, there is the requirement to implement the EU Birds[1] and Habitats Directives[2] to ensure protection and management of internationally important wildlife sites. The third theme unites the first two, which is what sort of research is needed to better understand bird-use of the Greater Solent and to ensure that 'site integrity' is maintained, against a background of 'natural' and 'man-made' changes.

---

[1] EU Council Directive of 2nd April 1979 on the conservation of Wild Birds (74/409/EEC).

[2] EU Council Directive (92/43/EEC) of 21st May 1992 on the Conservation of natural habitats and of wild fauna and flora.

In setting the scene and beginning to address these questions, this paper considers the ornithology of the Solent under the following headings:

- the range of bird species in the Solent;
- habitat requirements, numbers and distribution;
- conservation status and relevant designations;
- present monitoring; and
- issues, possible solutions, and information needs.

RANGE OF BIRD SPECIES IN THE SOLENT

The Solent is important for three key groups of birds, summarised as follows:

- breeding seabirds - such as little tern and black-headed gull;
- wintering and passage wildfowl - such as brent geese and wigeon; and
- wintering and passage waders - such as black-tailed godwit and dunlin.

Some of these species and/or species assemblages, occur in nationally and inter-nationally important numbers. These, in turn, support several of the key nature conservation designations affecting the south coast harbours and estuaries.

HABITAT REQUIREMENTS, DISTRIBUTION AND NUMBERS

Habitat requirements

The distribution of these species, across a range of broad habitat types, begins to highlight the importance of the mosaic of habitats to breeding, passage, and wintering waterfowl, both in front of and behind the sea wall. This is illustrated in general terms in Table 1 below.

Table 1 Broad habitat types and their use by key bird assemblages in the Solent

Habitat	Feeding	Roosting	Breeding
Mudflats	Wildfowl, waders	-	-
Saltmarsh	Wildfowl, waders	Wildfowl, waders	Gulls, terns
Grazing marsh	Wildfowl, waders	Wildfowl, waders	-
Lagoons	Wildfowl, waders	Wildfowl, waders	Gulls, terns
Shingle, islands	-	Waders	Gulls, terns
Agricultural land	Brent geese	-	-
Amenity grassland	Brent geese	-	-

Distribution

The overall distribution of birds within the Greater Solent is defined most easily by highlighting the principal sites. From east to west, these are:

- Pagham Harbour, West Sussex;
- Chichester Harbour, West Sussex, Hampshire;
- Langstone Harbour, Hampshire;
- Portsmouth Harbour, Hampshire; and
- the Solent and Southampton Water, Hampshire, Isle of Wight.

For general location and main sites referenced, refer to *Maps A* and *B* in the *Preface*. Geographically, these sites are relatively straightforward to define. The four harbours appear to be rather discrete bodies. Although in the case of the Solent and Southampton Water, the 'site' comprises a suite of coastal habitats extending from Hurst Spit to Calshot, much of the Southampton Water shoreline, and the estuaries on the north shore of the Isle of Wight.

In biological terms, however, the ornithological interest is dependent upon habitats both in front of and behind the seawall. Furthermore, observation of bird movements and the preliminary results of ringing studies are beginning to indicate a much more complicated relationship between the use of particular sites, or parts of sites, by certain species at various stages of the tidal cycle. It is helpful to consider the wider area as part of the relevant Important Bird Area (IBA) (Pritchard *et al.*, 1992) as this may extend beyond the boundary of currently designated sites.

Numbers of birds

The most obvious indicator of importance for a particular site is the numbers of birds using it. There are two complementary approaches for assigning levels of importance to sites:

- the total numbers of a species occurring on the site, expressed as a percentage of either that species' Great Britain (GB), or international populations; and
- sites regularly supporting over 20,000 waterfowl.

All regularly monitored species have agreed Great Britain and international qualifying thresholds, set at 1% of the respective populations. Thus, an estuary may support several species at either national or international levels of importance, and hold over 20,000 waterfowl. The thresholds are revised periodically in the light of new data (e.g. Waters *et al* 1998).

Two examples illustrate this approach. Table 2 shows the 5-year mean peak count for the top five brent goose sites in the Solent, whilst Table 3 presents the 5-year mean peak counts for the five most important sites for all waterfowl.

Table 2 5-year mean peak counts (1992/3 to 1996/7), for the top five brent goose sites in the Solent (Waters et al., 1998).

Site	5-year mean peak count 1992/3 to 1996/7
Chichester Harbour	10,616
Langstone Harbour	6,676
Portsmouth Harbour	2,862
Pagham Harbour	2,823

Note: International importance threshold for brent goose: 3,000
       Great Britain importance threshold for brent goose: 1,000

Table 3 5-year mean peak counts (1992/3 to 1996/7), for the top six Wetland Bird Survey (WeBS) count sites in the Solent (Waters et al., 1998).

Site	5-year mean peak count 1992/3 to 1996/7
Chichester Harbour	54,969
Langstone Harbour	44,416
Southampton Water*	20,389
Pagham Harbour	17,357
North-west Solent* (part of the Solent and Southampton Water complex)	14,988
Beaulieu Estuary*	13,858

*Note: all part of the Solent and Southampton Water Special Protection Area.

The data presented in these two tables (Table 2 and 3) illustrate the point that a site, such as Chichester Harbour, can be internationally important for a particular species, and internationally important for its assemblage of waterfowl. When this exercise is extended to all the key sites in the Solent, the 5-year (1992/3 to 1996/7) mean peak count for brent geese totals some 32,566 birds. A figure that is equivalent to about 35% of the UK wintering population and some 11% of the world population.

Similarly, the 5-year (1992/3 to 1996/7) mean peak count for all waterfowl on all sites amounts to about 178,093 birds; this makes the Greater Solent one of the top five estuarine complexes in the UK.

CONSERVATION STATUS AND RELEVANT DESIGNATIONS

Clearly, the numbers of birds recorded against accepted criteria give rise to the range of national and international designations that apply to at least the core areas of the six sites listed under *Distribution* (see above).

Many of the harbours, individual estuaries or sections of shoreline are designated as Sites of Special Scientific Interest (SSSIs), for a range of both biological and sometimes, geological features, under the Wildlife and Countryside Act of 1981. Where the appropriate criteria are met, these SSSIs are the basis for two key international designations, with particular relevance to birds: Special Protection Areas (SPAs) designated under the EU Birds Directive and Wetlands of International Importance under - the Ramsar Convention[3].

As noted above, those sites qualifying as SPAs may do so based upon a number of criteria; namely:

- sites regularly used by 1% or more of the Great Britain population of birds listed on Annex 1 of the Birds Directive - in these cases applying to the breeding tern and gull species;
- sites regularly used by 1% or more of the biogeographic population of a regularly occurring migratory species; and
- sites regularly used by over 20,000 waterfowl.

The latter two criteria overlap with those in the Ramsar Convention, which relate specifically to birds. There are others that refer to other taxa and physiographic features of wetlands. Pagham, Chichester, Langstone and Portsmouth Harbours, together with the Solent and Southampton Water are designated SPA/Ramsar sites. The SPAs overlap to varying degrees with the candidate for Special Areas of Conservation (SACs), to be designated under the EU Habitats Directive. Together, these will form the *Natura 2000*, network of European sites under the provisions of the Directive. In addition to pulling together the SPAs and the SACs, Article 6 of the Habitats Directive sets out the process for considering the impact of plans or projects on such European sites (DoE, 1994). This is referred to later (*Issues, possible solutions and information needs*).

In addition to these statutory designations and the obligations that accompany them, there are a number of key non-statutory frameworks referring either specifically to birds, or the wider environment, which affect the status, protection and management of Important Bird Areas within the Solent. Recent reviews of bird population data in Europe and the UK have led to a clear set of bird conservation priorities.

At the European level, this has led to the publication of *Birds in Europe; their conservation status*, by Tucker and Heath (1994). The analysis presented here prioritises birds into four categories as Species of European Conservation Concern (SPEC). Those species accorded SPEC1 status are globally threatened; those listed as SPEC4 have their global populations

---

[3] The convention on Wetlands of International Importance especially as Waterfowl Habitat (Cmnd 6465)

concentrated in Europe, but which currently enjoy favourable conservation status. Most waterfowl species fall within the SPEC2 (global populations concentrated in Europe, unfavourable conservation status in Europe) and SPEC3 (global populations not concentrated in Europe, unfavourable conservation status in Europe). Examples in each category would be redshank and sandwich tern (SPEC2), and dunlin and bar-tailed godwit (SPEC3).

This European-wide analysis provided information for a similar exercise in the UK, which led to the publication in 1996 of *Birds of Conservation Concern* (RSPB *et al.*, 1996). This prioritises UK bird species into red, amber and green lists according to a number of criteria regarding numbers, range, trends and SPEC status, with the red-listed species being those under greatest threat. Within a Solent context, species such as roseate tern are red-listed on account of their rarity as breeding birds, whilst many of the wildfowl and wader species are amber-listed because of the large proportion of biogeographic populations wintering at relatively few sites.

Lastly, the framework provided by Biodiversity Action Plans (BAPs) at the UK, Regional, and County levels will help to co-ordinate action for coastal habitats and their bird fauna. As a general rule the wintering waterfowl and breeding seabirds of the Solent will benefit from the delivery of habitat actions through the Hampshire and Sussex BAPs; they are unlikely to require individual species plans (Hampshire Biodiversity Partnership, 1998; Sussex Biodiversity Partnership, 1998).

CURRENT MONITORING

The Wetland Bird Survey

The key datasets on wintering waterfowl within the Solent's harbours and estuaries are collected as part of the national Wetland Bird Survey (WeBS), co-ordinated by the British Trust for Ornithology (BTO), the Wildfowl and Wetlands Trust (WWT), the Royal Society for the Protection of Birds (RSPB) and the Joint Nature Conservation Committee (JNCC). Volunteers collect the raw data. The background to the WeBS programme is given in the latest report for 1996-97 (Waters *et al*, 1998). In essence, however, the WeBS data comprises the core high tide counts for all key sites, principally carried out between September and March each autumn and winter. These data provide total numbers of all relevant species by site, UK species totals, and population indices.

The low tide counts comprise a seven-year rolling programme around UK estuaries, with the counts taking place between November and February. These counts provide a picture of waterfowl distribution at low tide and are especially helpful in highlighting the relative importance of individual sections of mudflat for feeding waterfowl.

Other studies

As noted above, a range of other monitoring studies are underway; these consist principally of ringing and marking studies carried out by the Farlington Ringing Group and the Solent Shorebird Study Group. The preliminary results of this work are showing a much greater degree of multiple site use by waterfowl species in the Solent. At the same time, there is a significant need for further information arising from the conservation issues referred to below.

ISSUES, POSSIBLE SOLUTIONS AND INFORMATION NEEDS

Clearly, there are a wide range of factors that can impact upon the bird populations of the Solent and their habitats. Some relate to questions of policy interpretation and practice; still, other factors such as sea level rise, will have a direct effect on much of the existing habitat resource. Development proposals are more likely to affect specific sites, either directly and/or indirectly, but the need for a better understanding of waterfowl ecology is equally important. The following paragraphs are not intended to be comprehensive, but rather provide a flavour of the range of issues currently under consideration. They suggest possible solutions and highlight further bird data needs where these would help in reaching solutions.

Designation boundaries

The currently designated suite of SPA and Ramsar sites in the Solent is based upon the existing SSSIs. Generally speaking, these encompass the key intertidal habitats and some habitats behind the sea wall, which are broadly considered to be semi-natural habitats. Inevitably, birds do not recognise these boundaries and significant numbers may roost or feed beyond the SPA boundaries. Thus a range of man-made structures, or more intensively managed habitats, may play a critical role in supporting birds for which the SPAs have been designated. These issues can produce serious conflicts with development proposals - as the case of the proposed football stadium at Farlington, outside Portsmouth, proved in 1994. In this instance, the development, if it had been permitted, would have destroyed amenity grasslands used by nationally and (occasionally) internationally important numbers of brent geese (GOSE, 1994). A principal reason for the refusal was the regular use of the site by brent geese from the Chichester and Langstone Harbour SPA.

Nevertheless, this event and other similar cases led to the establishment of Brent Goose Strategy Working Group (1998), comprising representatives of Hampshire County Council, Gosport, Havant and Fareham Borough Councils, Portsmouth City Council, English Nature, the Hampshire Wildlife Trust, the Solent Shorebird Study Group and the RSPB. The aim of the Group is to produce a strategy, which will provide an integrated approach to the management of brent goose feeding sites behind the sea wall. The key objective is to foster a pro-active approach to site specific issues and thus minimise conflicts between development and recreational pressure and geese, and to promote the value of the environment to local communities.

Whilst there may be cases where formal modification of designation boundaries is desirable, a second option is the better recognition, not least through the planning system, of the ecological value of areas outside SPA/Ramsar sites, which nevertheless are integral to the ecological function of those sites.

Sea Level Rise

The most serious threat to bird habitats in the Solent is sea level rise, and the 'coastal squeeze' of these habitats that will occur as a result. Sea levels are predicted to increase by 6 mm per year, leading to a 32 cm rise in levels by 2050 (Solent Forum, 1997). Recent research, undertaken by the University of Newcastle for the Environment Agency and English Nature, predicts a net loss of coastal habitats, within European sites, of some 377 ha between Portland Bill and Selsey Bill over this period, *if* the current proposals in the relevant Shoreline Management Plans (SMPs) are followed (Lee, 1998). The worst affected habitat types are saltmarsh and wet grassland, with predicted net losses of 360 ha and 442 ha respectively.

The predicted changes will clearly impact on the current mosaic of habitats, within and outside the SPA/Ramsar sites; and they could affect bird numbers, distribution, or usage. Implementation of the SMPs, as currently written, will need to take account of the requirements of the Habitats Directive in ensuring the protection of *Natura 2000* sites. Much work, both theoretical and practical remains to be done in this area, but the bird data for existing habitats will assist in setting objectives, modelling, and post-project monitoring of any habitat creation schemes.

Development pressure

Development proposals are, almost by definition, highly site-specific in location terms. However, any impact on important bird habitats, and the bird use of them, may extend beyond the confines of the site itself. The level of interest (local/national/international) of the site(s) affected also determines the type of treatment accorded to such proposals in the planning system. Given the significant areas of internationally protected habitats in the Solent, the application and implementation of the requirements of the Habitats Directive is critical.

Site-specific impacts, such as direct habitat loss, are perhaps easier to quantify than the indirect effects; these may be associated with additional disturbance, or changes to coastal processes. Furthermore, the cumulative effects of otherwise unrelated developments could have serious impacts on the habitat resource and its function over time. These impacts are difficult to predict and resolve, but clearly critical to maintaining the conservation status of the designated sites. If damaging developments do proceed on or adjacent to internationally important sites, the issues surrounding the suitability and likely success of habitat creation measures are similar to those noted under *Sea Level Rise*.

Recreation management

The Solent is a major recreational resource for a wide range of interests and activities, which, like industry, are critical to the economic health of the area (Solent Forum, 1997). There are potential conflicts between recreation and nature conservation. These may arise through either the direct impacts of recreational facilities on, or adjacent to, key habitats or the conduct of the activities concerned in sensitive areas such as breeding seabird sites or wader roosts. Generally, recreation groups, the Local Authorities, Harbour Authorities, and the conservation organisations have tried to take a pro-active approach to managing recreational impacts on nature conservation interests. These approaches rely on a mixture of controlling access, zoning and providing information to recreational users, such that they can avoid sensitive locations at specific times of the year or the tidal cycle. Clearly, new proposals for recreational developments would be dealt with through the planning system, as for any other project.

SUMMARY

The above presents only a short synopsis of the conservation importance of the Solent for birds, and some of the key issues affecting it. Nevertheless a number of key points emerge, these are summarised below.

- The ornithological importance of the Solent is acknowledged and described under accepted criteria. The individual harbours and/or estuary complexes are internationally important for their bird populations; taken together they constitute one of the top five sites in the UK.

- This status is founded on many years of detailed bird counts, which continue today under the Wetland Bird Survey (WeBS) programme of high and low tide counts.

- Recent bird ringing and marking work on bird movements is beginning to provide a better understanding of how the birds actually use these sites; and more work is needed on the ecology of waterfowl at the sub- and inter-site levels.

- The habitats used by the birds are threatened generally by sea level rise. Consequently, a proactive approach to habitat creation is needed urgently if the integrity, structure, and function of these intertidal and behind-the-sea wall habitats are to be maintained.

- This relates to the need for a better integration of land use planning policy and practice, at national and local levels, to ensure that our international obligations under the Habitats Directive are met.

- Finding solutions to the issues raised above requires a multi-disciplinary approach; this is already in progress on some topics. Nevertheless, a more proactive approach to habitat creation (in particular) is needed urgently. The common ground established at the Solent Science Conference will hopefully provide the necessary starting point.

REFERENCES

Brent Goose Strategy Study Working Group. 1998. *A 'draft scoping document' towards a Brent Goose Strategy*, English Nature, Lyndhurst, 20 pp.

Clark, J.M. and Eyre, J.A. (Eds.) 1993. *Birds of Hampshire*. Hampshire Ornithological Society,. 512 pp.

DoE. 1994. *Planning Policy Guidance* Note 9. Nature Conservation. Department of the Environment, HMSO, 59 pp.

GOSE 1994 *Secretary of State's decision letter on application by Portsmouth Football Club for an all-seater stadium and associated retail and leisure and railway station, Eastern Road, Farlington, Portsmouth*. Government Office for the South East, London, 95 pp.

Hampshire Biodiversity Partnership 1998 *Biodiversity Action Plan for Hampshire Volume 1.* Hampshire Biodiversity Partnership, 80 pp.

Lee, E.M. 1998. *The implications of future shoreline management on protected habitats in England and Wales.* Environment Agency and English Nature, 51 pp.

Pritchard, D.E., Housden, S.D., Mudge, G.P., Galbraith, C.A, & Pienkowski, M.W. (Eds.) 1992. *Important Bird Areas in the UK including the Channel Islands and the Isle Man.* RSPB/JNCC Sandy, 540 pp.

RSPB, Birdlife International, The British Trust for Ornithology, The Game Conservancy Trust, The Hawk and Owl Trust, The National Trust, The Wildfowl and Wetlands Trust, & The Wildlife Trusts. 1996. *Birds of Conservation Concern in the United Kingdom, Channel Islands and the Isle of Man.* RSPB, Sandy, 6 pp.

Solent Forum 1997. *Strategic Guidance for the Solent.* Hampshire County Council, 200 pp.

Sussex Biodiversity Partnership 1998. *Biodiversity Action Plan for Sussex Volume 1.* Sussex Biodiversity Partnership, Lewes, 185 pp.

Tucker, G.M. & Heath, M.F. 1994. *Birds in Europe; their conservation status.* Birdlife International, Cambridge, 600 pp.

Waters, R.J., Cranswick, P.A., Musgrove, A.J. & Pollitt, M.S. 1998. *The Wetland Bird Survey 1996-97: Wildfowl and Wader Counts. The results of the Wetland Bird Survey in 1996-97.* British Trust for Ornithology, The Wildfowl and Wetlands Trust, RSPB and JNCC, Slimbridge, 176 pp.

Fisheries of Southampton Water and the Solent

Antony Jensen

School of Ocean and Earth Science, Southampton Oceanography Centre, University of Southampton, European Way, Southampton, SO14 3ZH, U.K.

INTRODUCTION

The fisheries in the Solent and Southampton water are mixed, in terms of catch and methods used. Each fishery varies in importance throughout the year according to market demand, regulation and availability of target species. Every year varies, according to the dynamics of demand, weather and stock size. Consequently, any summary of such a coastal fishery must be read with such dynamics in mind.

The fisheries lie within an area that is heavily used by commercial, military and recreational interests. Associated British Ports (ABP) operate the large port of Southampton with its extensive car import/export trade and significant container traffic. To maintain shipping volume, the port requires a deep, dredged, navigation channel, maintained at 12.2 m below chart datum; this alters the natural hydrography of the estuary and restricts the areas that can be fished. In addition, large petrochemical works associated with the Fawley oil refinery and the Fawley oil fired power station (with its requirement for cooling water), add to the industrial pressures on the marine environment, where Southampton Water joins the Solent. Military shipping is concentrated in Portsmouth Harbour, but the Marchwood military base does contribute to the traffic in the area (for general location and main sites referenced, refer to *Maps A* and *B* in the *Preface)*. Pollution from shipping and oil-related activities is a constant concern. Additionally, the passage of large vessels with restricted ability to manoeuvre within marked shipping channels causes short-term loss of access and prevents the use of static gear in some parts of the system.

Recreational interest in the area is high, with the Solent being an important centre for yachting. The sheltered waters of the Solent are ideal for cruising and racing; they are used throughout the summer 'season', with peaks of activity at weekends and during special events such as Cowes Week. Winter activity is much reduced, but some clubs do organise winter racing (Shears, 1986). Recreational craft can interfere with fixed fishing gear, in some circumstances.

Although speculative, it is suggested that if it were not for industrial and recreational pressures that exist, the sheltered waters of the Solent and Southampton Water would provide prime locations for aquaculture initiatives.

FISHERY MANAGEMENT AND REGULATION

Southampton Water and the Solent lie within the Southern Sea Fishery District (SSFD), which extends from Hayling Island to the Dorset/Devon border. Local regulations are drawn up by the Southern Sea Fishery Committee (SSFC) and implemented by the Fisheries Officers of the SSFD. Any local regulations and bylaws are additional to national and EU fisheries legislation, the enforcement of which is the responsibility of both MAFF (Ministry of Agriculture, Fisheries and Food) and SSFD Fishery Officers working in the area. Such inshore fisheries management capability allows regional strategies to be developed, which are designed to maintain the fishery whilst taking into account local conditions and interests. The Sea Fishery Committee consists of a mixture of local government and MAFF appointees, providing a group consisting of locally elected representatives, fishermen and scientists. As examples of the type of regulation used, the SSFD bylaws (SSFC, 1996): (a) prohibit vessels over 12 m in length and not registered in the district before 1995 from fishing within 6 nautical miles of the coast; (b) place restrictions on some fishing seasons e.g. closure of oyster fisheries from March 1st to October 31st; (c) define some fishing gear specifications e.g. in the oyster fishery, the maximum blade length of a dredge is 1.5 m; and (d) set Minimum Landing Sizes (MLS) for some species e.g. for the oyster (*Ostrea edulis*), the MLS is set at 70 mm.

Current SSFC issues, relating specifically to the Solent and Southampton Water are:

- the possible loss of the 6 nautical miles coastal fisheries management derogation during the re-negotiation of the new Common Fisheries Policy, which will come into force in 2002, effectively removing local inshore fisheries management and opening up coastal fisheries to the EU fishing fleet; and

- the effective management of the oyster fishery, in particular controlling the problem of pre-season poaching from grounds within the regulated fishery.

In addition, part of Southampton Water is a MAFF designated protected nursery area for juvenile bass, a valuable fish to the inshore fishing industry.

LANDINGS

Landing data (MAFF) show 24 species are landed from boats working from ports within the area (Table 1). About 154 boats are registered with the SSFD, as working from ports in the Solent (Figure 1). However, caution must be exercised when looking at catch data for the Solent ports. The catch data provided by MAFF (1998) shows landings for Portsmouth only, whereas for Southampton Water and the Solent, landings from Southampton, Bursledon, Cowes, Lymington and the Isle of Wight are incorporated together with non-Solent ports such as Christchurch and Barton-On-Sea to the west (note: there is no port at Barton-On-Sea, just a beach; conjecture suggests this may refer to the port of Mudeford). This means that Solent landing figures cannot be totally isolated from other data. In addition, landing does not always imply local capture. Large harbours such as Portsmouth are the home port for some trawlers, which work almost exclusively outside the region.

Additionally, vessels under 10 m are not required to make catch returns to MAFF. These omission make the interpretation of official MAFF catch statistics almost meaningless, when trying to assess the total landings from such an area.

Table 1 Selected catch data for a variety of species landed at 'Solent ports' (Portsmouth, Southampton, Bursledon, Cowes, Lymington and the Isle of Wight) and including 'non-Solent ports' (Christchurch and Barton-On-Sea) in 1997 (MAFF, 1998).

Species	1997 Landing (tonne)	Approx. Value (£)
Bass	72	407,000
Crab	516	675,000
Lobster	30	280,000
Oyster	350	650,000
Skate/ray	50	48,000
Sole	59	364,000

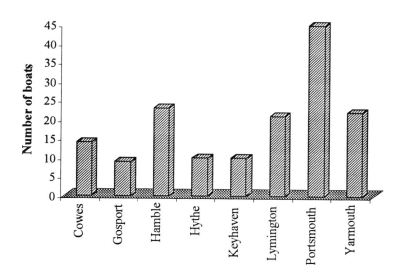

Figure 1 Number of boats registered with SSFD, at ports within the Solent and Southampton Water.

A local source of published information is the SSFD Chief Fisheries Officer's quarterly report (Whitley, 1997a, b; 1998a, b, c, d); these provide valuable qualitative information about local fishing activity, some of which is specific to the Solent and Southampton Water area. A summary of the past two years fishing activity within the Solent has been extracted from these reports.

Annual variability of fishing activity in the Solent

In 1997

- Oysters were still being fished until 21st March, the close of the second phase of the Solent Regulated Oyster Fishery, with good prices being obtained for the catch. Total oyster landings from the fishery in the 1996-7 season were estimated at about 285 tonnes.

- In April, the cuttlefish season started with catches fetching between £200 and £700 per day.

- The dominance of cuttlefish continued into May, with extensive effort focused on this species and the prices holding up. May saw the start of the bass season, with good catches in the West Solent (Whitley, 1997a).

- Bass fishing continued in June, as the cuttlefish catches declined. Mackerel shoals were present, but there was little commercial fishing as the market demand was minimal.

- In July, longliners in the Solent were still catching good quality bass.

- In August, bass fishing continued to prosper. Trawl fishermen, although hampered by drift weed, started to land increasing numbers of sole and skate, as well as dogfish, however; there was no market for the latter. Reports of poaching of oysters from the Solent Regulated Fishery were received by the SSFC (Whitley, 1997b) during this month. September was a poor month for fishing, as catches were low and unpredictable. Bass fishing continued in a sporadic fashion and netters were thwarted by large amounts of drift weed. Small bass (< 150 mm) were plentiful.

- October saw dogfish and codling landings develop and good quality sole were caught in the West Solent. During the month, bass landings were generally good for the time of year and netting improved as the weed problem disappeared by the end of the month.

- In November, netting continued and both sole and skate were landed in reasonable quantities, bass fishing continued. The oyster fishery opened on 3rd November, with 77 licences in operation. Catches were moderate, but demand low; this, together with the strong pound affecting export prices, kept first sale prices below those seen in 1996 (ranging between £1200 to £1500 per tonne). Stockpiling of oysters in bags was common around the Solent and concerns were expressed about the stress caused by poor storage conditions, increasing the susceptibility of these oysters to disease (Whitley, 1998a).

- Oyster prices were maintained in December with the 'hoped for' Christmas peak not materialising. With the poor price of oyster catches, many of the larger boats turned to other fisheries, with weather restricting many to the eastern Solent and east Wight. Flatfish and skate provided reasonable catches.

In 1998

- Bad weather in early January prevented fishing outside the Solent and the poor demand for oysters dampened effort in this fishery. In February, oyster prices improved to £1600 per tonne for a short while, and fishery activity increased, but by 13th March (end of season) oyster fishing had stopped (Whitley, 1998b).
- Poor weather in early March restricted fishing to sheltered waters and landings were poor. There was some sporadic activity in the Solent clam fishery.
- By April, sole were being caught in the Solent and bass catches were increasing. In contrast to 1997, no cuttlefish were caught. This situation continued into May (Whitley, 1998c).
- In June, fishing continued in a similar vein, bass numbers increased, but fish were generally small and attracting a lower price (about £5 per kg) than the rarer, larger bass (about £8 per kg). Cuttlefish were still absent from the fishery.
- Improvements were seen in July, when the East Solent yielded catches of very good quality plaice for a time, but the fishery, in general, continued at a low level of reward.
- In August, sole catches were increasing and the bass fishery improved, providing good catches in the East Solent (Whitley, 1998d).

At no time in 1998 did the 'hoped for' cuttlefish fishery develop. Because of depressed demand and price, the pre-season poaching on the regulated oyster fishery was minimal and licensed fishing effort was less than expected when the season opened in November (Whitley pers. comm.).

This 'thumb nail' sketch of two year's activity shows how variable the catches and rewards can be from an inshore area such as the Solent and Southampton Water. The poor catches in 1998 are not the only financial problem for inshore fishermen; many, working on 1997 experience, invested in new fishing gear to catch cuttlefish hoping for good returns in 1998. Such returns just did not materialise.

SOLENT OYSTER FISHERY

The Solent holds the largest remaining, naturally regenerating, fishery for the European oyster, *Ostrea edulis*, in Europe. Much of the fishery was brought under the control of the SSFC, in 1980, by the Solent Oyster Fishery Order. The regulated area stretches from a line between Hurst Spit and Fort Albert, in the west, to a line from Seaview to Southsea Castle (including the two Solent forts) in the east, excluding the estuaries of the Lymington, Beaulieu and Medina rivers, Newtown and Wootton Creeks, Southampton Water and Portsmouth Harbour and an area between Stansore Point and Hill Head.

The SSFC aims to manage the fishery within the regulated area, to sustain the resource. This management has generally taken the form of effort limitation (licences, gear control and restricted fishing season) and enforcement of a MLS for oysters (at 70 mm). In 1998-1999, only licensed skippers will be authorised to fish, in the region of 55 licences are expected to be 'taken–up'. The season will run from 2nd November – 11th December and re-open on 11th January until 12th March 1999; only oyster dredges of approved dimensions can be used. These regulations are enforced by the Fishery Officers of the SSFD, operating from SSFD patrol craft.

The management regime has altered over time and, of late, has tried to balance fishing pressure against the requirement to maintain a large enough brood stock of oysters in the Solent; this is to provide sufficient spat for settlement. One of the major problems in managing the fishery is a lack of understanding of the relationship between the density of mature stock, on the ground, and the successful spatfall and survival to reach MLS. However, since there is no definitive numerical basis for calculating total allowable catch, the fishery is managed using: empirical information from previous seasons; noting information from an annual 'stock on the ground' dredge survey by MAFF; and taking a precautionary approach when it comes to increasing fishing effort, as market demand is so variable as to be unpredictable year to year. This approach appears to work, as oyster catches have not been limited by stock levels in recent years.

The oyster fishery is of great local value, providing revenue at a time of year when other important species, such as bass and (recently) cuttlefish, are unavailable i.e. November through to March. Additionally the grounds are located in areas generally protected (to some extent) from bad weather. There is considerable debate regarding the value of the fishery to the local economy. Estimates for the 1997-98 season range upwards from £500,000, a year where demand was low and prices generally poor.

Areas other than the regulated fishery also contribute to the oyster yield. In the Solent, two 'several order' fisheries exist for oysters. MAFF have designated an area within Stanswood Bay (for sole use by Stanswood Bay Oystermen Ltd) and another close to Calshot Spit (for the sole use of Calshot Oyster Fishermen Ltd) as 'several orders'. Here, limited companies (formed by local fishermen) have the exclusive right to manage and harvest the oyster population. Beyond the several orders and the regulated fishery are the public fishery areas, such as Langstone Harbour and the eastern side of Southampton Water.

CLAM FISHERY

By 1965 and extending into the 1970s and early 1980s, there was extensive fishing for the American hardshelled clam, *Mercinaria mercinaria*, in Southampton Water and some parts of the Solent; this is a fishery that dates back to World War II (Mitchell, 1974). Various anecdotes suggest that this species was introduced into Southampton Water in the 1930s, either discarded from a cruise ship or 'escaping' from a failed clam farming enterprise. Mitchell (1974) reports that introduction took place in 1925, when clams were imported from America on the vessel 'Leviathan' to be tried as bait for eels. The trials were unsuccessful and the remaining clams (reported to be a few dozen) were laid in the Test estuary, close to where the Marchwood Power Station once stood. The clams reproduced successfully possibly assisted by the warm water emerging from the power station cooling system. The spat settled and grew throughout the Test and Itchen estuaries and along the eastern shore of Southampton Water, together with some parts of the Solent.

By 1971, annual catches had reached 100t per annum, estimated as about 1 million 50 mm clams live weight (Mitchell, 1974). Al-Sayed (1988) reports that the number of boats involved in the fishery increased, from about 10 in 1979 to over 40 by 1983, with some using hydraulic dredges, such equipment having a much higher fishing efficiency than the traditional dredge. A combination of poor recruitment to the population, over successive years, and heavy fishing pressure is thought to be the reason for the significant depletion of stock and collapse of the fishery. Al-Sayed (1988) estimated that the total Southampton sub-littoral stock had declined to about 205 tonnes, by 1985. In response to this decline, by the late 1990s, only a few boats were occasionally involved in the fishery.

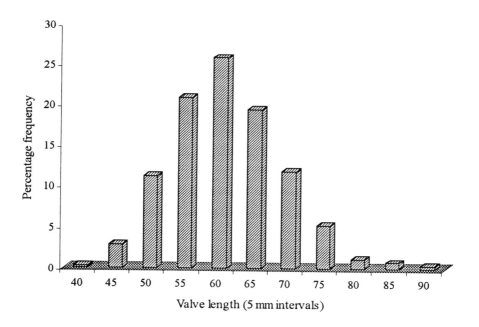

Figure 2 Size frequency distribution of clams (*Mercinaria mercinaria*) from Dibden Bay re-laid within Marchwood yacht moorings in 1998.

In March 1998, a joint initiative by the Waterside Fisherman's Association and Associated British Ports led to a period of intensive clam dredging in Dibden Bay; this area would be severely disturbed if plans for a new container port are approved. Some 2000 clams were collected and re-laid in the area of moorings used by Marchwood Yacht Club and the area declared a prohibited fishing zone by SSFC. It is hoped that the clams will settle into the sediments and act as brood stock, supplying spat to colonise Southampton Water and so re-establish a clam fishery in the fullness of time. The size frequency distribution of re-laid clams (Figure 2) suggests that most of the animals are of reproductive size. Mitchell (1974) reports that clams in Southampton Water were sexually mature at 3 years of age, an average size range of 20-21mm.

SUMMARY

Southampton Water and the Solent provide ports and sheltered fishing for a significant number of fishing boats. Whilst many boats will fish outside this area, there are times when the shelter afforded by the Isle of Wight means that the Solent is the only viable fishing location. The Solent oyster fishery is of local importance, providing an income at a time of year when other fisheries are very limited.

Throughout the area, catches vary with season and year, and a flexible approach to fishing technique and targeted catch is a requirement of those fishing the Solent, just as it is with all inshore fishermen. In the case of the Solent and Southampton Water, this flexibility has to be extended to accommodate the significant use of the area by shipping, coastal industry and recreation.

ACKNOWLEDGEMENTS

Thanks are due to Mark Whitley and Ian Carrier of SSFD, for helping with statistical data and other information; and to the Waterside Fisherman's Association and their colleagues, for measuring the clams as they were moved from Dibden Bay to Marchwood.

REFERENCES

Al-Sayed, H.A.Y. 1988. Population studies of a commercially-fished bivalve, *Mercinaria mercinaria* (L.) in Southampton Water. PhD thesis University of Southampton, Oceanography Department.

Shears, J. 1986. *Solent Environmental Studies: Solent Sailing and Water Use*. Report 16a/86 to Shell UK Exploration and Production by GeoData Unit, Faculty of Science, University of Southampton, 50 pp.

SSFC, 1996. *Southern Sea Fisheries District By-Laws*. Published by SSFD, 13p.

MAFF 1998. *Statistics of fish landings in England, Wales and Northern Ireland by Port. Annual Figures for 1997*. Report from MAFF Fisheries Statistical Unit, London.

Mitchell, R. 1974. Studies on the population dynamics and some aspects of the biology of *Mercinaria mercinaria*. Ph.D. thesis, University of Southampton, Oceanography Department.

Whitley 1997a. *Report of the Chief Fishery Officer 1 March 1997 to 31 May 1997.* Report to the Joint Quarterly Committee of the Southern Sea Fishery District, 17 July 1997, 8 pp.

Whitley 1997b. *Report of the Chief Fishery Officer 1 June 1997 to 31 August 1997.* Report to the Joint Quarterly Committee of the Southern Sea fishery District, 16 October 1997, 7 pp.

Whitley 1998a. *Report of the Chief Fishery Officer 1 September 1997 to 30 November 1997.* Report to the Joint Quarterly Committee of the Southern Sea fishery District, 22 January 1998, 9 pp.

Whitley 1998b. *Report of the Chief Fishery Officer 1 December 1997 to 28 February 1998.* Report to the Joint Quarterly Committee of the Southern Sea fishery District, 9 April 1998, 7 pp.

Whitley 1998c. *Report of the Chief Fishery Officer 1 March 1998 to 31 May 1998.* Report to the Joint Quarterly Committee of the Southern Sea fishery District, 9 July 1998, 8 pp.

Whitley 1998d. *Report of the Chief Fishery Officer 1 June 1998 to 31 August 1998.* Report to the Joint Quarterly Committee of the Southern Sea fishery District, 15 October 1998, 8 pp.

SECTION 4

Short Contributions

Underwater Light in Tidal Waters: Possible Impact on Macro-Algae Communities

S. Charrier[1], A. Weeks[1], S. Lewey[1] and I. Robinson[2]

[1]Maritime Research Centre, Maritime Faculty, Southampton Institute, East Park Terrace, Southampton, 8014 OYN, U.K.

[2]School of Ocean and Earth Science, Southampton Oceanography Centre, University of Southampton, European Way, S014 3ZH, U.K.

INTRODUCTION

The intensity and spectral composition of solar energy play essential roles in the development of aquatic photosynthetic organisms (Dring, 1992). In the intertidal zone, the light available to macro-algae is a combination of direct sunlight and underwater light. At low tide, exposed macro-algae are physiologically affected by the intense direct sunlight; they rapidly stop photosynthesis as they dry out (Dring & Brown, 1982). Therefore, it can be assumed that most of the light available to intertidal macro-algae for photosynthesis is underwater light.

In the intertidal zone, the underwater light reaching attached macro-algae depends upon the combination of various factors such as solar and lunar cycles, tidal characteristics and processes, and water optical properties. This paper presents both measured and simulated downwelling underwater irradiance, and shows the variability in the light field available to intertidal macro-algae growing at different levels on the shore.

METHODS

Profiles of the underwater light field have been measured over semi-diurnal tidal cycles at three stations, on the south coast of England (UK). Optical sensors (Satlantic Inc., Ca.) were deployed at the edge of the intertidal zone in the Hamble estuary, a turbid water site, and as close as possible to the intertidal zone at Swanage bay, a relatively clear water site. Measurements were also carried out at a subtidal site in Southampton Water. The three stations, shown in Figure 1, were chosen to allow for comparisons between turbid and clear waters, intertidal and subtidal environments, and small and large tidal range. A simple mathematical model has also been developed, to simulate the relative underwater light field in tidal waters (Charrier *et al.*, 1998).

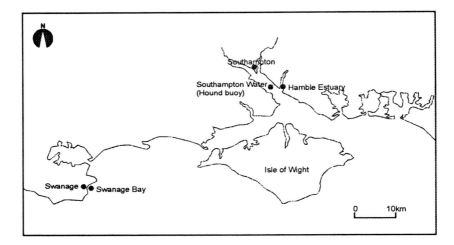

Figure 1 Map showing the sampling stations.

RESULTS

The percentage of surface downwelling irradiance at different levels above the seabed shows important variations throughout a semi-diurnal tidal cycle (Figure 2). These variations are particularly important in the Hamble estuary and Southampton Water, as a result of the combination of the tidal range and the turbidity of the water column. Figure 3 shows the predicted seasonal and spring-neap variations of daily integrated downwelling irradiance.

This diagram shows daily amounts of blue light reaching macro-algae growing at different levels above the seabed (i.e. different height above Chart Datum). In the Hamble estuary maxima occur at neap tide, at about 2 m above Chart Datum and between neap and spring tides around 4 m above Chart Datum; near Chart Datum, the light remains constantly low. At Swanage Bay, the maxima also occur at neap tide and at about 1 m above Chart Datum.

DISCUSSION AND CONCLUSIONS

Both measured and predicted underwater downwelling irradiance reveal the high variability and complexity of the underwater light field available to intertidal macro-algae. Measurements show the semi-diurnal variations, while simulated irradiance suggests complex semi-diurnal and seasonal patterns of temporal and vertical distribution of the daily integrated irradiance.

Therefore, intertidal macro-algae experience a highly variable light supply, depending on their position on the shore (i.e. above Chart Datum); this could contribute to the species distribution patterns commonly observed.

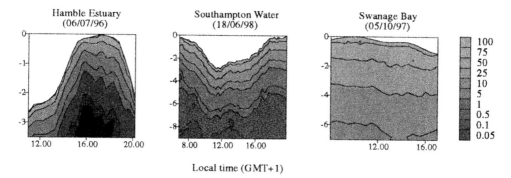

Figure 2 Percentage of downwelling irradiance (412 nm) on the water surface, remaining at different depths.

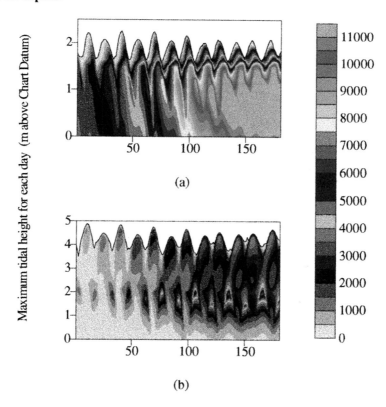

Julian day

Figure 3 Predicted seasonal and spring-neap variations of daily integrated downwelling underwater irradiance (relative units; Hamble estuary, Kd(443 nm) = 1.8 m^{-1}; Swanage Bay, Kd(443 nm) = 0.28 m^{-1}).

ACKNOWLEDGEMENT

The authors would like to express their thanks to the School of Ocean and Earth Science of the University of Southampton for helping with the collection of data in Southampton Water.

REFERENCES

Charrier, S.C., Weeks, A., Robinson, I. & Lewey, S. 1998. The optical characteristics of tidal waters. *Proceedings of the XIV Ocean Optics conference,* SPIE, USA.

Dring, M.J. 1992. *The Biology of Marine Plants,* Cambridge University Press, 199 pp.

Dring, M.J. & Brown, F.A. 1982. Photosynthesis of intertidal brown algae during and after periods of emersion: a renewed search for physiological causes of zonation. *Marine Ecology Progress Series,* **8**: 301-308.

Phytoplankton - Annual Sequences in the Hamble Estuary

J. O'Mahony and A. Weeks

Maritime Faculty, Southampton Institute, East Park Terrace, Southampton, SO14 0RP, U.K.

INTRODUCTION

Knowledge of the species composition of a habitat, as well as the processes that determine the presence of those species, is fundamental to the prediction of system production at all levels. The particular species of phytoplankton present at any one time determine the type of herbivore population and the herbivore niche; therefore, information on phytoplankton assemblages is essential. Phytoplankton composition and concentrations in the Hamble Estuary are affected by not only a seasonal pattern, but also by processes occurring on much shorter time-scales. Evidence indicates that tidal cycles and diurnal variations are important (Wright *et al.* 1998). An investigation of the timing of the phytoplankton spring bloom and subsequent community development was undertaken in the Hamble Estuary. The data presented here is a time-series of species composition and concentrations, from March - December 1996.

METHODS

The sample site was located at the Southampton Institute pier, Warsash, close to where the river Hamble enters Southampton Water (Figure 1). Temperature, conductivity and downwelling irradiance profiles, as well as meteorological data were recorded. Surface water samples were taken at high water springs and neaps, for suspended particulate matter (SPM), chlorophyll and quantitative phytoplankton analysis. Phytoplankton samples for quantitative analysis were preserved in Lugol's Iodine solution; these were analysed using sedimentation techniques and phase-contrast inverted microscopy, using a modified Utermohl's method (Hasle, 1978).

RESULTS AND DISCUSSION

Sea surface temperature

Surface water temperatures ranged from 4.79°C in mid March (JD[1] 74) to 23.29°C in mid June (JD 164, Figure 2). Temperatures were less than 10°C until mid April (JD 102) and then increased rapidly to reach 15.87°C by the end of that month (JD 121). Surface temperatures then fell during May to a minimum of 12.05°C (JD 142). Another rapid increase followed the annual maximum surface water temperature (23.29°C, JD 164). Following the annual maxima, water temperatures remained steady between 17 and 18°C until the end of July (JD 212), and then rose to 19-20°C until the beginning of September (JD 247). From September onwards temperatures began to decrease steadily towards winter values (approximately 5°C).

---

[1] The conversion table for Julian Days (JD) is given in Table 1.

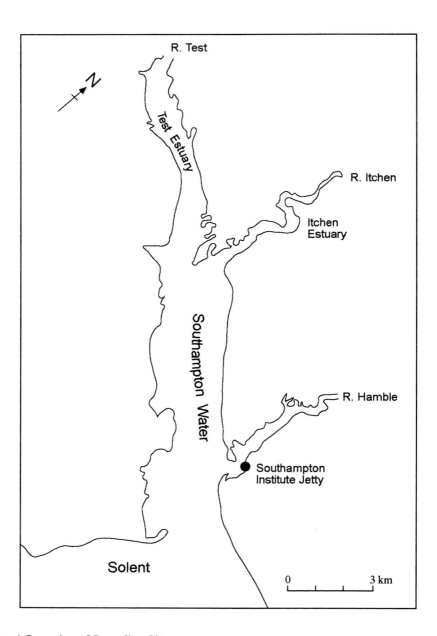

Figure 1 Location of Sampling Site

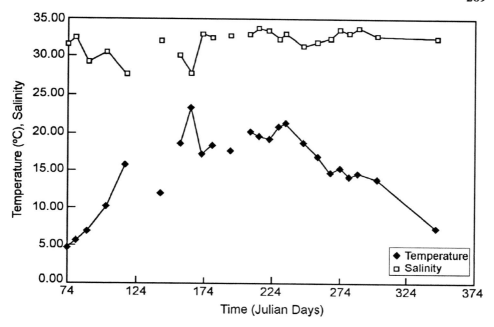

Figure 2 Temperature and salinity data. The break in line indicates that temperature and salinity data were not recorded due to mechanical breakdown.

Table 1 Conversion table for Julian Days (JD) mentioned in text.

JD	Date	JD	Date	JD	Date
74	14.03	164	12.06	234	21.08
79	19.03	172	20.06	247	3.09
88	28.03	179	27.06	257	13.09
102	11.04	186	4.07	267	23.09
116	25.04	193	11.07	274	30.09
128	7.05	200	18.07	281	7.10
137	16.05	208	26.07	288	14.10
142	21.05	214	1.08	302	28.10
149	28.05	222	9.08	-	-
156	4.06	229	16.08	346	11.12

Sea surface salinity

Salinity values, at high water, ranged from 27.66 to 33.82 throughout the survey (Figure 2). These are within the ranges recorded by other researchers in Southampton Water (Howard et al., 1995; Iriarte & Purdie, 1994). Highest salinities were recorded in July and August.

Suspended particulate matter

Levels of suspended particulate matter (SPM), both organic and inorganic, ranged from a minimum of 11.72 mg/l in early October (JD 281) to a maximum of 38.93 mg/l in late May (JD 149, Figure 3). The SPM composition was principally inorganic and contributed more than 70% of the total observed on each sample.

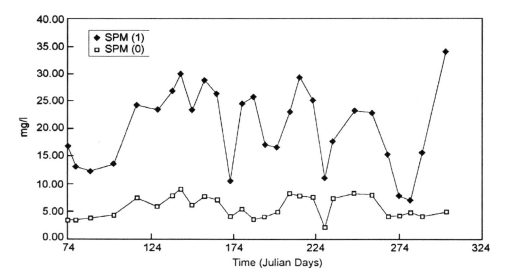

Figure 3 Total organic and inorganic SPM.

Phytoplankton biomass and composition

Biomass

Chlorophyll concentrations ranged from 0.5 to 7.37 µg/l during the period of sampling. (Figure 4). The phytoplankton biomass began to increase around mid March (JD 79). Four peaks in chlorophyll levels were recorded: 4.24 µg/l at the end of June, 7.37 µg/l at the end of July, 4.26 µg/l, and 3.71 µg/l at the beginning and end of August (JDs. 179, 208, 222 and 234, respectively). From early September (JD 247), levels decreased to less than 2 µg/l.

Chlorophyll values appeared to show a spring-neap variability, as most of the chlorophyll peaks were recorded on neap tides. Higher chlorophyll levels generally were found in association with a water temperature greater than 18°C and a salinity of more than 32.68.

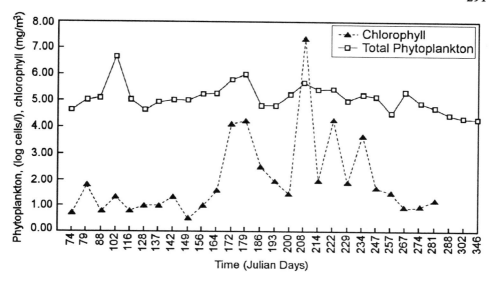

Figure 4 Phytoplankton and Chlorophyll Data.

There was an apparent delay in the spring peak, in terms of biomass; the spring bloom usually develops in late April/early May (Howard *et al.*, 1995, Iriarte & Purdie, 1994)). This delay was most likely due to the colder and duller weather recorded in January-March of that year (Brugge, 1996). It was unusually cold in May with the maximum mean temperature about 2°C below average. These conditions would all have affected phytoplankton composition and growth. Chlorophyll levels did not increase significantly until June, when the weather improved.

Species diversity

Over 143 species of phytoplankton from 8 algal classes were recorded during the survey. Diatoms comprised 64% of the species present, dinoflagellates 23% with the other classes making up the rest (Figure 5). Both diatoms and microflagellates were the most important groups in terms of cell concentrations (Figure 6). Fluctuations in both species composition and cell concentrations were observed between all groups, throughout the sampling period, but this was more marked among the microflagellates and the dinoflagellates than the diatoms (Figure 6).

Species diversity, in terms of the numbers of different species recorded, varied throughout the survey (not shown). In general, the diversity was greatest throughout July and August (>50 spp.). Diversity was also high in early April (JD 102, 52 spp.). The lowest diversity was found in mid May (JD 137, 15 spp.) and early December (JD 346, 23 spp.). Higher species diversity (>50 spp.) was recorded generally on neap tides.

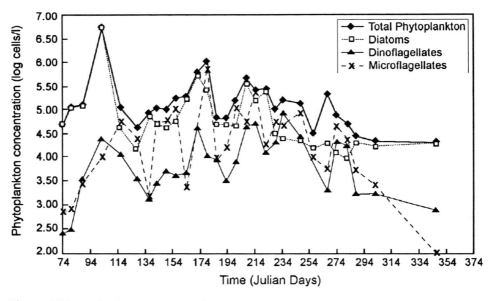

Figure 5 Phytoplankton concentrations.

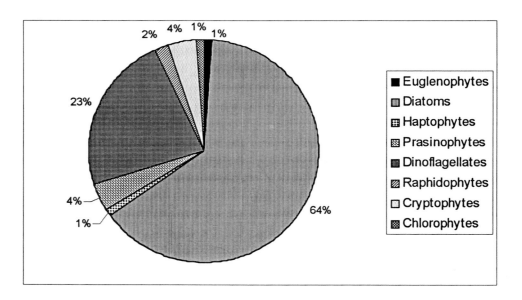

Figure 6 Phytoplankton Group Composition.

Species composition

The species composition differed throughout the survey period; similarly, the number of different species per sample varied. Different species also dominated the flora at different times.

The diatom *Skeletonema costatum* and an unnamed cryptophyte species were important components of the phytoplankton flora during 1996. One or the other was numerically important, on each sampling date, with the exception of early August and December when other species dominated (JD 222 and 346 respectively). In early April (JD 102) the diatoms *Asterionellopsis glacialis* and *Thalassiosira* spp. were also present in large numbers.

Cryptophytes became important, from late April through to early June, and from late July onwards. The diatom *Chaetoceros* spp. became dominant from mid-late June and from mid-July to early August. The only chrysophyte that was numerically dominant during the survey was *Apedinella spinifera,* at the end of June (JD 179).

Dinoflagellates were only present in significant numbers on a few occasions throughout the summer: *Heterocapsa triquetra* and *Scrippsiella trochoidea* (JD 172); *S. trochoidea* (JD 208); and *S. trochoidea* and *Prorocentrum micans* (JD 234).

Examination of the phytoplankton group composition (Figure 6) shows that fluctuations in numerical dominance between the different groups occurred frequently, indicating a continually changing floral assemblage.

CONCLUSIONS

The data that have been collected from the Hamble Estuary, during this ongoing survey, constitute a valuable resource for monitoring the diversity of phytoplankton communities and population changes, both in terms of species composition and concentrations over time. This has relevance, as recent research from the Continuous Plankton Recorder Survey, indicates that the distribution and abundance of phytoplankton appears to be responding to climate variability, on a decadal scale (Reid *et al.,* 1998).

The value of long-term biological monitoring cannot be underestimated. It is the composition of the dominant species present within phytoplankton assemblages that may be undergoing great changes; this is not detected by monitoring biomass alone. With increased concern today about toxic algal blooms and the introduction of non-indigenous species, we need to monitor the "hidden flora" currently present in small numbers in the phytoplankton assemblages - they may be the toxic blooms of tomorrow.

REFERENCES

Brugge, R. (Ed.) 1996. Weather Logs, January - December 1996. In: *Weather*, **52**: parts 1-12. Royal Meteorological Society.

Hasle, G.R. 1978. Using the Inverted Microscope. In: *Phytoplankton Manual.* Sournia, A. (Ed.) UNESCO, 191-196.

Howard, A.G., Comber, S.D.W., Kifle, D., Antai, E.E. & Purdie, D.A. 1995. Arsenic speciation and seasonal changes in nutrient availability and microplankton abundance in Southampton Water. *Estuarine Coastal and Shelf Science*, **40**: 435-450.

Iriarte, A. & Purdie, D.A. 1994. Size distribution of chlorophyll a biomass and primary production in a temperate estuary: the contribution of photosynthetic picoplankton. *Marine Ecology Progress Series*, **115**: 283-297.

Reid, P.C., Planque, B. & Edwards, M. 1998. Is observed variability in the long term results of the Continuous Plankton Recorder a response to climate change? *Fisheries Oceanography*, **7**: 282-288.

Wright, P.N.W., Hydes, D.J., Lauria, M-L., Sharples, J. & Purdie, D.A. 1998. Results from data buoy measurements of processes related to phytoplankton production in a temperate latitude estuary with high nutrient inputs: Southampton Water U.K. *Deutsches Hydrograficshes Zeitschrift*, **49**: 201-210.

Truncatella subcylindrica (Mollusca: Prosobranchia) in the Solent Area: Its Distribution, Status, and Conservation

J. M. Light[1] and I. J. Killeen[2]

[1] *Department of Geology, Royal Holloway University of London, Egham, Surrey, TW20 OEX, U.K.*

[2] *Malacological Services, 163 High Road West, Felixstowe, Suffolk IP11 9BD, U.K.*

INTRODUCTION

Truncatella subcylindrica (L., 1758), sometimes known as the 'looping snail' is a small marine prosobranch mollusc, whose geographical distribution reaches its northern limit in Britain. It is one of a group of molluscs that are considered to be part of the crevice fauna of the tidal zone, which forms the interface between intertidal animals of terrestrial origin and those who are truly marine. The animals inhabiting the uppermost region of the intertidal zone of a shore are often very small, cryptic and subjected to ephemeral and extreme conditions imposed by both the diurnal and monthly tidal regimes.

Truncatella subcylindrica lives in muddy places at the level of high tide, where it may be only occasionally inundated by seawater. It is associated with fully saline conditions and commonly associated with the plants *Suaeda maritima* and *S. fruticosa* (sea blite) and *Halimione portulacoides* (sea purslane). Where it is to be found living interstitially in shingle, it is usually found at about 15 cm depth, amongst rotting saltmarsh vegetation and moist fine sediment. At other sites, it can be found living between unmortared but firmly embedded slabs, which form sea walls. The animals are found amongst the mud and detritus infilling the joints between the slabs. Searching for the species requires detailed examination of its cryptic habitat. Such sites and habitats are often overlooked by conchologists and marine biologists alike and, although *T. subcylindrica* is rare, it may well be under-recorded.

The adult shell of *Truncatella subcylindrica* is nearly cylindrical and reaches a height of 5 mm. Ribbed and smooth-shelled forms occur. It has 3-4 whorls with a blunt tip that is considered to be a spiral turn rather than a true protoconch. The tip marks the point at which the juvenile snail sealed the opening with a calcareous plate and the apex then became truncated during maturation (Fretter & Graham, 1978). Juvenile shells have 6-7 whorls with a narrow and tapering spire; the aperture is more pointed apically. The animal has a cylindrical snout, ending in a round expanded disc on which the mouth lies, and 2 short and rapidly tapering tentacles with an eye at the base of each tentacle. The name 'looping snail' refers to the method of locomotion which resembles that of a caterpillar and involves both foot and snout (Morton, 1964). The sexes are separate and the females attach their eggs singly to particles or stones, with the young hatching as small snails with no waterborne dispersal phase (Fretter & Graham, 1978).

Truncatella subcylindrica is associated frequently with lagoonal habitats that are recognised as important habitats within the UK Biodiversity Action Plan (BAP) and the European Habitats & Species Directive. *Truncatella* is listed as Category 2 (rare) in the British Red Data Book (Bratton, 1991) and is also listed in the Joint Nature Conservation Committee (JNCC) Rare Species Directory (Sanderson, 1996). It was not included on the BAP short or middle lists (HMSO, 1995).

EUROPEAN DISTRIBUTION

Truncatella subcylindrica is a southern species occurring mainly in the Mediterranean, and is found on Madeira and the Azores; its distribution extends north to southern England. It was formerly known from sites ranging from Porthcurno, on the Land's End Peninsula, along the south coast of England and extending as far as the Deben and Orwell estuaries in Suffolk; however, few of these records refer to living populations (Seaward, 1992). There is some evidence that *T. subcylindrica* is in decline at the northern limit of its range, e.g. the species no longer inhabits the Suffolk estuaries (Killeen, 1992). Presently, living populations are known only from seven sites in the British Isles, all of which are located on the south coast of England: the Fleet in Dorset, Pagham Harbour, West Sussex; a recently discovered site in Cornwall (Killeen & Light, 1998) and four sites in the Solent area.

DISTRIBUTION & HABITAT IN THE SOLENT AREA

In the Solent area, the four currently known sites are: Warsash in the Hamble estuary; Eastney/Langstone Harbour; Freshwater Gate, River Yar, Isle of Wight; and King's Quay, Isle of Wight. Of the four, only King's Quay has a relatively long historical record (Morey, 1909). The other three Solent sites were discovered in the late 1980's and early 1990's, as a result of systematic molluscan surveys in the area (Light, 1994).

At Kings Quay, *Truncatella* was described as living "in considerable numbers" Morey (1909). The species could not be located after the 1940s and was presumed to have died out. King's Quay is a small tidal inlet, approximately 0.5 km in length, on the northeast coast of the Isle of Wight. The opening to the Solent is narrow and is protected by a shingle bar. *Truncatella* was rediscovered by the present authors in 1992. Initial searching for the species was focused in the shingle, below the plant, *Halimione portulacoides,* at the seaward end of the estuary. Eventually, snails were located behind slabs of an old wall that runs across the saltmarsh and into the estuary, some 250 m from the shingle bar.

The second Isle of Wight site lies at the upper reaches of the river Yar estuary, at Freshwater Gate. The snails live within the muddy crevices, between the loose slabs of the low wall of a causeway carrying a minor road. The micro-habitat at this site is very similar to that at Warsash, on the river Hamble. *Truncatella* was first recorded at the site based on dead shells found around High Water Mark, in the region of the ferry landing. The living animals were subsequently located at places along a loosely-mortared wall beneath the causeway, which runs up the Hamble River. The habitat at Eastney closely resembles that found in the Fleet. A small population of *Truncatella* was discovered at the western end of Langstone Harbour, living interstitially in shingle at approximately 15 cm depth, beneath and around the roots of *Halimione portulacoides* and *Suaeda* spp. This site is close to a housing estate and receives significant quantities of domestic rubbish.

The most frequent molluscan associates of *Truncatella subcylindrica*, in interstitial shingle and in the crevice habitats, are the pulmonates *Ovatella myosotis* and *Leucophytia bidentata*.

CONSERVATION

Our present state of knowledge indicates that the Solent populations of *Truncatella subcylindrica* are considerably smaller and less extensive than those in the Fleet and in Pagham Harbour. There is also evidence that some populations are in decline. Morey's (1909) comments on the Kings Quay population suggest that the species was formerly more abundant and widespread, whereas it is now restricted to a very small area of occupation. The reasons for this decline are unclear, particularly as King's Quay lies in a relatively inaccessible, undisturbed area that is unaffected by tourism, development or marinas.

The habitat at the other Solent sites is vulnerable, as it occurs in places that may be subject to development proposals, particularly from the marine leisure industry - new marinas and beach tourist attractions, for example. There is evidence from the recently discovered site in Cornwall (Killeen & Light, 1998) that removal of upper shore slabs of rock, to create hearths for beach fires and barbecues, could cause local habitat destruction. Two of the sites (Freshwater Gate and Warsash) are beneath causeways that are likely to receive periodic maintenance and repair by Local Authorities. Excessive use of tarmac could effectively seal up the habitat. It is not known how sensitive the species might be to pollution. There is evidence that the site at Eastney experiences rubbish dumping from the neighbouring housing estate.

The above threats are potentially all the more damaging because, lacking an aquatic dispersal phase during development, the species is vulnerable to loss of habitat. Since its powers of re-colonisation are assumed to be poor, any fragmentation of its habitat would be detrimental.

Within the Solent area, *Truncatella subcylindrica* may be considered 'Conservation Dependent'. Although the snail itself receives no direct protection, the sites are all within existing or proposed Sites of Special Scientific Interest (SSSIs), Special Areas of Conservation (SAC's) /Special Protection Areas (SPA's) or Ramsar areas.

Further information on the species' distribution at the known sites and within the Solent area is urgently required to enable an informed strategy for conservation and habitat management to be developed.

REFERENCES

Bratton, J.H. (Ed.) 1991. *British Red Data Books: 3. Invertebrates other than insects.* Joint Nature Conservation Committee, Peterborough, 253 pp.

Fretter, V. & Graham, A. 1978. The prosobranch molluscs of Britain and Denmark. Part 3 - Neritacea, Viviparacea, Valvatacea, terrestrial and freshwater Littorinacea and Rissoacea. *Journal of Molluscan Studies,* Supplement **5**:101-152.

HMSO, 1995. *Biodiversity: the UK Steering Group Report.* HMSO, London. 103 pp.

Killeen, I.J. 1992. *The land and freshwater molluscs of Suffolk.* Suffolk Naturalists' Society, Ipswich. 171 pp.

Killeen, I.J. & Light, J.M. 1998. A discovery of *Truncatella subcylindrica* living in Cornwall. *Journal of Conchology,* **36**: 50-51.

Light, J.M. 1994. *The marine Mollusca of Sea Area 15 (Wight): Provisional Atlas* (Revised edition). Godalming, 73 pp.

Morey, F. 1909. *A guide to the natural history of the Isle of Wight.* The County Press, Newport, 560 pp.

Morton, J.E. 1964. Locomotion. In: *Physiology of Mollusca.* Wilbur, K.M. & Yonge, C.M. (Eds.), Academic Press, New York & London, 473 pp.

Sanderson, W.G. 1996. Rare benthic marine flora and fauna in Great Britain: the development of criteria for assessment. *Joint Nature Conservation Committee Report* No.240, Peterborough, 36 pp.

Seaward, D.R. 1990. *Distribution of the marine molluscs of north-west Europe.* Nature Conservancy Council, Peterborough. 53 pp + maps.

Evolution and Current Status of the saltmarsh grass, *Spartina anglica*, in the Solent

A.F. Raybould, A.J. Gray and D.D. Hornby

Institute of Terrestrial Ecology, Furzebrook Research Station, Wareham, Dorset, BH20 5AS, U.K.

INTRODUCTION

Spartina anglica is a perennial grass that grows on saltmarshes and mudflats. It appeared in the Solent at the end of the 19th century. By 1918, *S. anglica* had spread as far as Poole harbour in the west and Pagham harbour in the east, as well as to the Isle of Wight and the northern coast of France. Between the Wars, *S. anglica* was planted throughout the British Isles to stabilise mudflats. It now covers thousands of hectares and has had a major impact on the topography and ecology of the British coast, as well as in several other countries where it has been introduced (Gray *et al.*,1991).

THE ORIGIN OF *SPARTINA ANGLICA*

Several pieces of evidence show that *S. anglica* evolved by hybridisation between a native species, *S. maritima*, and a North American species, *S. alterniflora*. The latter species was introduced accidentally into the Solent early in the 19th century. Chromosome doubling in the hybrid (*S. x townsendii*) produced *S. anglica*. The evidence of hybridisation is outlined below:

- Morphological: *S. x townsendii* and *S. anglica* are morphologically intermediate between *S. alterniflora* and *S. maritima* (Marchant, 1967).

- Cytological: *S. x townsendii* has 62 chromosomes, rather than 61 as expected if it is a hybrid between *S. maritima* and *S. alterniflora*. However, chromosome pairing at meiosis is irregular, indicating a hybrid origin. Some clones of *S. anglica* have 124 chromosomes, double the number of *S. x townsendii*. Clones with 120 or 122 are assumed to have arisen by loss of chromosomes during reproduction of the 124 chromosome form (Marchant, 1968).

- Biochemical: *S. maritima* and *S. alterniflora* both have protein variants that are not found in the other species. *S. x townsendii* and *S. anglica* contain all the protein types found in the putative parents. These data show that *S. anglica* did not arise directly from chromosome doubling, in either *S. maritima* or *S. alterniflora* (see Plate 1; and Raybould *et al.*, 1991a).

- Molecular: Chloroplast DNA in *S. x townsendii* and *S. anglica* has the same DNA sequence as *S. alterniflora* and is different from *S. maritima*. This pattern shows that *S. alterniflora* is the female parent as chloroplasts are inherited maternally in most plant species (Ferris *et al.*, 1997).

THE HISTORY OF *SPARTINA ANGLICA* AND ITS PARENTAL SPECIES IN THE SOLENT

S. alterniflora was first recorded in about 1816 from the River Itchen, its centre of distribution, and it spread as far as Thorney Island in Chichester Harbour. The first records of *S. maritima* are from Hythe and the mouth of the Itchen, in 1826. The distributions of *S. alterniflora* and *S. maritima* overlapped in the Itchen and, probably, at Hythe, during the middle of the 19th century (Marchant, 1967).

The first confirmed record of *S. x townsendii* is from 1870, at Hythe; during the 1870s and 1880s, it spread slowly along the shores of Southampton Water. In the early 1890s, the new *Spartina* began to spread more rapidly. This sudden expansion is believed to signal the production of *S. anglica*, from *S. x townsendii*. The chromosome doubling turned the sterile hybrid, able to spread by vegetative means only, into a fertile plant able to spread by seed as well. *S. anglica* was first recorded in 1892, at Lymington (Marchant, 1967).

S. anglica spread naturally into mudflats around the Solent, between 1892 and 1918; this coincided with the beginning of the decline of both *S. maritima* and *S. alterniflora*. By 1930, *S. maritima* was "practically exterminated" in Hampshire (Hall, 1934) and 1933 is the last record of *S. alterniflora* from the Itchen (Marchant, 1967). By the 1960s, *S. alterniflora* was restricted to a single site at Marchwood (Marchant, 1967); by the 1980s, *S. maritima* probably occurred on Hayling Island and the Isle of Wight only, although it was still reasonably frequent at several sites in Essex and Suffolk (Raybould *et al.*, 1991b). Saltmarsh erosion and habitat loss, because of development, are the main reasons for the decline of the parental species.

THE CURRENT STATUS OF *SPARTINA* SPECIES IN THE SOLENT

The Institute of Terrestrial Ecology (ITE) has surveyed the status of *Spartina* species in the Solent, for English Nature. On the northern shore, *Spartina maritima* is probably restricted to a single site, at Northney on Hayling Island (Plate 2). Several plants were present in 1986 (Raybould *et al.* 1991b); however, in 1998, we found just a single plant, which was confirmed as *S. maritima* by isozyme electrophoresis (Plate 1 and 2). Many places on Hayling Island (e.g. on the Gutner Peninsular) have species-rich high-level saltmarshes that appear very similar to the characteristic habitat of *S. maritima* on the east coast: absence of the species, from these Solent marshes, is puzzling. On the Isle of Wight, there is a large population of *S. maritima* on the high-level saltmarsh below Walter's Copse. *S. alterniflora* is still present at Marchwood, but it is being lost to erosion and through invasion by *S. anglica*. The extinction of both parental species in the Solent seems inevitable in the face of relentless sea-level rise.

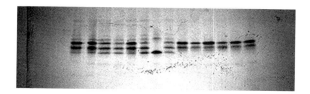

Plate 1 Electrophoresis (esterase profiles) of Spartina species from different locations: *S. anglica* from Northney (lanes 1-6 and 8); *S. maritima* from Northney (lane 7); *S. alterniflora* from Marchwood (lanes 9-14). This gel illustrates that *S. maritima* and *S. alterniflora* have unique esterase proteins that are all present in *S. anglica*. Lane 7 confirms that *S. maritima* still occurs on Hayling Island.

Plate 2 High level saltmarsh at Northney, Hayling Island – possibly the last site for *S. maritima* on the northern shore of the Solent. For general location and main sites referenced, refer to Maps A and B in the *Preface*.

There is a large population of *S. x townsendii* at Hythe; this seems stable and in no immediate danger from erosion. This species is probably scattered throughout the Solent and elsewhere on the south and east coasts, at the landward fringe of *S. anglica*-dominated saltmarshes (Gray *et al.*, 1991).

Spartina anglica is still very common in the Solent. However, since the 1930s populations throughout the Hampshire Basin have declined because of a process called 'die-back'. The exact mechanism of die-back is not known; however, it is widely accepted that *S. anglica* causes its own downfall, by altering the drainage and sedimentary processes of saltmarshes and creating anaerobic soils (Goodman & Williams, 1961). There is debate about whether the loss of *S. anglica* is beneficial to nature conservation. For example, areas in the UK that have lost *S. anglica* have shown an increase in wading bird populations, although the cause is not established (Raybould, 1998). However, in the Solent, wading bird numbers have declined despite the loss of *Spartina* from their potential feeding grounds (Tubbs *et al.*, 1992). In the Solent, therefore, reduced *Spartina* cover may accelerate saltmarsh erosion, causing loss of species-rich saltmarsh, without any compensatory beneficial effects on other species. A research priority should be how we manage the loss of *Spartina* saltmarsh, to prevent significant loss of biodiversity.

REFERENCES

Ferris, C., King, R.A. & Gray, A.J. 1997. Molecular evidence for the maternal parentage in the hybrid origin of *Spartina anglica* C.E. Hubbard. *Molecular Ecology*, **6**: 185-187.

Goodman, P.J. & Williams, W.T. 1961. Investigations into 'die-back' in *Spartina townsendii* agg. III. Physiological correlates of 'die-back'. *Journal of Ecology*, **49**: 391-398.

Gray, A.J., Marshall, D.F. & Raybould, A.F. 1991. A century of evolution in *Spartina anglica*. *Advances in Ecological Research*, **21**: 1-62.

Hall, P.M. 1934. A note on the genus *Spartina*. *Report of the Botanical Society and Exchange Club of the British Isles*, **10**: 889-892.

Marchant, C.J. 1967. Evolution in *Spartina* (Gramineae): I. History and morphology of the genus in Britain. *Botanical Journal of the Linnean Society*, **60**:1-24.

Marchant, C.J. 1968. Evolution in *Spartina* (Gramineae): II. Chromosomes, basic relationships and the problem of the *S. x townsendii* agg. *Botanical Journal of the Linnean Society*, **60**: 381-409.

Raybould, A.F. 1998. The history and ecology of *Spartina anglica* in Poole Harbour. *Proceedings of the Dorset Natural History and Archaeological Society*, **119**: 147-158.

Raybould, A.F., Gray, A.J., Lawrence, M.J. & Marshall, D.F. 1991a. The evolution of *Spartina anglica* C.E. Hubbard (Gramineae): origin and genetic variability. *Biological Journal of the Linnean Society*, **43**: 111-126.

Raybould, A.F., Gray, A.J., Lawrence, M.J. & Marshall, D.F. 1991b. The evolution of *Spartina anglica* C.E. Hubbard (Gramineae): genetic variation and status of the parental species in Britain. *Biological Journal of the Linnean Society*, **44**: 369-380.

Tubbs, C.R., Tubbs, J.M. & Kirby, J.S. 1992. Dunlin *Calidris alpina alpina* in the Solent, southern England. *Biological Conservation*, **60**: 15-24.

Solent Science – A Review
M. Collins and K. Ansell (editors)
© 2000 Elsevier Science B.V. All rights reserved.

Saltmarsh Monitoring Studies Adjacent to the Fawley Refinery

S. May

Cordah Environmental Management Consultants / Oil Pollution Research Unit, 3 Dolphin Court, Brunel Quay, Neyland, Pembrokeshire, Wales SA73 1PY, U.K.

Present address: 73, High Street, Pembroke Dock, Pembrokeshire, Wales, SA72 6PB, U.K.

INTRODUCTION

The Oil Pollution Research Unit (OPRU), now a division of Cordah Environmental Management Consultants, has been undertaking saltmarsh studies on behalf of, and funded by, Esso Petroleum Co. Ltd. and Exxon Chemical Ltd. in the vicinity of the Fawley Refinery, since 1969. The Solent and Southampton Water system is characterised by the mix of biological, recreational and industrial activity, the latter particularly associated with oil refining, chemical manufacture and their associated jetty activities on the western shore at Fawley. The Fawley saltmarshes are important for their extensive coverage of *Spartina anglica* (Marchant, 1975) and for feeding and roosting areas provided for waders and wildfowl (Tubbs, 1980a and 1980b; Hampshire County Council Planning Department, 1982) and have been designated a Site of Special Scientific Interest (SSSI). Having been subjected to damage by effluent discharges from 1951 to 1970 (Baker, 1976; Baker *et al.*, 1990) and also being vulnerable to accidents such as oil spills, Esso Petroleum Co. Ltd. and Exxon Chemical Ltd. initiated a major effluent quality improvement programme; this was designed to aid saltmarsh recovery together with monitoring of the marshes, mudflats and the adjacent sea bed.

METHODS

Monitoring work was carried out to assess whether improved effluent quality had beneficial effects on the saltmarsh vegetation. A vegetation mapping method was used, which has remained consistent from its inception. The same area of saltmarsh is walked and the distribution and abundance of saltmarsh plant species recorded using four categories of Abundant (most plants less than 50 cm apart), Common (most plants between 50 cm and 1 m apart), Rare (individual plants more than 1m apart and may be widely scattered) and Present (parts of high saltmarsh where plants are widely scattered). The monitoring programme gathered pace in 1972; it currently includes a vegetation survey carried out on a biennial basis.

In 1979, saltmarsh sediment infauna was also included in the monitoring programme. This infauna could also be affected by refinery effluent quality; hence, regular studies were undertaken to assess if the infauna were also making a recovery along with the vegetation. The infaunal studies were originally conducted twice per year, until 1991, when a triennial programme was established. Eighty fixed stations were established over the saltmarsh, with samples being obtained with a 6.5 cm diameter coring tube to a depth of 15 cm.

General sublittoral surveys, throughout the length of the waterway were initiated in 1976 and 1977 to assess the status of the macrobenthic sediments and animal communities. As a result, a more intensive sublittoral monitoring programme was started close to the saltmarsh which included the jetty area. Thirty-five sampling stations were established. The sampling methodology involved obtaining two grab samples (0.1 m^2), for analysis of the macrobenthic infauna, and sub-samples from other grabs for determination of sediment characteristics.

RESULTS

By 1981, a large part of the denuded area of saltmarsh was re-colonising with plants, but there were still areas of bare mud where growth had not occurred and other areas where growth of *Spartina* was poor. Further resources were put into improving effluent quality, in 1980 and 1981, and a *Spartina* transplant programme, originally started in 1975, was extended in the 1980s. Planting trials were undertaken to establish the most successful techniques. Surveys have shown that the denuded saltmarsh has progressively re-colonised, augmented by the *Spartina* transplant programme and improvement in effluent quality.

There has been a marked increase of *Spartina* vegetation cover from 1970 to 1996. Annual species such as the glasswort (*Salicornia* spp.) and annual seablite (*Suaeda maritima*) have moved into suitable areas by seeding and, together with *Spartina*, have managed to maintain populations closer to the outfalls than in earlier years. It should be noted that fluctuations in density and distribution do occur naturally but the long-term trend has been one of extensive re-colonisation.

The infaunal surveys have shown that since 1979, sediment invertebrates have spread across the saltmarsh from healthy areas, in tandem with the spread of saltmarsh plants, as effluent quality has improved. Oligochaete worms are an example of animals that have spread with the plants. They are the commonest organisms found in the sediments, in every survey; this is probably because the saltmarsh provides good conditions for oligochaetes to thrive, with its relatively high levels of organic matter and stable sediment. There have been encouraging signs also of an increase in faunal diversity, over time with the appearance of greater numbers of ragworms (*Hediste diversicolor= Nereis diversicolor*) and nematode worms. Surface-dwelling animals such as mud snails (*Hydrobia ulvae*), shore crabs (*Carcinus maenas*) and velvet swimming crabs (*Liocarcinus* spp.) have increased also. Furthermore, the number of stations where no infauna has been recorded has decreased markedly, between 1985 and 1997. It is clear, however, that numbers of certain species have fluctuated over some years. Oligochaetes are a good example where, in 1997, the numbers were less than half those found in 1994. This change has been attributed to a natural cause brought about by a dry summer in 1996.

The early sublittoral monitoring surveys initiated in 1975 and 1976 to assess the range of macrobenthic sediments and animal communities, showed that the sediments on the western shore adjacent to the saltmarshes had communities of low diversity and species richness. The most common species in the fine silts bordering the saltmarsh were the bivalves *Abra nitida* and *Cerastoderma edule* (common cockle) and the polychaete worm

Nephtys hombergii. The muddy, deeper water was dominated by the cirratulid polychaete *Chaetozone gibber*. Monitoring surveys carried out in 1979, 1980, 1982 and 1984 showed little change in the community composition, compared to 1975 and 1976. However, the survey of 1986 showed encouraging changes. Before 1986, the opportunist polychaete *Capitella capitata*, characteristic of organically-polluted sediments, dominated at the point of discharge of the two outfalls to the sea; its density decreased at stations farther from this area. By 1986, *C. capitata* was confined mainly to the seabed close to the discharges. The common estuarine polychaete *Hediste diversicolor* (= *Nereis diversicolor*), which is less tolerant of organically-enriched sediments, became commoner closer to the discharge mouth in Cadland Creek while still maintaining its presence further away (for general location and main sites referenced, refer to *Maps A* and *B* in the *Preface*).

DISCUSSION

The impressive results of the saltmarsh recovery project and the long-term monitoring programme for the Fawley saltmarsh have shown clearly that an industrial presence on Southampton Water is compatible with statutory regulation governing effluent quality. Equally important, however, is the commitment of a refining company to maintaining and enhancing the environmental quality of the shoreline and seabed adjacent to its installation (Lemlin, 1981). Management of the Solent and Southampton Water system will depend crucially on an understanding of the ecology of the region and in particular, that of its extensive saltmarshes (Adam, 1990). Rational decisions of engineers and involvement of the public in the planning process can only be undertaken in the light of scientific knowledge, obtained from programmes such as that established at Fawley.

REFERENCES

Adam, P. 1990. *Saltmarsh Ecology*. Cambridge University Press, Cambridge, 461 pp.

Baker, J.M. 1976. Investigation of refinery effluent effects through field surveys. In: *Marine Ecology and Oil Pollution*. Baker, J.M. (Ed.), Applied Science Publishers, Barking, 201-226.

Baker, J.M., Oldham, J.H., Wilson, C.M., Little, D.I. & Levell, D. 1990. *Spartina anglica and oil: spill and effluent effects, clean-up and rehabilitation*. In: Spartina anglica-a research review. Gray, A.J. & Benham, P.E.M. (Eds.), ITE Research Publication No. 2. HMSO, London, 52-62.

Hampshire County Council Planning Department. 1982. *Solent Sailing Conference*. Report of Working Party. Hampshire County Council, Portsmouth.

Lemlin, J.S. 1981. The value of ecological monitoring in the management of petroleum industry discharges: experience in Esso Petroleum Company UK refineries. *Water Science and Technology*, **13**: 437-464.

Marchant, C.J. 1975. The introduction and spread of *Spartina* in the UK. In: *Spartina in the Solent*. Stranack, F. & Coughlan, J. (Eds.), Proceedings of The Rothschild Symposium, Exbury, Hampshire, 1-4.

Tubbs, C.R. 1980a. Bird populations in the Solent, 1951-77. In: *The Solent Estuarine System*. NERC Publications Series C, **22**: 92-100.

Tubbs, C.R. 1980b. Processes and impacts in the Solent, 1951-77. In: *The Solent Estuarine System*. NERC Publications Series C, **22**: 1-5.

Use of the Dog-Whelk, *Nucella lapillus*, as a Bio-Indicator of Tributyltin (TBT) Contamination in the Solent and Around the Isle of Wight

R.J.H. Herbert[1], S. Bray[2] and S.J. Hawkins[2]

[1] *Medina Valley Centre, Dodnor Lane, Newport, Isle of Wight PO30 5TE, U.K.*

[2] *Centre for Environmental Science, Southampton University, Building 46, Highfield, Southampton SO17 1BJ, U.K.*

INTRODUCTION

Dog-whelks are predatory molluscs characteristic of rocky shores, breakwaters and pier piles; they feed mainly on barnacles and mussels and can survive at least seven years. Eggs are internally-fertilised and laid in crevices or beneath boulders. There is no planktonic larval stage and young 'crawlaways' emerge from the egg capsules, in late spring. Around the Isle of Wight, fully grown animals have a shell length of between three and four centimetres (Bray & Herbert, 1998).

In the mid-1980's, there was a noticeable decline in the populations of dog-whelks in the vicinity of ports and marinas around the British Isles (Bryan *et al.*, 1986). Moreover, many female whelks had grown a penis and *vas deferens* - a phenomenon known as 'imposex' (Bryan *et al.*, 1986; Gibbs *et al.*, 1987). Experimental work implicated exposure of these populations, to water contaminated with tributyltin (TBT) antifouling paints. Exposure of TBT at 1 ng/l can initiate imposex and, at 4-5 ng/l, imposex can cause blockage of the oviduct and the resultant build-up of unlaid eggs may cause the death of the female. Concentrations above 500 ng/l were recorded in Solent estuaries in the 1980's (Langston *et al.*, 1994).

In 1987, a survey undertaken around the Isle of Wight revealed that, while still common on the south coast, populations on the Solent shores had either become extinct or reached dangerously low levels. Moreover, the degree of imposex was higher in populations closer to the Solent (Herbert, 1989).

In July 1987, TBT paints were banned for use on small crafts less than 25 m in length. Ten years on, in the summer and autumn of 1997, dog-whelk populations around the Island were re-surveyed. Fieldwork revealed evidence of a rapid recovery in some populations, but a decline in others (Bray & Herbert, 1998). The degree of imposex (measured as relative penis size (RPSI); see Gibbs *et al.*, 1987) at Bembridge has fallen dramatically and the dog-whelk population has risen as a result of higher recruitment. However, at locations on the south coast of the island, imposex values are dropping surprisingly slowly, and populations between Ventnor and St. Catherine's Point appear to be decreasing (Figure 1).

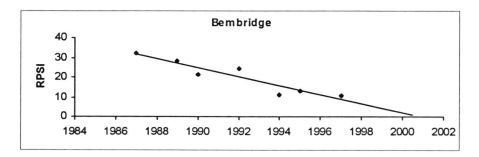

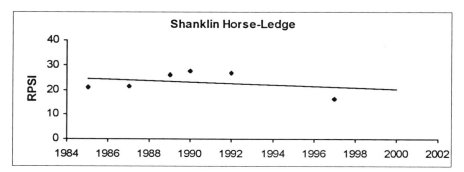

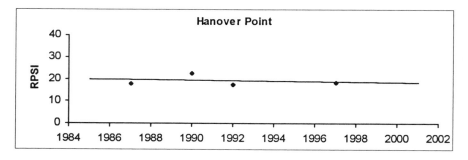

Figure 1 RPSI values, with regression lines, for *Nucella lapillus* from Isle of Wight populations: Ethel Point, Bembridge, Horse Ledge at Shanklin and Hanover Point. For sources, see Bray & Herbert (1998).

There are two possible reasons for this:

- Non-TBT related factors, such as poor food supply. Recruitment of the barnacle *Semibalanus balanoides* on the south coast has been relatively poor in recent years (*pers. obs)*.
- Contaminated sediments transported out of the Solent might expose populations to pulses of TBT that are responsible for the observed fluctuations in RPSI values.

DISCUSSION

There is no doubt that the presence of TBT in Solent waters over the past 30 years has had a detrimental impact upon our local marine environment. The use of the dog-whelk as a bioindicator of its presence and biological effect has been as valuable here, as elsewhere in the UK, where similar monitoring has been carried out (Evans *et al.*, 1995). The dog-whelk has been the most intensively studied species, but the lethal and sub-lethal effects of TBT on other marine organisms have been reported by Evans *et al.* (1995). In many areas, there appears to be a good recovery of dog-whelk populations since the 1987 legislation: further improvements might be expected over the next decade, should international legislation be implemented. However, such is the toxicity of TBT, that the anomalies concerning the rate of recovery of local populations require further investigation. This approach should be combined with anatomical studies and tissue analysis, to determine the extent to which TBT might be implicated. There must also be a simultaneous monitoring of TBT in Solent sediments and a study undertaken of its transport; both within and out of the Solent, through natural processes, or by the dumping of dredge spoil. Areas where there is a concentration of contaminated sediments need to be identified.

Of additional interest is the process of population recruitment. How can a species which has no pelagic stage re-colonise sites? *Nucella* young emerge from their egg capsules fully formed, with their movements limited to a few metres only (Day *et al.*, 1994); they seek refuge in crevices and empty barnacle shells. With no pelagic stage of the juveniles recorded, the possibility of individuals re-colonising sites which have been severely affected by TBT pollution is an interesting prospect. One suggestion is that intertidal species such as *Nucella* may use passive transport e.g. 'rafting' as a method of dispersal (Martel & Chia, 1991; Day *et al.*, 1994; and Ingolfson, 1995). This mechanism may be achieved by juveniles using large clumps of floating seaweed as a raft or by drifting in water currents using mucus threads (Martel & Chia, 1991). Direct evidence of such dispersal capabilities is difficult to identify, but the authors have noted a site where this may have recently occurred. At Highcliffe in Dorset, limestone groynes have been positioned to trap sediments as part of coastal management scheme. There are twelve groynes at the site, constructed between 1987 and 1992 (Christchurch Borough Council, 1994). The rock used is of terrestrial origin from a quarry in Dorset, and the structures are several kilometres from other *Nucella* populations. The groynes have been found to support populations of variable size consisting of large individuals. These animals are young in physical appearance with no 'teeth' in the shell aperture, indicating that they are below 'normal' breeding age; however, small numbers of juveniles have also been found. Further investigation is required into interactions between the individual groyne populations, and dispersal and arrival mechanisms at the site. In addition, it has been shown that *Nucella* populations are often genetically distinct from one another (Day *et al.*, 1994), and further work to establish the origin of the Highcliffe and new Solent populations is planned.

Insights derived from these data will enhance existing knowledge into recovery patterns of this important bio-indicator species around the Solent. With this in mind, it is hoped to assess further the effectiveness of current UK legislation against the use of TBT and the ability of *Nucella* to re-colonise formerly denuded sites.

ACKNOWLEDGEMENTS

The authors acknowledge financial assistance from the Environment Agency towards the field and laboratory work.

REFERENCES

Bray, S. 1996. An investigation into the effects of tributyltin anti-fouling paints, using the bioindicator species *Nucella lapillus*. B.Sc. Thesis, University of Southampton. Department of Environmental Sciences, unpublished, 53 pp.

Bray, S. & Herbert R.J.H. 1998. A reassessment of populations of the Dog-whelk *Nucella lapillus* on the Isle of Wight following legislation restricting the use of TBT antifouling paints. *Proceedings of the Isle of Wight natural History and Archaeological Society,* **14**: 23-40.

Bryan, G.W., Gibbs, P.E., Hummerstone, L.G. & Burt, G.R. 1986. The decline of the gastropod *Nucella lapillus* around south-west England: evidence for the effect of tributyltin from antifouling paints. *Journal of the marine biological Association, U.K.,* **67**: 525-544.

Christchurch Borough Council 1994. *Keeping the sea at bay: Cliff and coast protection in Christchurch*. Christchurch Borough Council, 55 pp.

Day, A.J., Leinaas, H.P. & Anstensrud, M. 1994. Allozyme differentiation of populations of the Dog-Whelk *Nucella-lapillus* (L.) – the relative effects of geographic distance and variation in chromosome-number. *Biological Journal of the Linnean Society,* **51** (3): 257-277.

Gibbs, P.E., Bryan, G.W., Pascoe, P.L. & Burt, G.R. 1987. Reproductive failure in populations of the dog-whelk *Nucella lapillus* caused by tributyltin from antifouling paints. . *Journal of the marine biological Association, U.K.,* **66**: 767-777.

Evans, S.M., Leksono, T., & McKinnell, P.D. 1995. Tributlyltin Pollution: A diminishing problem following legislation limiting the use of TBT-based anti-fouling paints. *Maine Pollution Bulletin,* **30** (1): 14-21.

Herbert, R.J.H. 1989. A survey of the Dog-whelk *Nucella lapillus* (L.) around the coast of the Isle of Wight. . *Proceedings of the Isle of Wight natural History and Archaeological Society,* **8** (3): 15-21.

Ingolfsson, A. 1995. Floating clumps of seaweed around Iceland – natural microcosms and a means of dispersal for shore fauna. *Marine Biology,* **122** (1): 13-21.

Langston, W.J., Bryan, G.W., Burt, G.R. and Gibbs, P.E. 1990. Assessing the impact of tin and TBT in estuaries and coastal regions. *Functional Ecology.* 4: 433-443.

Langston, W.J., Bryan, G.W., Burt, E.R., & Pope, N.D. 1994. *Effects of sediment metals on estuarine benthic organisms*. National Rivers Authority R&D Note 203. 141 pp.

Martel, A. & Chia, F.S. 1991. Drifting and dispersal of small bivalves and gastropods with direct development. *Journal of Experimental Marine Biology and Ecology,* **150:** (1), 131-147.

The Biology and Distribution of the Kelp, *Undaria pinnatifida* (Harvey) Suringar, in the Solent

P. Farrell and R. Fletcher

The Institute of Marine Sciences, School of Biological Sciences, University of Portsmouth, Ferry Road, Eastney, Hampshire, PO4 9LY, U.K.

INTRODUCTION

The arrival of the seaweed *Undaria* on the south coast of England was not unexpected; it is thought to have arrived in the Solent via the hull of a small boat (probably from Brittany) either as microscopic gametophytes or as young sporophytes. Hay (1990) predicted such a pattern of spread, via boats, using ports in the English Channel.

Undaria pinnatifida is a laminarian kelp, native to Japan, Korea and parts of China. This adventive *seaweed* has considerably extended its world-wide distribution over the past three decades. There are reports of its introduction into regions as far apart as the Mediterranean coast of France, the Adriatic, the Atlantic coast of France (Brittany), New Zealand, Tasmania and, more recently, Argentina, the Venice Lagoon and the Channel Islands (publication in preparation). In the majority of these introductions, the seaweed arrived accidentally, with imported shellfish or shipping identified as the most likely vectors; its introduction into Brittany was, however, deliberate and made for commercial reasons.

Undaria is economically important as a food crop (Wakame) in all of the above countries: the current world harvest exceeds 500,000 tonnes fresh weight (Yamanaka and Akiyama, 1993). *Undaria* has also been cultivated on the Atlantic coast of France since, 1983. The first introduction of *Undaria* into Europe was discovered in l'Etang de Thau (France) and the vector for this introduction was thought to be the imported spat of the Japanese oyster *Crassostrea gigas* (Perez *et al.*, 1981; Boudouresque *et al.*, 1985).

Hence a study of this large adventive kelp was carried out with the aim of determining the present distribution and to provide information on its ecology and biology in British coastal waters. Particular attention was given to studying a population of plants attached to marina pontoons in the Hamble estuary where it was originally first introduced; the main findings from this study are presented here.

METHODS

All the marinas and ports in and around the Solent region were visited in the early stages of the project. To assist in the search for other possible introductions around the British Isles, a descriptive poster was prepared; it was sent to all marinas and ports along the south coast, and to strategic locations around the British Isles.

For the biological studies, monthly, destructive sampling of plants attached to marina pontoons in the Hamble was carried out, during 1996; this was to determine density, seasonality, growth rates and reproductive patterns. Also, a random set of 50 plants was tagged *in-situ*; these were measured, every two weeks.

RESULTS

Distribution

Since its initial discovery in the Hamble in 1994, (Fletcher & Manfredi, 1995), *Undaria* has greatly increased its distribution, both within the Hamble and the Solent area (Figure 1).

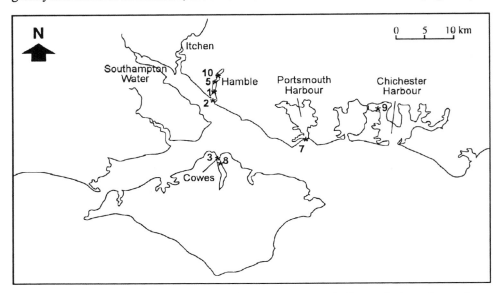

Figure 1 Distribution of *Undaria* in the Solent. Numbers indicate the locations where *Undaria* has been identified (south coast of England). The numbers are in chronological order of introduction. Numbers 4 and 6 are outside the Solent area, but are still mentioned in the following key:

1) Port Hamble Marina, Hamble Estuary (June 1994); 2) Hamble Point Marina, Hamble Estuary (March 1996); 3) West Cowes Marina, Isle of Wight (August 1996); 4) Torquay Marina, Devon (June 1996); 5) Mercury Yacht Harbour, Hamble Estuary (March 1997); 6) Brighton Marina, Sussex (June 1997); 7) Haslar Marina, Portsmouth Harbour (December 1997); 8) East Cowes Marina, Isle of Wight (April 1998); 9) Northney Marina, Chichester Harbour (April 1998 (boat), July, 1998 (pontoons)); and 10) Swanwick Marina, Hamble Estuary (June 1998).

In June 1994, when *Undaria* was first discovered in the Hamble, it was only present at Site 1, and its distribution was restricted to a few floating pontoons. *Undaria* has since spread along the estuary, in both directions. The upper limit of distribution is at Site 10, where just a few plants were found in the 1998 survey. Hydrological data shows that there is very little difference in water temperature between the two sites, with fairly consistent seasonal changes.

There are, however, some differences in salinity, with Site 10 being consistently lower in salinity than at Site 1, during some parts of the year. This difference undoubtedly provides a relatively narrower window of opportunity for *Undaria* to colonise the marina floats and would explain the late arrival of the plants at Site 10, together with the low abundance compared with site 1.

Life History

Undaria exhibits a different seasonality in reproduction and growth, to that observed in its native habitats in the Far East. In Japan, Korea and China, the sporophytes largely disappear by late summer: *Undaria* overwinters as the gametophyte stage, with the young sporophytes reappearing in early spring. In the UK, sporophytes also largely disappear by late summer, but the young sporophytes reappear just a month later (in September) forming a 'new crop' of mature and fertile plants as early as December. Mean densities of individual sporophytes were found to vary seasonally, between 56.12 m^{-2} (April) and 8.5 m^{-2} (September). Nevertheless, spore release has been observed for most of the year from May-December under laboratory conditions, when fresh field samples were examined. Therefore, plants may be recruited over an extended period, with two rather than one crop of plants per year.

Tagging of plants (from April to September) revealed a maximum growth rate for an individual plant of 39 cm in 14 days (2.4.96-16.4.96), with a mean growth rate of 19.89 cm (sample size, n = 11) during the same period of 14 days as above (see Figure 2). The highest mean growth rate was recorded at 23.74 cm (1.4.96 to 30.4.96). Although many of the plants disappear after September, some remain throughout the year.

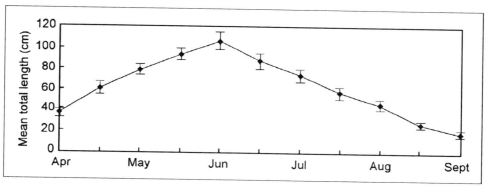

Figure 2 Mean Length (±Standard Error) of tagged *Undaria* sporophytes, from a random sample of 50 plants at Site 2 in the Hamble, during 1996.

DISCUSSION

With the now, seemingly inevitable colonisation of the coasts of Britain by *Undaria*, it will be necessary to gauge the behaviour of the plants towards the native flora and fauna; similarly, to determine whether its introduction into indigenous settlements will generate any major environmental problems. This introduction presents a unique opportunity to closely study the establishment, spread and ecological impact of a recently introduced, large, adventive, alien kelp on the coasts of the British Isles. The ecological consequences of its introduction, especially in relation to the native flora, and fauna are likely to be considerable and it is very important that these are fully monitored and understood.

On the shore, *Undaria* is able to compete with the native kelp species in occupying the shallow sublittoral/infralittoral zone. Kelps are major primary producers in neritic ecosystems (Breen & Mann, 1976), providing a rich food source for organisms at several trophic levels, and hence play an important role in fisheries and marine ecology. These plants are effective also in providing habitats, nursery areas and protective cover for many species.

In addition, *Undaria* can tolerate a lowered salinity and grows well in estuarine conditions, unlike many of the native kelp species. Estuaries are crucially important areas for many reasons e.g. for fish and shellfish fisheries, wildlife and nature reserves and as nursery areas for many species of fish. Any effect, therefore, upon the productivity of the latter areas will have profound environmental and commercial impacts.

REFERENCES

Boudouresque, C. F., Gerpal, M. and Knoepffler-Peguy, M. 1985. L'algue japonaise *Undaria pinnatifida* (Phaeophyceae, Laminarlales) en Mediterranee. *Phycologia,* **24**: 364-366

Breen, P. A. & Mann, K. H. 1976. Changing lobster abundance and the destruction of kelp beds by sea urchins. *Marine Biology,* **34**: 137-142.

Fletcher, R. L. & Manfredi C. (1995) The occurrence of *Undaria pinnatifida* (Phaeophyceae, Laminariales) on the South Coast of England. *Botanica Marina,* **38**: 355-358.

Hay, C. H. 1990. The Dispersal of Sporophytes of *Undaria pinnatifida* by Coastal Shipping in New Zealand, and Implications for further Dispersal of *Undaria* in France. *British Phycological Journal,* **25**: 301-313.

Perez, R., Lee, J. Y. & Juge, C. 1981. Observations sur la biologie de l'algue japonaise *Undaria pinnatifida* (Harvey) Suringar introduite accidentellement dans l'etang de Thau. *Science et Peche,* **315**: 1-12.

Yamanaka, R. & Akiyama, K. 1993. Cultivation and utilisation of *Undaria pinnatifida* (Wakame) as food. *Journal of Applied Phycology,* **5**: 249-253.

The Sussex *SEASEARCH* Project

R. Irving

Sussex SEASEARCH Project, 14 Brookland Way, Coldwaltham, Pulborough, West Sussex, RH20 1LT, U.K.

INTRODUCTION

SEASEARCH is a national project, which uses the skills of volunteer divers to record seabed types in the near-shore zone (typically, to a maximum depth of 30 m or to approx. 6 km offshore). The recording methodology was devised by the Marine Conservation Society (MCS), in association with the Joint Nature Conservation Committee (JNCC). Essentially, it is a 'Phase 1' survey of benthic habitat and community types. The presence of human activities and impacts is also recorded. Additional objectives include: educating volunteer divers in basic survey methods; fostering an interest and commitment to marine conservation; and providing guidance on possible policies for marine conservation, to be incorporated into future coastal zone planning strategies.

The Sussex project began in 1992 and covers the near-shore zone from Chichester Harbour in the west to Rye Harbour in the east, a total length of coastline of about 140 km. The western part of the area, including Bracklesham Bay and Selsey Bill, lies within the Solent region (for general location and main sites referenced refer to *Map A* and *Map B* in the *Preface*). The Sussex project is funded by a consortium of organisations including: English Nature; The Environment Agency; the county and district councils; Standing Conference on Problems Associated with the Coastline (SCOPAC); World Wildlife Fund U.K. (WWF); Sussex Wildlife Trust; Chichester Harbour Conservancy; and the Sussex Downs Conservation Board.

METHODS

Any diver with suitable qualifications (minimum of BSAC Sports Diver standard) is able to take part in the project. Training is given at the start of the diving season, but if newcomers are unable to attend these sessions, inexperienced recorders are paired within experienced ones on their dives. Once on the seabed, divers are asked to note on their recording slates the following:

- the depth of the seabed (for each habitat encountered) - later converted to below Chart Datum;

- the type of seabed - divided into percentages of each category (i.e. bedrock, boulders, cobbles, pebbles, gravel, sand, mud, clay & artificial substrata); and

- the dominant, conspicuous or characterising marine life associated with any one habitat type.

To assist with the recording process, a laminated Prompt sheet has been designed; this can be taken under water and used as an *aide memoire*. The area of 'search' that any one pair of divers can cover depends upon a number of variables. For example, whether they are diving at slack water or having to contend with a current, underwater visibility (obviously, far less can be recorded in poor visibility conditions), and the maximum water depth of the dive, hence the time available within non-decompression limits. Typically, the maximum duration of dives is set at 30 minutes. Once back on dry land (and, preferably, on the same day that the dive is carried out), each pair of divers transcribes their underwater notes to a standard recording form. The location of dive sites, as determined by GPS, is also noted.

RESULTS

Since 1992 until the end of 1998, 636 dives have been completed on behalf of the project by 218 volunteer divers. The distribution of these sites is given in Figure 1. Over 1000 habitat records have been entered onto the project's database from these dives.

At the western end of the region (i.e. that part included within the area of the Solent), the presence of a number of features of interest has been observed. In brief, these include:

- areas of moorings within Chichester Harbour, which avoid the ravages of the oyster dredges and have rich mixed sediment communities associated with them;
- the entrance to Chichester Harbour, which experiences strong tidal flows;
- the Bracklesham 'Balls' - spherical boulders of concretions of fossil shells;
- numerous reefs of sandstone and limestone to the south of Selsey Bill; and
- the Mixon Hole, which has a particularly impressive 20 m high clay cliff, on its northern side.

DISCUSSION

Quality of recording

It is accepted that there will be a wide range of recording capabilities among the volunteers taking part in the Project. In general, greater reliability can be placed on the descriptions of seabed habitats than that on the identification of species, especially where these have been recorded by inexperienced surveyors, who may have a tendency to guess identifications. Those records completed by experienced surveyors, in particular, are likely to provide more reliable information on species presence and abundance. However, inexperienced surveyors are quite capable of providing acceptable descriptions of the seabed, even if few species are positively identified.

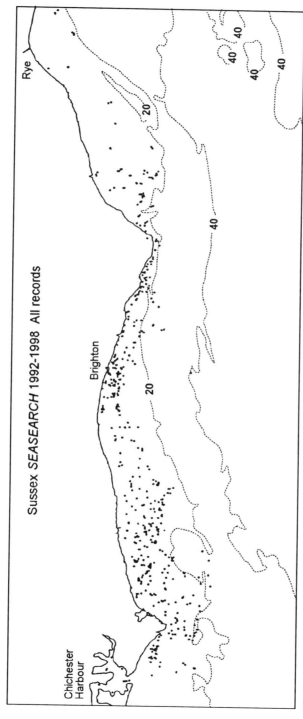

Figure 1 The distribution of SEASEARCH dive sites to the west of Beachy Head, 1992-1998. Please note that there are additional sites to the east of Beachy Head, which are not shown on this map.

Outcomes

The Project has led to a number of 'end products':

- A promotional and training video aimed at publicising the project and for assisting dive clubs embarking on the project.

- An A2-size colour poster entitled *Undersea Sussex*, which depicts examples of benthic habitats and communities present off the Sussex coast; this has been distributed free of charge to all schools within East and West Sussex.

- A report on an initial list of marine Sites of Nature Conservation Interest (Irving, 1996). Twelve sites have been highlighted for their physical or marine biological importance. The 'quality' of these is comparable to terrestrial Sites of Special Scientific Interest (SSSI) standard. These sites are now being incorporated into relevant statutory plans such as Shoreline Management Plans.

- A species identification guide, *Sussex Marine Life* (Irving, 1998).

- A sublittoral biotope manual, identifying the variety of seabed types off the Sussex coast, is in preparation.

- A detailed report of the project to date (Irving, 1999).

CONCLUSIONS

The Sussex *SEASEARCH* Project has demonstrated that accurate descriptions of near-shore seabed types can be obtained, using volunteer divers at relatively low cost. However, checks on the accuracy of this information are essential, for any reliability to be placed on the data. Information on habitat types is likely to be of a higher quality than that of species recognition.

REFERENCES

Irving, R.A. 1996. *A dossier of Sussex Marine Sites of Nature Conservation Importance*. An unpublished report compiled on behalf of the Sussex Marine SNCI Steering Group. Coldwaltham, West Sussex, 52 pp.

Irving, R.A. 1998. *Sussex Marine Life - an identification guide for divers*. Lewes, East Sussex County Council, 178 pp.

Irving, R.A. 1999. *Report of the Sussex SEASEARCH Project, 1992-1998*. Lewes, English Nature, 120 pp.

Analysis of the Numbers and Distribution of Wildfowl and Waders, as an Aid to Estuarine Management

A. de Potier

Chichester Harbour Conservancy, Itchenor, Chichester, West Sussex, P020 7AW, U.K.

INTRODUCTION

Chichester Harbour is recognised as internationally important for wintering and migrating shorebirds. As managing authority, Chichester Harbour Conservancy requires information on which to base decisions and recommendations, and from which to assess the effectiveness of its management. Analysis of the data arising from regular counts, which contribute to a national and international scheme, provides management information not only on the birds themselves but also on wider issues and concerns. The scheme, now known as the Wetland Bird Survey (WeBS), is a partnership between the British Trust for Ornithology, the Wildfowl and Wetlands Trust, the Royal Society for the Protection of Birds and the Joint Nature Conservation Committee. This is not a novel exercise; it is recognised internationally as an essential aid to the conservation of the various species and their habitats (Smit, 1989). Much of the work on the Solent, and sites within it, was carried out by the late Colin Tubbs (see for example Tubbs, 1977; Kirby & Tubbs, 1989).

In broad terms, information is available on: trend analysis of population size; the use of different areas (roosting and feeding); variation through the year; the functioning of the harbour ecosystem, and signposts to further research; and the effects of management or other activities and potential effects of proposed action.

THE COUNTS

Monthly counts at high tide, from September to March, have been made at Chichester Harbour since 1964. The harbour is divided into 13 sectors; each counted by a volunteer who also notes the positions of the birds on an outline map. This mapping was introduced in 1971 as a local project, and has proved invaluable. Data exists on birds present in the Harbour between April and August over several years. This paper is based on work carried out in 1994. It is important to note that comparisons between local and national data are based on an index derived from January numbers. This was a crude system which since 1995 has been replaced by the 'Underhill Index' (Underhill and Prys-Jones, 1994). It is also worth remembering that the data give total numbers, not densities. The majority of data analysed are taken from the archives of The Sussex Ornithological Society, augmented by Chichester Harbour records. Other sources, such as wildfowl records, are sometimes available (Tubbs, 1991), but have not yet been explored for Chichester Harbour. National data are from the annual WeBS reports, the most recent at the time of writing being Cranswick *et al*. (1997). A sequence of thirty years may seem long, but ecologically this is only a short span of time; hence, some apparent trends may only be part of a cycle. Further potential problems with the dataset include: the influence of the weather (on both birds and counters); the skill of the counters; and gaps due to missed counts. Nonetheless, the system has stood the test of time.

EXAMPLES OF FINDINGS

Care must be taken when interpreting data. Local data for Redshank (Figure 1) show a peak (in 1983) followed by a decline; comparison with the national data shows that it is the peak that is the unusual feature. This local rise did not occur elsewhere in the Solent (Kirby & Tubbs, 1989). More worrying is the apparent steep decline in 'Autumn' numbers (Figure 2); however, 'Autumn' for the main count series includes September and October only. By looking at data from July and August, it emerges that the peak can occur earlier. From 1964-1969 (the only sequence of early years for which adequate data for this species exist), two out of the 6 annual peaks were recorded in July or August; from 1988-1993, five peaks were recorded in these months, although in the last 3 years the peak has occurred in October. The trend is, therefore, not as severe as it appeared, and recently seems to be halting.

These data illustrate not only the value of counts throughout the year, but also the importance of the harbour for birds even during the peak summer holiday period. There are, therefore, implications for disturbance management: disturbance is a complicated subject, research into which, is best addressed nationally or internationally; a useful general review is given in Hill *et al.* (1997).

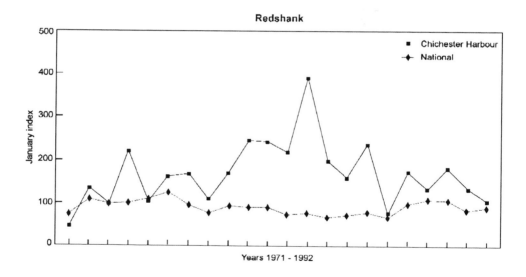

Figure 1 Redshank: comparison of local and national 'January index' (1971-1992).

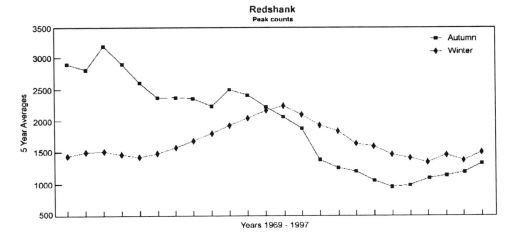

Figure 2 Redshank: comparison of 'autumn' and 'winter' average peak counts (1969-1997).

Identification of the factor(s) behind fluctuations is essential, in order to determine the nature of management action required. Variation in numbers of Bar-tailed Godwit (Figure 3) can mostly be attributed to cold winters. Cold weather affects different species in different ways, causing both influxes and exoduses, but it is a factor beyond the control of the site manager. There is no doubt that numbers of Brent Geese are increasing locally and nationally (Figure 4); the Figure shows also a close correlation between local and national fluctuations. This is because numbers of wintering geese are linked to their breeding success, which is generally cyclical and depends on a number of factors (Summers et al., 1994). This is clearly also beyond the control of the site manager. However, the local trend also shows a slower increase than nationally; this could be caused by farmers scaring the geese efficiently from crops. Comparison with data from neighbouring harbours would help interpret this trend: whether the birds' choice of habitat is widening, or more birds are concentrating on neighbouring sites. This has implications for the managers of those more urban sites, and for Chichester Harbour, where more refuges are needed.

Management information can be derived from within-site comparisons. Table 1 shows the proportions of the harbour population of Ringed Plover roosting in various locations. In the 1980s, considerable use was made of East Head, an area popular with walkers; ten years later the birds were concentrated on to Pilsey Sands. This concentration makes the birds more vulnerable to adverse influences, such as disturbance - the factor that is likely to have caused the shift from East Head in the first place. Research in northwest England has studied the effects this kind of disturbance can have (Mitchell et al., 1988). Secure roost sites in Chichester Harbour are at risk, not only from disturbance but also from sea level rise and associated coastal erosion and squeeze; similarly, from changes in land use and agricultural practice. Loss of roosts in Chichester and neighbouring harbours, whether physical loss or through disturbance, could be a serious threat to the viability of populations as the extra energy involved in flying greater distances could be critical to survival (Rehfisch *et al.,* 1996). By studying the changing usage of the harbour by shorebirds, the problems associated with particular sites can be identified and tackled, by working to prevent the problem and provide alternative sites.

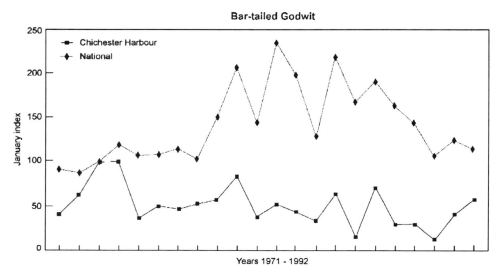

Figure 3 Bar-tailed Godwit: comparison of local and national 'January index' (1971-1992).

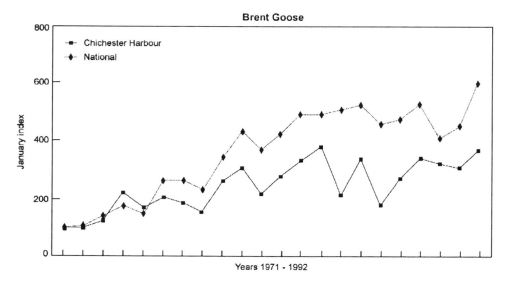

Figure 4 Brent goose: comparison of local and national 'January index' (1971-1992).

Table 1 The proportions of the harbour populations of Ringed Plover roosting in different locations.

Mean % of total counted				
1980/1-1983/4	*Nov*	*Dec*	*Jan*	*Feb*
East Head	34.9	40.6	32.9	24.8
Pilsey	23.4	11.7	44.5	32.0
Rest of harbour	41.7	47.7	22.6	43.2
1990/1-1993/4				
East Head	0	0	0	1.3
Pilsey	91.4	72.9	82.1	62.5
Rest of harbour	8.6	27.1	17.9	38.2

LOW TIDE COUNTS

The examples so far have been derived from counts of birds at high tide. These can be complemented by counts and maps of the distribution of feeding birds. In Chichester Harbour, these have taken place between November 1990 and February 1994, and November 1996 and February 1999. Counts are based on the unit of a 'mudflat'; Chichester Harbour is divided into 53 of these, which vary considerably in size but try to reflect local topography and bird usage. The value of these counts, especially in conjunction with detailed maps, is to show the relative importance of different areas of the harbour for different species of feeding and roosting waterfowl. Some species are distributed throughout the harbour whereas others are more specialist, favouring particular substrate types and their associated invertebrates. In addition, the maps are especially helpful when assessing applications for development, dredging and similar activities, as they provide particularly fine detail.

MOVEMENTS BETWEEN EAST SOLENT HARBOURS

With caution, one can compare totals derived from high tide counts and those from low tide counts. Care is needed because the aims and methods of the two surveys are different, but it was noted that in Chichester Harbour there were markedly fewer Dunlin and Ringed Plover at low tide than high. The suggestion was that while Langstone and Portsmouth Harbours were able to support feeding birds at low tide, there were insufficient roost sites at high tide, and the birds needed to use Chichester Harbour. The results of a pilot survey were inconclusive (Badley, 1998), showing birds moving in both directions. The surveys will be continued, and augmented by ringing and colour marking data.

FURTHER RESEARCH

The work achieved so far is only the beginning of what is necessary to understand the position of these birds in the harbour ecosystem and, hence, the threats that might affect them. There are many other issues yet to be pursued, such as the application of the theory of carrying capacity (see for example: Moser, 1988; and Tubbs, 1991): is Chichester Harbour becoming 'full' in relation to any species? Answers to this might help define priorities and parameters for habitat creation. The effect on birds of eutrophication and macroalgal mats is a complicated issue summarised in Green *at al*. (1992). Although some research results from elsewhere may be transferable to Chichester Harbour, specific information is required. A start has been made on background research, such as the correlation of feeding maps with sediment, intertidal vegetation and invertebrate data. Similar analyses of data from other Solent estuaries would help set the Chichester data in context. Regional indices have been produced by the British Trust for Ornithology; a comparison between these and data from individual sites would also be useful.

CONCLUSIONS

Careful study of the data from regular counts is an invaluable aid to monitoring the well-being of the internationally-important intertidal ecosystem. It helps raise awareness of particular issues, such as the threats to roost sites; hence, can assist prioritising research and practical action. Therefore, it is essential not only that these counts continue and form part of an adequately funded national partnership, but also that other research prompted by them is carried out. Likewise, that the information gathered is transferred to practical positive action.

ACKNOWLEDGEMENTS

I acknowledge with gratitude the help given, now and in the past, by the volunteer ornithologists who contribute their time and skill. Other volunteers helped process the sector data. The majority of the summer data were provided by C.B. Collins and the late Mrs J.H.M. Edom. Help with the original analysis and its interpretation was given by A.J. Prater, R.D.M. Edgar and the late C.R. Tubbs. My colleague P.J. Couchman helped with advice, data handling, and the production of Figures. I am grateful to them all.

REFERENCES

Badley, J. 1998. *Movements in East Solent Harbours*. Unpublished report to Solent Shorebird Study Group, 3.

Cranswick, P.A., Waters, R.J., Musgrove, A.J. & Pollitt, M.S. 1997. *The Wetland Bird Survey 1995-6: Wildfowl and wader counts*. BTO/WWT/RSPB/JNCC, 165.

Green, P.T., Hill, D.A. & Clark, N.A. 1992. *The effects of organic inputs to estuaries on overwintering bird populations and communities*. BTO Research Report 59. ETSU/ Department of Energy, 166 pp.

Hill, D.A., Hockin, D., Price, D., Tucker, G., Morris, R. & Treweek, J. 1997. Bird disturbance: improving the quality and utility of disturbance research. *Journal of Applied Ecology*, **34**: 275-288.

Kirby, J.S. & Tubbs, C.R. 1989. Wader populations in the Solent 1970/71 to 1987/88. *Hampshire Bird Report 1988*: 83-104.

Mitchell, J.R., Moser, M.E. & Kirby, J.S. 1988. Declines in midwinter counts of waders roosting on the Dee estuary. *Bird Study*, **35**: 191-198.

Moser, M.E. 1988. Limits to the number of Grey Plovers *Pluvialis squatarola* wintering on British estuaries: an analysis of long-term population trends. *Journal of Applied Ecology*, **25**: 473-485.

Rehfisch, M.M., Clark, N.A., Langston, R.H.W. & Greenwood, J.J.D. 1996. A guide to the provision of refuges for waders: analysis of 30 years of ringing dada from the Wash, England. *Journal of Applied Ecology,* **33**: 673-687.

Smit, C.S. 1989. Perspectives in using shorebird counts for assessing long-term changes in wader numbers in the Wadden Sea. *Helgolander Meeresuntersuchungen,* **43**: 367-383.

Summers, R.W., Underhill, L.O., Syroechkovski Jr., E.E., Lappo, H.G., Prys-Jones, R.P. & Karpov, V. 1994. The breeding biology of Dark-bellied Brent Geese *Branta b.bemicla* and King Eiders *Somateria spectabilis* on the northeastern Taimyr Peninsula, especially in relation to Snowy Owl *Nyctea scandiaca* nests. *Wildfowl,* **45**:110-118.

Tubbs, C.R. 1977. Wildfowl and waders in Langstone Harbour. *British Birds,* **70**: 177-199.

Tubbs, C.R. 1991. The population history of Grey Plovers *Pluvialis squatarola* in the Solent, Southern England. *Wader Study Group Bulletin,* **61**:15-21.

Underhill, L.G. & Prys-Jones, R.P. 1994. Index numbers for waterbird populations. 1. Review and methodology. *Journal of Applied Ecology,* **31**: 463-480.

SECTION 4

Workshop Findings

Findings of the Biodiversity and Conservation Workshops

Session Leader: Sarah Fowler *(Nature Conservation Bureau)*

INTRODUCTION

Six Workshops were held to discuss the various issues surrounding biodiversity and conservation in the Solent region, as outlined below.

1. Bird conservation and sustainable development
2. Sea level rise and nature conservation
3. Special Area of Conservation, management and information needs
4. Archaeology
5. Fisheries
6. Habitat creation

There is a need to collate existing work and baselines. Many of the studies, to date, have led towards the establishment of an inventory. Now there is a need to improve greatly the understanding of processes, if the challenge of managing dynamic systems and migratory species is to be met. There is an order of magnitude difference between knowing where the resource is, whether it is important, and its management.

Moving into this new field, as discussed within these Workshops, will drive the research agenda significantly in the future. Researchers should understand the policy and legislative context, when planning research programmes; likewise, managers should have more of input into helping to establish research priorities.

Summary of Workshop findings

Integration

- Information and management systems.
- GIS – opportunities, to use existing Solent area Institutes.
- Research and policy-making communities, managers and users.
- Consideration to be given to the role of the Solent Forum.

Regional and National Considerations

- In the sense that the Solent *is not* unique – we need to assess best practice and experience, from other areas.
- In the sense that the Solent *is* unique – we cannot replace lost Solent habitats at other sites e.g. East Anglia, where there is a different climate, different species and different communities.

Strategic Approach

- Understanding sea level rise (SLR) in the Solent and the need to identify an associated habitat creation strategy.
- Modelling of sea level rise in the Solent.
- The need to take a long-term view (over 25, 50, 75+ years), including economic factors and the land use planning context.
- Setting of priorities and targets, together with the need to monitor progress.

The need for a single Authority?

- Too many authorities, lack of a holistic approach.
- The expense of modelling and other initiatives; these are not likely to happen unless commissioned.
- Central database of information – possibly accessed from the Internet (Solent Forum to act as a broker).
- Role for the Inter-Agency Committee on Marine Science and Technology (IACMST). Nevertheless, would this really address the issues of conservation and management?

BIRD CONSERVATION AND SUSTAINABLE DEVELOPMENT

Rapporteur: Dave Burges *(Royal Society for the Protection of Birds)*

Main Issues

Ecology

Research needs:

- interactions between sites;
- relationships to international flyway populations; and
- 'health' of bird populations – indicators/bioaccumulators.

Planning and policy

Collaboration:

- links between researchers (volunteers/professionals) and planners;
- translation of results into actions;
- desk studies of existing works;
- opportunities for managed retreat; and
- Solent Shorebird Conference.

Habitat management and creation
Research

- Sediment/invertebrates/bird studies

Management

- Roost disturbance.
- Bait digging.
- Wild fowling/refuges.
- Buffer areas.

Education and public affairs
Collaboration

- Effective communication, emphasising the importance of the Solent.

Quantifying the impacts of development

- Research.
- Baseline studies.
- Before/after comparisons of the impacts of development: examples from the Solent and elsewhere.

SEA LEVEL RISE AND NATURE CONSERVATION

Rapporteur: Robert Page *(Hampshire & Isle of Wight Wildlife Trust)*

Main Issues

- A requirement for modelling of the Solent for the effects of sea level rise (SLR) based on the present situation. It should be noted that no action would be taken until coastal communities realise that their homes are at risk.

- There is a need for the modelling of visions on the future potential for habitats/species creation, within the context of the prevailing economic situation i.e. land purchase, planning realignment, bearing in mind the time-scale for habitat recreation.

- The main objectives need to be considered within an UK and international context, but local biodiversity must be maintained. Habitat creation in other regions is not an adequate replacement for Solent losses.

- There is a need for a single national authority, charged with the above requirements: the present functions are too diffuse. Further, modelling will not be initiated unless it is commissioned.

SPECIAL AREA OF CONSERVATION (SAC) MANAGEMENT AND INFORMATION REQUIREMENTS

Rapporteur: Graham Bathe *(English Nature)*

There is a need for more information, to meet statutory requirements.

Baseline

- Long-term data sets.
- Understanding of processes (natural and man-modified changes and impacts).

Monitoring and reporting

- Success of management.
- Conflict resolution.
- Impact of proposed development/unplanned events, drawing up of a management scheme.

What information is needed?

- Collation of existing data – baseline.
- Broad-scale habitat mapping.
- Review of historical processes/activities.
- Vulnerability/sensitivity of biotopes/features.
- Assessment of the operation of natural driving forces and their modification by man.
- Human uses – physical and potentially damaging activities.
- Long-term prognoses, in relation to sea level rise.

Finally, consideration should be given to how the SAC can stimulate economic development.

ARCHAEOLOGY

Rapporteur: Garry Momber *(Hants & Wight Trust for Maritime Archaeology)*

Introduction

Since 1992, new initiatives have been undertaken by the Isle of Wight Council, Hampshire County Council, English Heritage and the Hampshire and Wight Trust for Maritime Archaeology. These initiatives have revealed a wide range of palaeo-environmental evidence such as: identifying submerged forests, earlier shorelines, wave-cut platforms and deep sediments associated with former configurations of the Solent estuarine system. At the same time, Government planning policy guidance on coastal planning (PPG20) and the new Shoreline Management Plans for the United Kingdom have highlighted the need for coastal planners and engineers to achieve a firm understanding of local and regional coastal processes, together with the overall scale and rate at which coastline changes are proceeding.

Main Issues

- The age of the Solent, as an open seaway separating the Isle of Wight from the mainland, is still unknown. Without this critical piece of evidence, the nature, scale and pace of present coastal erosion cannot be viewed, within a context that is sufficiently secure to permit reliable predictions of future trends and coastal protection needs.
- Recent archaeological surveys undertaken of the intertidal and sub-tidal zones have identified submerged palaeo-landscapes. Evidence offered by the study of pollen, diatoms and tree-ring chronology could identify and fix environmental changes on the Solent coast, over the past 8,000 years. A suite of radiocarbon dates, obtained mostly for archaeological purposes, has helped to fix some specific events.
- Preliminary results of the new surveys in the Solent demonstrate that many of the archaeological and palaeo-environmental sites in the coastal zone should be viewed as critical sources of evidence, concerning long-term coastal processes. At present, there has been a noticeable tendency to view such sites simply as a conservation issue (concerning cultural resources or, at best, the historic environment).

Research Priorities

Since the completion of the Shoreline Management Plans, the Government (through Agenda 21) has expressed new concerns over the need to understand and measure long-term trends in climate change and sea-level rise. In coastal areas, like the Solent, where deep deposits of silt and mud are present, 'sediment archives' have accrued which represent long-term changes in the Holocene coastline and the local and regional environment. These 'archives,' together with the tree-ring sequences from the submerged forests, should be seen as potential pieces in the reconstruction and understanding of European and global climate history.

The main research priorities in this case are:

- to identify these scientific sites;
- to include them in the Sites and Monuments Records and Environmental Records, which inform Central Government and guide the decisions of local authorities; (hence, averting the loss of these non-renewable scientific sites); and
- to implement steps to ascertain the full scientific potential of the site.

The Way Forward

Continued research is necessary, to identify and study these invaluable sites. Members of the Workshop, including the commercial sector, have agreed to pool their resources; likewise, with the assistance of SCOPAC, to pursue European partners with similar interests.

FISHERIES

Rapporteur: Antony Jensen *(SOES, University of Southampton)*

Aquaculture Potential

- Constraints (water quality, vandalism/theft, exotics, planktonic food, skills and motivation of fishermen).
- Possibilities/opportunities (Fawley outflow – fluctuates, collaboration with academics on pre-feasibility studies, artificial reefs for protection, Southampton Institute/CEFAS biofilms study).
- Existing experiences (bivalve fattening and Manila clams in Poole Harbour, oyster farming in Newtown and Beaulieu, Bonamia Poole, Emsworth oyster poisoning).

Fisheries Management

- Need for local initiatives (no coherent industry/community).
- The example of the Hythe clam relaying demonstrates what can be achieved.
- Spanish vocational training initiative, as a model?

Gravel extraction and crab fishing

- Six years of data are available from Hastings.
- There is a need for a Solent study on female crab migration.
- Some difficulties have arisen associated with collaborative research, as there is a perception that data may be used against the industry.

Lack of quality fin fishery data

- Improved links with fishermen need to be established.
- Mutual suspicion between regulators and fishermen should be reduced.
- Anonymity of data collecting/handling requires improvement.
- Voluntary logbook system, with follow-up feedback, should be instigated.
- Winch activity monitoring needs to be undertaken.

HABITAT CREATION

Rapporteur: Philip Couchman *(Chichester Harbour Conservancy)*

Main issues

- The definition of priorities, objectives and opportunities.
- Insufficient knowledge.
- Obtaining the necessary land and financial support whilst, at the same time, marketing the idea to the public.

Research priorities

- Greater understanding of processes.
- Experimentation linked to monitoring.
- Intertidal recharge, with fine-grained sediment.

Collaborative research opportunities

- Solent habitat creation strategy.
- Register of projects.
- Publications and liaisons.

Other concerns

- Concerns about the loss of naturalness.
- Policies of retreat.
- Dissemination of existing experience.

SECTION 5

Integrative Solent Case Studies

Aggregate Extraction

Elizabeth Dower

Oakwood Environmental Ltd, The Limes, Combe Lane, Wormley, Godalming, Surrey, GU8 5SX, U.K.

INTRODUCTION

Oakwood Environmental Ltd (Oakwood) are leading experts in the preparation of Environmental Assessments (EA) for marine aggregate extraction licences; they have undertaken 14 Environmental Statements (ES) and their associated consultation programmes. Of this total, 8 are now complete and have either achieved a favourable Government View, or await a decision by Department of the Environment and Transport and the Regions (DETR). Experience gained in the preparation of these ESs has been based upon extensive detailed scientific studies set within the context of EU Environmental Impact Assessment legislation.

This paper addresses the legislation covering the licensing process, the scientific methodology utilised during the preparation of marine aggregate EAs, the range of scientific studies commissioned and the consultation process, together with the way in which scientific information is collated and illustrated within the final ES. Short and medium term trends in environmental assessments are covered, together with an up-to-date review of the *Strategic Cumulative Effects of Marine Aggregates Dredging* research project currently being finalised by Oakwood, on behalf of the US Minerals Management Service (MMS). This project will have far-reaching implications for the Solent, in terms of the Cumulative Effects Assessments (CEA) of all activities; these include maintenance and harbour dredging, dumping applications and extensions to existing port and harbour facilities and the further development of the marine aggregate industry.

LEGISLATION

Licence Application and the "Government View" Procedure

The process of consent for mineral prospecting, together with subsequent exploitation of marine aggregates, is different from that for land-based projects. For planning purposes, local planning authorities have powers to control development of land, including the coastal zone down to Mean Low Water Mark (MLW) under the Town and Country Planning Act, 1990. Decisions on development proposals below MLW are generally outside the scope of this planning system. All mineral rights, with the exception of coal, oil and gas, are vested in the Crown, which owns most of the seabed around the coasts of the British Isles, from Mean Low Water out to the limits of the territorial waters. The management of these resources is the responsibility of the Crown Estate, who issues two types of licence: a Prospecting Licence and a Production Licence.

Production Licences are controlled at present in England and Wales by an interim non-statutory "Government View" procedure, as issued in May 1998 by the DETR. On receipt by the Crown Estate of a formal application (usually accompanied by an ES) for an aggregate production licence, DETR initiates the "Government View" procedure. Should these extensive formal consultations proceed without overriding objections from consultees, then a favourable "Government View" should be forthcoming, and an aggregate extraction licence is issued by the Crown Estate. The licence frequently incorporates conditions including, for example, measures to mitigate the environmental impact of the dredging.

These interim non-statutory procedures are presently being observed in England and Wales, prior to the implementation in the UK of statutory procedures (on 14th, March 1999). The implementation of these statutory procedures is to meet the requirements of the EC Environmental Impact Assessment Directive (85/337/EEC) supplement and amendment (EC Directive 97/11/EC) of March 1997 (see below). Previously, licensing was controlled by the (then) Department of the Environment (DoE), under a non-statutory procedure issued in May 1992; this modified the original procedures, issued in May 1989. In July 1992, the Crown Estate issued a "Project Management" procedure for Production Applications seeking a "Government View". Until the introduction of the recent interim procedures, this remained the basis for applications for production licences for the marine aggregates sector.

Under the present interim procedures, only one ES is produced rather than a draft and a final version, as was the case with the previous non-statutory procedures. Instead of informal and formal rounds of consultation under the previous procedures, the ES is now subjected to a single Consultation Stage. Any relevant comments on the ES, or further work or analysis identified as necessary during the consultation stage, are included in an accompanying Consultation Report, rather than in a 'final' ES.

Existing and Future Requirements for Environmental Impact Assessment

In the UK, European Community (EC) Directives on the environment are incorporated into UK law and, therefore, affect the consent procedure (as outlined above). The key legislation comprises EC Environmental Impact Assessment Directive 85/337/EEC "The assessment of the effects of certain public and private projects on the environment", which came into effect in July 1988.

The EC Environmental Impact Assessment Directive requires an EA to be carried out, before consent is granted for certain types of major projects (listed in Annexes) likely to have significant environmental effects. Exploitation of mineral resources is identified in the Directive as an activity that might, under certain circumstances, require a supporting EA.

The original EC Directive (85/337/EEC) was supplemented and amended by a new Directive (97/11/EC), in March 1997. The changes within are to be transposed into the domestic legislative regime of Member States by 14th March 1999. The main aims of the amendments are to clarify and supplement the original Directive to achieve a more uniform application in all Member States; likewise, to improve the quality and scope of the information provided. The increased scope of the amendments added 12 new classes of projects to Annex I (for which an EA is always required) and 8 new classes of project to Annex II (where an EA is required where the project is likely to have significant environmental effects). It also clarified that modifications to both Annex I and Annex II projects are considered to be Annex II projects, in their own right.

As a response to the new EC Directive, DETR issued an initial Consultation Paper (in July 1997), setting out the general principles for implementing in the UK the amendments of the new Directive. Essentially, the main effects of the amendments identified in the DETR Consultation Paper were:

- increased coverage - more projects require an EA;
- changes to procedures - clarification of the way in which Member States can decide whether Annex II projects (those which are likely to have significant environmental effects, but which do not always require an EA) require an EA;
- competent authority must publicise its decision, on whether or not an EA is required;
- competent authority must, if requested by a developer, give advice on the content ('scoping') of any proposed EA;
- competent authority is required to give reasons for consent decisions; and
- improved arrangements for consultation with other Member States, for projects likely to have significant trans-boundary environmental effects.

For the purposes of the "Government View" Procedure, most recent marine aggregate licence applications have already required a supporting EA. With the introduction of the new EC Directive, which classifies the extraction of minerals by marine dredging under Annex II, all new operations to extract minerals by dredging, or any increase in extraction from existing dredging operations, must be considered for EA. In a subsequent DETR Consultation Paper (of 19th December 1997), comments were invited on proposals by the Government to determine the need for an EA in the UK. It was proposed that, with regard to the 'extraction of minerals by marine or fluvial dredging' an:

> "EA may be required where it is expected that more than 100,000 tonnes will be extracted per year. EA will generally be required for any major extraction activities which are proposed in or close to sensitive locations such as spawning grounds, Marine Nature Reserves, Sites of Special Scientific Interest etc."

As extraction rates for marine aggregate licences generally exceed 100,000 tonnes per year, and coastal deposits are sometimes close to spawning areas for commercially-exploited fish and shellfish stocks, or other areas of conservation interest, it was assumed that an EA would be required for most Production Licence Applications under the current and proposed EC Directives.

In July 1998, DETR issued draft Regulations for consultation, which will implement the requirements of the amended EC Directive. Unlike the earlier Consultation Papers, which set out the general principles for implementation, this consultation exercise related only to amendments to the Town and Country Planning system in England and Wales. Implementation would be under the Town and Country Planning (Assessment of Environmental Effects) Regulations 1988 (Statutory Instrument (SI) No 1199). Under these draft Regulations, and as a result of the previous consultation exercise, the Government propose to continue to rely primarily on a case-by-case judgement, detailing criteria under Schedule 2 by which the need for an EA is considered. However, with regard to the 'extraction of minerals by marine or fluvial dredging,' an EA was now considered mandatory for 'all development'.

In line with these requirements, DETR and the Scottish Office issued draft Regulations for consultation which should have introduced proposed statutory procedures for marine dredging. These Environmental Assessment and Habitat (Extraction of Minerals by Marine Dredging) Regulations bring together the requirements of the EC Environmental Impact Assessment Directive and EC Directive 92/43/EEC (the Conservation of Natural Habitats and of Wild Fauna and Flora). Commonly known as the Habitats Directive, the introduction of these Regulations will mean that the importance of marine habitats and the protection of sensitive areas from the impact of marine dredging, will be recognised in line with the requirements of the EC Environmental Impact Assessment Directive.

Whilst the practical approach to the application for marine aggregates extraction will not be fundamentally changed, in that EAs are already routinely undertaken, the introduction of a statutory regime does mean that all applications will be more carefully considered (through a scoping exercise), regulated and licensed. The potential route to public inquiry or informal hearing, provided by the interim procedures, will continue under the statutory procedures; this is consistent with public inquiries for any other terrestrial planning application. With Oakwood having, in effect, undertaken the new requirements for some considerable time (Oakwood published the first scoping document for an onshore terrestrial application), little will, in fact, change. It is hoped that all EAs will now reach the standard that Oakwood strives for.

Importantly, the Regulations recognise the significance of fully investigating the potential for trans-boundary impact, where relevant. If a proposal is likely to have a 'significant effect upon the environment' of another Member State, it will be necessary to fully consider, inform, consult, advertise and notify that Member State during the EA process. This exercise will be critical in resolving both the individual and cumulative effects of projects impacting on neighbouring EU countries.

EA PROJECT TEAM AND STUDIES

In order to undertake such EAs (to include the scoping study, field surveys, production of the ES and all consultation) the Oakwood in-house EA multi-disclipinary consultancy team is supplemented by a range of specifically-skilled associates who regularly contribute to Oakwood projects around the world. EA teams typically comprise: a strategic/project manager; marine environmental scientists; marine biologists; experts in coastal processes, fisheries biologists and commercial fisheries; as well as marine archaeologists and ornithologists.

EAs have to address a broad range of studies including: bathymetric and geophysical surveys; coastal processes studies; sediment plume mobility and dispersion studies; benthic surveys; commercial and recreational fisheries assessments; and crab and shell fish studies. Further studies include: marine archaeological reviews; recreational use studies; assessments of contaminants at sea; (together with) a review of the actual aggregate resource; and the possibility of use, in some instances, of the resource as a secondary aggregate.

The Terms of Reference developed by Oakwood for benthic surveys have now been adopted as a standard requirement of MAFF[1]/CEFAS[2], reflecting EAs based upon technically-robust field information.

KEY SCIENTIFIC ELEMENTS OF AN ENVIRONMENTAL ASSESSMENT

To undertake an EA, the full range of scientific experts is required to identify data sources and specific issues, relating to the proposal through a scoping study.

In many cases, very little scientific data exists covering fisheries, water quality, turbidity, coastal processes, seabed conditions, benthic communities and other issues. At the same time, ongoing research, as part of monitoring programmes, often does not provide meaningful scientific information in isolation. Project promoters are required, therefore, to prepare Terms of Reference and commission detailed baseline studies, to supply the necessary data.

The scientific experts on the project team then analyse data and report on specific subject areas. On completion of the description of the project details and the baseline environment, it is possible to undertake an impact assessment and to consider mitigation and monitoring, where appropriate. The interaction or the cumulative impact of proposals on subject areas is also now an important issue.

The interpretation and display of often hard scientific facts, for non-scientists to appreciate, requires technical writing skills. Consultations are directed to specific scientific interest groups and those with wider responsibilities such as DETR, MAFF, CEFAS and English Nature, as well as the public at large. Specific commercial or interest groups, such as fisheries or amateur divers, follow specific agendas without regard to the wider issues.

PREPARATION OF THE ENVIRONMENTAL ASSESSMENT

Scoping Study

The EA firstly consists of a scoping study that will identify specific issues relating to the proposal. It will also identify where incomplete datasets exist and the need for additional surveys, such as:

- benthic baseline surveys;
- commercial fisheries assessments; and
- coastal impact studies.

---

[1] Ministry of Agriculture, Food and Fisheries
[2] Centre for Environment Fisheries and Aquaculture Science

Key baseline environmental issues within the Environment Assessments

The key physical, biological and human issues describing the environmental baseline are then compiled within the EA process, as outlined below.

- Coastal processes - sediment transport, wave climate, turbidity and hydrodynamics, and the impact of the dredging plume i.e. fines (clays/silts) released into the water column.

- Benthic communities – species and community composition, biological diversity and significance as a food resource.

- Commercial and non-commercial fisheries - demersal and pelagic species, species diversity, productivity, feeding and nursery areas, spawning and migration routes and predator/prey relationships.

- Ornithology - seabird movement and feeding patterns.

- Sea mammals.

- Commercial fisheries - fishing effort patterns, landings and net income.

- Marine archaeology - recorded marine archaeological sites and assessment of national and international importance.

- Other human issues - military exercise areas; waste disposal sites (dumping); submarine cables and pipelines; oil and gas exploration and extraction sites; leisure yachting and navigation.

- Conservation status of the coastline - within international, European, national, local or non-regulatory policy frameworks.

Assessment of impacts in the EA

Impact assessment of physical, biological and human effects is based upon knowledge of the marine aggregate dredging companies proposed operations and their scientific area of expertise. Scientific experts will discuss, at project meetings, the assessment of impacts and (particularly) any potential for cumulative effects of the proposed project.

Physical impact assessment is undertaken through the analysis of: hydraulic and hydrodynamic effects; effects on transportation of coastal sediments; coastal erosion; beach drawdown; modification to offshore sandbanks; and any effects from the release of fines, from within the dredge plume.

Biological impact assessment includes: analysis of the effects on benthic communities (both direct and indirectly resulting from dredging), including sensitivity to disturbance; the effect of dredging plumes; and rates of re-colonisation by the benthic communities. Biological impact assessment also includes an analysis of effects on commercial and non-commercial fisheries species, taking into account species diversity, productivity, spawning and migration routes, nursery areas, predator/prey relationships and the indirect effects of dredging and plume dispersion on fisheries. Finally, an assessment of effects on other marine species; such as seabirds and sea mammals, is undertaken.

Human impact assessment includes analysis of effects on commercial fisheries, navigational issues, marine archaeology, leisure activities, pipelines, cables and disposal grounds.

Mitigation and monitoring in the EA

If any impact has been identified within the EA, mitigation measures and monitoring programmes may have to be developed. Mitigation measures are defined as those steps taken to make the predicted effects of dredging activity less severe: monitoring is the repeated measurement of an environmental variable, to ensure any impact is within acceptable predicted levels.

Typical mitigation measures may be:

- hours and seasons of operation;
- zoning of proposed licence area, to take account of fishing patterns;
- setting aside areas for conservation, or safeguarding, marine archaeological sites; and
- licence sharing.

Typical proposals for monitoring programmes may be:

- Crown Estate requirement for dredges to use an Electronic Monitoring System (EMS), to identify areas of operation;
- monitoring of benthic communities, to identify rates of recovery; and
- monitoring of fish and shell-fisheries, which can provide relevant data when used in isolation, but should, ideally, be viewed within the context of the wider biological habitat.

Therefore, monitoring can provide the evidence necessary to demonstrate that activities, with mitigation measures as appropriate, have not resulted in, or are likely to cause, individual, or cumulative impacts creating lasting damage to the marine environment. Identified mitigation measures and monitoring techniques will result, however, from not only the potential for direct and indirect individual impacts, but from the impact of cumulative effects. The most likely causes for potential impact from cumulative effects, as defined within a Cumulative Effects Assessment, have recently been studied by Oakwood.

CUMULATIVE ENVIRONMENTAL ASSESSMENT

In order to address the need for the assessment of cumulative effects, Oakwood are currently finalising a 'best practice' methodology for Cumulative Effects Assessment (CEA), appropriate to the marine aggregate dredging industry. This research project, undertaken for the US Minerals Management Service (MMS), has taken the form of a literature review and an investigation of possible methodologies. A strategic regional assessment of a pilot study area, off the south-east of the Isle of Wight, has been undertaken as a model for a 'best practice' methodology. It is this type of strategic environmental assessment that the Solent Forum has suggested is suitable for 'the assessment of cumulative impacts from dredging activity'.

A literature review of CEAs world-wide, in both the terrestrial and marine environment, has identified some 130 scientific papers, which have been reviewed and assessed. The literature review has identified the key CEA methodologies, together with tools and techniques; these include the need for determining spatial and temporal boundaries, and impact thresholds for each resource. The study has been accepted and approved by MMS in its final form[1]. It is hoped that a Best Practice Handbook can be produced subsequently, in order to provide guidance to practitioners. This research project should provide state-of-the-art methodology, for Oakwood and practitioners alike undertaking EAs for marine aggregate licence promoters.

CONCLUSION

Throughout this short paper on the preparation of Environmental Statements for marine aggregate extraction licences, the need for a high level of scientific knowledge and training has been demonstrated, coupled with the availability of good scientific data collected over a relevant time-scale. All are needed in order to achieve a credible Environmental Assessment of the proposed project. The interaction between scientists, Government Departments, agencies and scientists is vital to providing a wider understanding of the proposed project. It is now recognised that there is a need for those reviewing major applications within the marine environment to take account of a wide range of scientific data including an assessment of cumulative effects prior to making decisions on new projects.

---

[1] Note: This study is available on the MMS website, address www.mms.gov.

Main Channel Deepening - Port of Southampton - 1996/7

Colin Greenwell[1]

Associated British Ports Southampton, Ocean Gate, Atlantic Way, Southampton, SO14 3QN, U.K.

INTRODUCTION

Containerisation has become the common method for the movement of all types of cargo around the world. This type of trade commenced in the late 1960's when vessels were capable of carrying around 300 twenty foot (Panamax) boxes (6.1 m long, 2.9 m high and 2.4 m wide). It progressed through the Panamax size (incorporating 4250 boxes) of the 1980's, to the present day where the largest vessels can carry over 6,600 boxes.

Today, the largest vessels tend to ply their trade on long circular routes using the 'bus stop' principle. Typically, a vessel might make two northern European calls before moving down to the Mediterranean or South Africa, and from there to, for example, Asia or Australia. At each major 'bus stop,' containers are discharged/loaded both for inland distribution and to feeder ships, for onward short sea movement. Located within the middle section of the English Channel, Southampton (see Figure 1 for plan of the Port of Southampton) is ideally situated to serve the largest vessels, without creating major additional passage to their itinerary.

Virtually all Southampton's major customers started to build "post Panamax" ships (see above) and, for the Port to remain competitive, it needed to cater for such vessels. At the same time, Southampton is one of only three ports in the UK in contention to serve post-Panamax, 'round-the-world ships;' the other two being Felixstowe and Thamesport. If Southampton couldn't retain its interests, a significant proportion of the UK's trade would need to be handled through feeder traffic from the continent, adding to the UK cost of living.

Until the recent channel deepening, Southampton had a channel that was 10.2 m deep at low water, with berths a minimum depth of 12.8 m; this allowed a ship drawing 12.2 m to enter or leave the port, over a 5 to 6 hour window during each tide. New ships are drawing up to 13.5 m and were calling for an even greater window of opportunity, as it is expensive to maintain a vessel waiting for the tide. On the basis of the above requirements, Associated British Ports (ABP) had to undertake both land-based and marine developments. The land-based development involved deepening 206 Berth, building 207 Berth and extending the container-park. This contribution relates to the marine works - the main channel dredge.

---

[1]Present address: Associated British Ports Teignmouth, The Old Quay, Teignmouth, Devon TQ14 8ES, U.K.

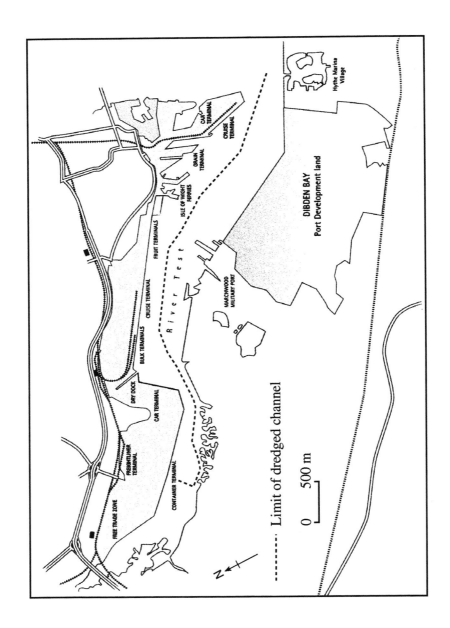

Figure 1 Plan of the Port of Southampton.

MAIN CHANNEL DREDGE

Background

Vessels approach Southampton from the Nab, where the depth of available water is approximately 13.3 m at low tide (for general location and main sites referenced, refer to *Maps A* and *B* in the *Preface*. From the Nab, there is at least 12.6 m clear depth to Fawley and 10.2 m maintained deep water to the container port. ABP elected to dredge to 12.6 m, right up the container port. Factors for this decision include: the tidal movement during the passage from the Nab inwards; the required underkeel clearance; and an operating window to remain attractive to the Port's customers. Thus, the channel had to be deepened by some 2.4 m; this was equivalent to the removal of 6.6 million cubic metres of material.

It is difficult to perceive such a volume of dredged material. Whilst maintenance dredging is a fairly unobtrusive activity, the main channel dredge would move enough material to bury a 500 acre farm to a depth of 3 m. At the same time, the area for removal was not only bordered by local communities, but is also surrounded by conservation designations (such as Sites of Special Scientific Interest and a proposed Special Protection Area). Additionally, both the Test and Itchen have established reputations as salmon rivers and there are recognised shell-fisheries in the area.

The first problem was the type of dredger to use; however, this was something ABP were unable to specify. There are a limited number of suitable international dredging contractors, capable of carrying out such work; they all had different equipment and ideas on what would be the most suitable method. Particularly, as for the most part, the material to be dredged was extremely hard Greensand, gravel, and solid clay. Opinion in the industry was divided, as to whether the Greensand could be suction-dredged economically.

The second problem was the requirement (also proposed by others) of the conditions of the Ministry of Agriculture Food and Fisheries (MAFF) dumping licence to examine beneficial uses for the dredged material. Some 21 sites were considered for the utilisation of the dredged material; however, apart from ABP, no one expressed an interest in the Greensand. Hence, the only material for which a beneficial use was identified was the gravel; some of this was landed ashore for future construction use, with some going to Gosport to recharge the Lee-on-the-Solent beach.

Investigation Undertaken

Virtually every aspect related to the project was studied and included in an environmental statement. Recommendations from this statement were incorporated into the contract conditions. Firstly, detailed soil investigations were undertaken; these included the collection of vibrocores, cone penetration tests and seismic profiling. Measurements were also obtained of heavy metals, tributyltin and organophosphate levels within the dredged material.

A detailed study of vessel characteristics, including their squat and low speed manoeuvrability was carried out. The results of this modelling work provided confidence in the selection of the planned dredged depth, as well as confirming that the Port could survive with a narrower channel, at full depth. If the channel had to be maintained at full width, at the new depth, the extra volume would have increased the size of the dredge by some 35%. On the basis of knowledge available, experience gained and measurements obtained, the conclusion was reached that the channel side slopes would be stable at a 1:6 gradient. Hence, it was decided to extend this profile from the toe of the existing slopes, to produce the new (slightly narrower) channel.

Studies were undertaken into the possible effects of a change in the geomorphology. Since the tidal prism would change, some effect on flow rates might be expected, but this proved to be negligible.

A considerable amount of effort was put into researching the effects of noise, both above and below water. It was known that bucket dredging would be far too slow to be a realistic option, likewise, other dredgers would be unlikely to cause problems. However, it was not possible to specify in advance the type of dredgers to use, a major exercise into investigating dredger noise was initiated. In the event, there were only two serious complaints received relating to noise during the dredging operations; one of these turned out to be unconnected with the dredge. Generally, the noise arising from the dredge went unnoticed.

In terms of fish population, however, the scale of the operation meant that there was the potential for serious (if only temporary changes) in the suspended sediment and oxygen levels within the water column. Furthermore, the Environment Agency was extremely concerned to investigate the effect of underwater noise on the migration of salmon, both adult and smolts. Research was undertaken on adult fish, which concluded they would not be affected. A significant contribution was made towards an Environment Agency tagging exercise, which tracked smolt movement down the Test. The results did not produce any evidence to show that they would be affected, but the Environment Agency was, nevertheless, concerned and wished to adopt the precautionary principle. As a result, certain restrictions (as to when and where dredging would not be permitted) were incorporated into the dredging contract.

Detailed studies of the effects of different types of dredging were undertaken to examine how the turbidity would be affected. The pre-contract advice was that most of the 'fall-out' would be retained fairly close to the dredger, remaining within the line of the main channel.

Lastly, but certainly not least, the effect on dumping such a large volume at sea needed to be carefully considered. A major exercise, including the use of fluorescent tracer material, was undertaken to confirm the capacity of the dump grounds, as well as to predict the effects on the marine environment at the Nab.

The Consultation Process

ABP's basic powers were already provided in the *Southampton Harbour Act*, 1911. However, consent was needed from the Crown Estate Commissioners, as the owner of a significant proportion of the channel bottom. The outcome of this process resulted in licence fee payments to dredge and dump, together with a royalty payment for any material beneficially-used. A dumping licence was required from MAFF and Department of the Environment Transport and Regions (DETR) (Department of Transport, DoT as it was at the time); their consent was also required.

In order to make certain that all the relevant parties were properly consulted, a Steering Group was set up, comprising representatives from the statutory agencies - English Nature, MAFF, Environment Agency and Hampshire County Council. To retain the group at a manageable level, each agency was asked to consult with other interested bodies within their sphere of activity e.g. Hampshire County Council represented the district authorities.

Towards the end of the above process, direct discussions were held with the various district authorities and the Royal Society for the Protection of Birds, before the formal public consultation process was undertaken. The public consultation was completed to DETR requirements and, as expected, a number of objections were raised from fisheries and nature conservation interests. The majority of these were satisfied, following discussion, and the DETR then issued its consent.

Safeguarding the Environment

When the Contract documents were issued, there were a number of safeguards incorporated to ensure that the requirements of the environmental statement were met: (a) physical and time constraints were made obligations of the Contractor; (b) monitoring exercises were undertaken by ABP Research and Consultancy; and (c) MAFF insisted also on the appointment of a Contract Liaison Officer, who kept an independent and careful watch on the Contractor's movements and sorted out complaints with fishermen (also overseeing the Nab monitoring works).

Examples of the safeguards were as follows:

- filling hoppers to below capacity (this allowed the spoil room to settle out and helped control turbidity);
- monitoring lobster pots near to the Nab (to check sediment movement);
- a restriction on night-time bucket dredging close to Hythe, or the SSSI; and
- the requirement not to dredge up-river of 203 Berth, during March to August.

Dredging operation/Siltation

Jan de Nul (UK) were awarded the dredging contract and commenced the dredge in August 1996. To overcome the problem of the hardness of the Greensand, they employed the 'Marco Polo', one of the world's largest cutter-suction dredgers. This vessel pre-cuts the Greensand, making it easier to remove than by conventional trailer suction dredgers. The clay section was removed directly by the Marco Polo; it was loaded then into a fleet of small barges. The gravel was trail-dredged by the trailing-suction dredger, the *Christophoro Columbo*, which pumped it through pipelines into No. 6 Dry Dock and on to the Lee-on-the-Solent beach.

Although the dredge itself was carried out efficiently, a siltation problem arose affecting the areas adjacent to the up-river section. The worst affected areas were Marchwood Yacht Club, Slowhill Copse, Marchwood Wharf/Husbands, Marchwood Military Port, Eling Marsh and (to a lesser extent) Hythe and the mouth of the River Itchen refer to *Maps A* and *B* in the *Preface* for locations. There remains an on-going legal debate with the contractor's insurers as to liability, rights to depth of water, etc. However, ABP adopted the simplistic view that it had caused a problem for our neighbours and set about cleaning it up. ABP's own maintenance dredgers undertook emergency work and Marchwood Yacht Club was relocated to the Bury Swinging Ground. This approach provided Jan de Nul with the time to modify a small shallow-drafted cutter dredger, which completed the clearance, by loading directly on to barges. The clam fishermen's complaint was complicated, as the fishery was already in terminal decline. This issue was resolved by funding a clam clearance operation, which repositioned all live clams found into the Yacht Club area (which is regarded now as a clam nursery area).

The most difficult area to deal with was the SSSI, up-river of the channel where re-dredging would be extremely detrimental. ABP commissioned a one-year monitoring exercise, undertaken by Ecological Planning and Research (EPR), to study the effects both up-river and at Hythe. The EPR report made several findings, which are summarised below:

- there were no adverse long-term changes at any of the sampling sites that could be attributed to sedimentation effect;

- the newly-deposited Greensand (which can bind metals) may have reduced the toxicity of the sediment and hence made it suitable for colonisation by a wide range of species (e.g. at Bury); and

- although sediments at Hythe remain highly contaminated, they appear to have become more attractive to a wider range of estuarine species over the course of the year.

In conclusion, due to the nature of the uncontaminated material, it would appear that the dredge had done more 'good' than 'harm', perhaps even temporarily arresting the natural erosion.

The findings of the above report are important, as it would have been difficult to correct any damage. It also gives encouragement to the use of the precautionary principle, as agreed with English Nature, prior to the commencement of the dredge. ABP promised to conduct a trial intertidal mud flat recharge experiment, using maintenance dredgings, following completion of the main dredge. Arrangements to undertake this recharge are presently being made; it is hoped that a small site at Hythe will be utilised for this activity.

Another monitoring requirement relates to Southampton Water. Changes in the flora and fauna of the intertidal mudflats are being monitored over a ten-year period. This work is being undertaken by ABP Research and Consultancy, using both site measurements and aerial photography.

The research undertaken to find out the cause of the siltation problem indicated that the overall power of the pre-cutting process caused a fluid mud layer to become established, within a metre or two of the bottom of the channel. The double-handling of the material by the trailer suction dredger, as well as the continual stirring by passing ships' propellers, raised the material in the water column. Further, the exceptionally high tides in February 1997 dispersed it across adjoining areas. It has taken some time to develop an understanding of the processes, but these have now been recreated in a modelling exercise.

It is perhaps worthy of note that the problem arose between December 1996 and February 1997, when several very large dredgers were working 'at capacity' in the upper reaches of the dredge area, prior to the Environment Agency's smolt deadline (of 1st March, 1997). It may be unfortunate but, by the time the problem arose, the precautionary principle measures had become embodied in both the consents and contract conditions. Had re-scheduling been possible, measures could have been taken to reduce the effects. Perhaps this is a lesson for the future?

CONCLUDING REMARKS

In conclusion a few personal and constructive criticisms of the Environmental Assessment process may be summarised as follows:

- there is a need for flexibility and ongoing review, to enable a proper response to be made to events as a project progresses;

- there is the tendency for single interest groups to attempt to divert the statutory consultees with unreasonable demands, in order to delay or frustrate a project, or (indeed) to extract a ransom, this needs to be resisted;

- the whole process is too time-consuming; and

- there needs to be better liaison between the statutory consultees, e.g. the requirements of birds can be the opposite to those of fish.

Evaluating the Intertidal Wetlands of the Solent

David Johnson

Maritime Faculty, Southampton Institute, East Park Terrace, Southampton SO14 0YN, U.K.

INTRODUCTION

This case study is an example of the need for science to inform coastal planning and management. It also links science with engineering, economics and politics in an effort to prescribe sustainable integrated coastal zone management (ICZM) or, in other words, coastal management policy for future generations.

The problem addressed is the need to reverse loss of intertidal wetland habitat in the Solent region. For the purposes of this paper, intertidal wetlands are defined as both the intertidal terrestrial wetlands of coastal saltmarshes (or saltings) and associated soft mobile sediments of intertidal marine muds; these are found generally between the mean high and mean low water mark (Adam, 1990).

Historically, intertidal wetlands have been lost principally as a result of reclamation and a fundamental "undervaluing" of indirect benefits, such as the ability of wetlands to enhance water quality. Latterly, global environmental change, in particular relative sea-level rise and increased storm frequency, combined with construction of hard sea defences has produced "coastal squeeze"; this is effectively the drowning of intertidal wetland areas, which are unable to migrate landwards.

Hitherto, conservation priorities accorded to intertidal wetlands have been based largely upon nature conservation criteria alone. The resulting mix of site acquisition, site designation and site management agreements, whilst influential, is failing to protect the Solent resource. The way forward, suggested in this paper, is: to evaluate intertidal wetlands by quantifying the broad range of beneficial functions they perform; to assess the cumulative pressures threatening the resource; and, then, by combining these, to establish a regional target for rehabilitation and re-creation.

FUNCTION BENEFIT INDICATORS

Intertidal wetlands perform a range of primary beneficial functions. These functions are important for nature conservation, coastal defence, water quality, commercial production, recreational activities, aesthetic qualities and as an educational/research resource. The relative importance of each of these functions is an informed subjective judgement.

In order to make that judgement, a panel of twenty-four experts (selected through a process of nomination and review), were asked to weight each of the functions using a Delphi technique (Silverman, 1997). The results, produced by asking each expert to apportion 100 points between the different functions, are presented in Figure 1; the experts highlight the relative importance of nature conservation and coastal defence functions. An analysis of the disciplinary and professional backgrounds of the experts suggested that the academics, scientists and environmental managers who participated in this exercise were more objective than engineers and conservation agency staff; this underlines the importance of using multidisciplinary teams for coastal management.

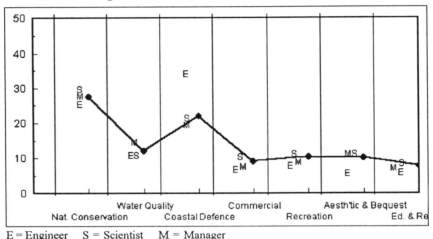

E = Engineer   S = Scientist   M = Manager
Function benefit weighting differences - academic disciplines

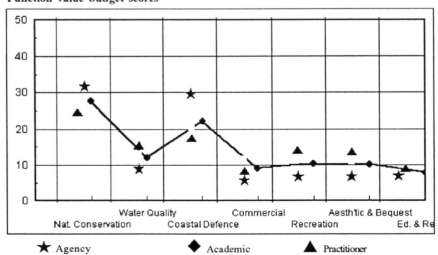

Function benefit weighting differences - professional groups

Figure 1 Function Benefit Weighting.

Having weighted the functions, their relative importance at different intertidal wetland sites must be calculated. 'Function benefit indicators' should be relevant, measurable, comparable, easy to interpret and targetable. The indicators selected for this exercise, which are described below, were chosen on the basis of structured interviews with subject experts (the secondary functions were weighted in the same way as the primary functions), literature review and data accessibility and repeatability. To reflect the weighting of primary functions, three indicators for nature conservation and two for coastal defence were selected.

The 'nature conservation indicators' adopted were waterfowl and wader abundance, numbers of soft substrate communities and network effectiveness (similar sites within a defined distance). Bird counts provide the most cost-effective indicator of nature conservation quality. Bird numbers were also selected on the basis of their scale of influence on the total biosystem. Migratory/resident bird support is more far-reaching, in terms of habitat importance, than fisheries support (which is regionalised), and endangered species support (which is localised). Diversity of benthic plant and animal biotypes (intertidal and subtidal) were chosen, as identified by the Marine Nature Conservation Review (MNCR) in its efforts to identify sites and species of nature conservation importance (Hiscock, 1996). Finally, the importance of the regional concentration and inter-connected value of intertidal wetlands was recognised, by attributing a network score based upon the proximity of neighbouring intertidal wetland resources.

For coastal defence, intertidal wetlands are important for both wave height reduction and shoreline stabilisation. Coastal engineers rate the absorption of energy by intertidal wetlands (dampening the attenuation of waves) as their key function. Gradient, extent, creek diversity and intertidal range are all influential in this respect (Allen & Pye, 1992). Consequently, mean intertidal width and creek density were selected as indicators. For the latter, on the basis that an undissected intertidal wetland presents a better coastal defence than a dissected one, ranking scales based on aerial photograph comparisons was considered to be a practical and useful measurement.

Largely on the basis of nutrient transformation, percentage saltmarsh cover was taken as the indicator for water quality. For commercial production, records of juvenile fish numbers were selected, as there is evidence of a linkage between the area of nursery grounds and the recruitment strength of 0-2 year group flatfish (whose diet is dependent on polychaetes and copepods, associated with the fine sediments of intertidal wetlands). To reflect participation, a percentage of local population density and total numbers of moorings were combined as the recreation indicator. As an aesthetic indicator, the length of visible undeveloped shoreline behind the intertidal zone was chosen and, to reflect scientific importance, a count was made of intertidal wetland specific scientific references.

THE SOLENT RESOURCE

The geographical setting for this study was the Solent and Poole Bay Marine Natural Area (English Nature, 1997). Sources of indicator data were the Shoreline Management Plans for the region, together with regional inventories (JNCC, 1996; Buck, 1997). Not all the data could be drawn from secondary sources. Creek density was calculated from aerial photographs, juvenile fish numbers obtained from the Centre for Environment Fisheries and Aquaculture Science (CEFAS) survey records and research records were quantified using the Standing Conference On Problems Associated with the Coastline (SCOPAC) database.

Data sets were compiled for each of the 13 different intertidal wetlands within the region and a regional mean figure for each indicator calculated. To compare the different data sets, a radial graph technique was used (Brink *et al.*, 1991). These graphs, or AMOEBAs, provide a useful visual comparator and also enable, within the regional data set, production of a total score for each site and for the region as a whole. The regional indicator means, as represented by a circle on the radial graph, and each of the scaled values represents a score for each hectare of intertidal wetland. Examples of AMOEBAS calculated for Langstone Harbour and the Yar Estuary are shown in Figure 2.

The evaluation confirmed the relative importance of the four largest intertidal wetlands (Chichester Harbour, Poole Harbour, Langstone Harbour and Southampton Water); these are all greater than 1,000 hectares in extent. These four intertidal wetlands scored highest, in terms of function benefit per hectare, with Lymington/Keyhaven registering as the next most valuable resource.

When size was taken into account as a multiplier, the impact of the first four intertidal wetlands was further emphasised. The four combined, accounted for 77% of the regional function benefit score. As single exceptional intertidal wetlands, Chichester Harbour and Poole Harbour, which scored similarly, are the region's most valuable resources. As a unit however, Chichester and Langstone harbours represented over 40% of the total function benefit score for the region.

The relative unimportance of the intertidal wetlands of the northern coastline of the Isle of Wight was also highlighted. Together, the five Isle of Wight intertidal wetlands represented less than 6% of the total function benefit score.

Whilst it is important not to detract from the importance of the regional intertidal wetland resource as a whole, and to recognise uniqueness (e.g. the unspoiled character of Newtown), it is also clear that the results of this evaluation can help establish management priorities.

ENVIRONMENTAL RISK ASSESSMENT

Coastal processes within the Solent region have been reviewed in detail by the Shoreline Management Plan exercises. The plans draw on studies of the geological and geomorphological background, littoral transport regime, bathymetry, seabed sediment transport, water levels and wave climate. Accretion/recession evidence over the past 100 years suggests that all the region's intertidal wetlands have been receding, with the coastal plain estuaries of the Western Solent experiencing the most rapid erosion.

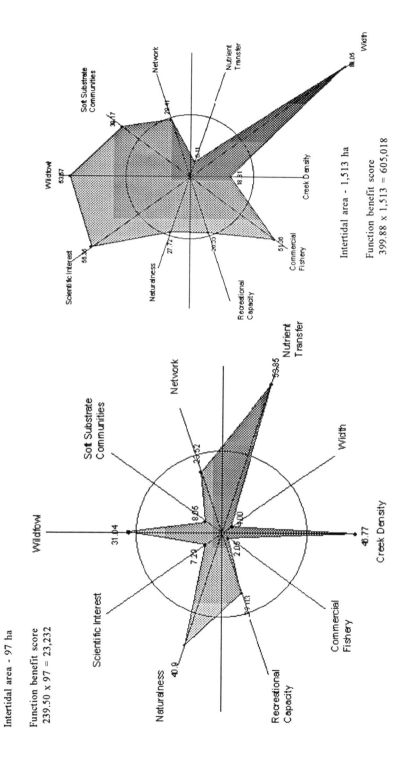

Figure 2 Radial graphs showing intertidal evaluations for Langstone Harbour and Yar Estuary.

However, to assess the level of future threat to the range of intertidal wetland function benefits (i.e. quality, as well as quantity of wetland), an environmental risk assessment was considered appropriate. At the same time, it was recognised that environmental risk assessment (estimation and evaluation or significance) and risk management (the process of implementing decisions about accepting or altering risks), within the context of sustainable development, is still in its infancy (DoE, 1995).

For the purposes of this study, to determine the cumulative pressures on each intertidal wetland, (and, thus, on the regional resource) wetland specific pressure-stress-function matrices were constructed. The matrices incorporated a generic list of pressures and summarised potential stresses and their effect on wetland functions. The pressures, or impacts, considered were: relative sea-level rise; sediment chemistry; reclamation for development; pollution; dredging; management intervention; and ship movements.

Some stresses are common to the whole region, whilst some are 'wetland-specific'. The matrices also took into account the present efforts to restore intertidal wetlands; thus incorporating both function gains and losses, when identifying consequences of impacts.

Estimations of magnitude (i.e. the level of effect, if consequences are realised) suggested that nature conservation and coastal defence function benefits (those of most importance with the highest weighting) are most at risk. For example, migratory/resident bird support is at risk from rising sea-level, both in terms of waves over-topping nests and, because of a diminishing intertidal wetland resource, adult birds who relegate juveniles to wetlands with less abundant food supply. When magnitude was combined with a probability estimation, the overall environmental risk assessment pattern within the Solent region was one where stresses on function benefits confirmed a probable continuing and exacerbated rate of loss of the intertidal wetland resource.

Although the risk/impact predictions did not change the regional intertidal wetland ranking, achieved by quantifying function benefit indicators, they served to illustrate which intertidal wetlands are vulnerable within a regional context. Lymington/Keyhaven and Southampton Water are considered to be most at risk. In the case of Lymington/Keyhaven, this vulnerability is related mainly to predicted natural change. Southampton Water is under pressure from a combination of anthropogenic-related risks, as well as the pressures of natural change. Chichester harbour was considered to be least at risk.

Based on the current mean function benefit score for the region a projected loss, and thus a restoration target, of 1,200 hectares of intertidal wetland was calculated, working on a risk estimation period of twenty years (i.e. 1998 - 2018).

DISCUSSION

Work undertaken for English Nature by Pye and French (1993), established targets for coastal habitat re-creation. However, there is still no National Strategy for the conservation of intertidal wetlands. For the Solent region, the Solent Forum's Strategic Guidance aims to conserve intertidal wetland areas and re-create them wherever possible. To this end, monitoring and protection must be combined with pro-active rehabilitation and re-creation. In practical terms, however, any surface and seaward restoration will only be a temporary solution. Long-term landward-managed retreat/realignment techniques need to be planned and implemented.

In this respect, as Fowler (1995) has highlighted, the Solent and Poole Bay Natural Area has a significant problem. Very few potential retreat/realignment sites are of little nature conservation value. Unlike the east coast of England, where many intertidal wetland reclamations have become agricultural land and there is no great biodiversity to lose, biodiversity behind the seawall in the Solent region is generally greater than in front. Of particular importance are the extensive grazing marshes and brackish lagoons.

An opportunistic approach to regional restoration of intertidal wetlands is likely, but preserving them in the wrong geomorphological setting is clearly unrealistic. Thus, whilst sentiment might favour restoring the most valuable, or perhaps the most threatened areas, in practice the most easily rehabilitated/re-created areas should be the priority, i.e. where the likelihood of success is greatest.

Furthermore, landowners and the public will need to support the elimination of development value and development rights along the coastal strip, at least on a selective basis. This approach will require a combination of designation, mitigation, land purchase, compensation and education or social marketing. Without an incentive for landowners and a major shift in public attitude, even the most attractive scientific arguments will fail to produce the scale of change required.

The recommendation of this integrative case study is, therefore, that intertidal wetland re-creation effort within the Solent region should be concentrated on projects within Chichester Harbour. This intertidal wetland is not only the most valuable and least at risk, in function benefit terms, but it also:

- has the necessary land available behind the sea wall. This is illustrated dramatically by the false colour composite of three Compact Airborne Spectrographic Imager (CASI) images produced by the National Centre for Environmental Data and Surveillance (NCEDS, 1997);

- is the most favourable geomorphological setting, in terms of sediment movement, within the Solent region; and

- is socio-economically well placed to achieve greatest net benefit from investment in intertidal wetland re-creation (because of the constitution of the Harbour Conservancy, with its policy to actively purchase land within the Area of Outstanding Natural Beauty).

CONCLUSIONS

The appropriateness of ranking information for decision-making purposes, as suggested in this contribution, is disputed within the scientific community. The most common criticisms are: (a) that summarising information loses detail and can be misleading; (b) some evaluations are necessarily subjective; and (c) the complex ecological interactions gives rise to mathematical complications (Dalby, 1987).

Although (in some cases) it is easy to appreciate why this argument applies, it can also be said to be a narrow view; this is unhelpful for decision-makers, who have to establish coastal conservation restoration priorities. Quantifying set parameters or indicators allows greater objectivity, and the identification of knowledge gaps; it is perhaps the only realistic way of communicating scientific and technical information, to policy makers and planners.

It is the contention of this case study, therefore, that the relative value of the Solent and Poole Bay Natural Area's intertidal wetlands (saltmarshes and mudflats) can be expressed by comparing the different functions they perform. Functions have been weighted using expert opinion and measurable indicators for each function were identified.

Comparison of secondary data sets for each of the indicators confirmed the relative importance of the largest intertidal wetlands, which are all over 1,000 hectares in extent. As a unit, Chichester and Langstone harbours are exceptional. By contrast, the intertidal wetlands of the north coast of the Isle of Wight are relatively unimportant.

An environmental risk assessment, considering potential pressures on the region's intertidal wetlands over the next 20 years, identified Lymington/Keyhaven and Southampton Water as the wetlands whose functions are most at risk. For both sites, the relatively high potential function loss (quality) is compounded by the prospect of significant erosion (quantity).

For the Solent region, the agreed policy is to conserve intertidal wetland areas and re-create them where possible. The trend, however, over the past 100 years, has been a steady reduction of this resource. In order to halt and/or reverse this trend, and retain the present level of intertidal wetland functions, a 20 year restoration target of 1,200 hectares is proposed.

A suggested future research agenda, based upon the evaluation of the Solent's coastal wetlands, should include:

- further research into physical and biological processes which control the formation and 'healthy' functioning of the region's intertidal wetlands to improve scientific certainty;
- scoping of additional managed retreat/realignment opportunities in Chichester harbour;
- intertidal wetland specific management plans, which foster short-term rehabilitation solutions, particularly for wetlands such as Lymington/Keyhaven which are under most pressure;

- consensus building, to achieve local ownership of intertidal wetland restoration projects; and

- research to identify a funding mechanism, potentially as part of the redirection of Common Agricultural Policy subsidies, to enable positive public/private sector collaboration to achieve practical solutions.

REFERENCES

Adam, P. 1990. *Saltmarsh Ecology*, Cambridge University Press, Cambridge, 461 pp.

Allen, J.R.L. & Pye, K. (Eds.) 1992. *Saltmarshes - morphodynamics, conservation and engineering significance*. Cambridge University Press, Cambridge, 184 pp.

Brink, B.J.E., Hosper, S.H. & Colijn, F. 1991. A quantitative method for description and assessment of ecosystems: the AMOEBA-approach. *Marine Pollution Bulletin*, **23**: 265-270.

Buck, A.L. 1997. *An Inventory of UK estuaries. Volume 6: Southern England*, JNCC, Peterborough, 234 pp.

Dalby, D.H. 1987. Salt Marshes. In: *Biological Surveys of Estuaries and Coasts*. Baker, J.M. and Wolff, W.J. (Eds.), Cambridge University Press, Cambridge, 38-80.

DoE 1995. *A Guide to Risk Assessment and Risk Management for Environmental Protection*. HMSO, London, 92 pp.

English Nature 1997. *Solent and Poole Bay Natural Area: A Nature Conservation Profile*. Confidential Draft, October 1997 (Unpublished), 47 pp.

Fowler, S.C. 1995. *Review of Nature Conservation features and information within the Solent and Isle of Wight Sensitive Marine Area*. Contract F80-18-09 on behalf of the Solent Forum Nature Conservation Topic Group. Nature Conservation Bureau Ltd., Newbury, 120 pp.

Hiscock, K. (Ed.) 1996. *Marine Nature Conservation Review: Rationale and Methods (Summary Report)*. JNCC, Peterborough, 12 pp.

JNCC 1996. *Coasts and Seas of the United Kingdom. Region 9 Southern England: Hayling Island to Lyme Regis.* Barne, J.H., Robson, C.F., Kaznowska, S.S., Doody, J.P. & Davidson, N.C. (Eds.). JNCC, Peterborough, 247 pp.

NCEDS 1997. *Aerial Surveillance of Fourteen Estuaries in England and Wales*. The National Centre for Environmental Data and Surveillance, Environment Agency Internal Report, 213 pp.

Pye, K. & French, P.W. 1993. Targets for coastal habitat re-creation. *English Nature Science* **13**. English Nature, Peterborough, 85 pp.

Silverman, D. (Ed.) 1997 *Qualitative Research: Theory, Method and Practice*. Sage Publications, London, 262 pp.

SECTION 6

Concluding Remarks

Developing a Research Agenda for the Future[1]

Ian Townend

ABP Research and Consultancy, Pathfinder House, Maritime Way, Southampton, SO14 3AE, U.K.

INTRODUCTION

Three areas of activity emerged from the Plenary Sessions, which provide a basis for summarising future needs:

- monitoring and measurement;
- processes and ecosystems; and
- management, training and promotion.

MONITORING/MEASUREMENT

- Improve upon the description of sediments in the Solent; for example, it may be possible to use existing models to focus the fieldwork on areas of greatest potential scientific value.
- Long-term measurements of flows, to identify patterns of water movement.
- Long-term measurements of nutrients, to identify inter-annual variability - this could be achieved by ensuring that the "ferry box" experiment is maintained for several years (see Water Quality and Chemistry sections).
- Monitor levels of copper and TBT, to assess the impact of changing legislation.
- Fieldwork to be undertaken, to identify sources of entroviruses and other pathogens.
- Seek better measures or surrogates, to measure and monitor the links between water quality and public health.
- Establish a catalogue of sites where the habitat has been changed; then monitor the manner in which they have colonised and changed, as new habitats have become established.
- Develop new methods for measuring impacts, with the aim of isolating cause and effect – particularly with respect to new developments.

---

[1] Note: This contribution summarises the research agenda identified from the Plenary Sessions; as such, it needs to be considered in association with the Workshop findings (see text).

- Consider the potential for the establishment of a strategic programme of monitoring - extensive monitoring is likely to be required, in response to a number of forthcoming directives and developments in licensing arrangements, notably:
- Habitats Directive;
- Water Framework Directive;
- Shellfish Waters Directive;
- Aggregate Licences; and
- Dredging Licences.

Historically, the various agencies have been very protective of their own data collection programmes. Given the scale of the above monitoring requirements and the limited funds available, this is no longer acceptable. There is an imperative requirement for a co-ordinated programme that meets the needs of individual agencies (to a reasonable degree), whilst, at the same time, offers added value by providing data sets of which a wide range of user groups can take advantage

PROCESS/ECOSYSTEMS

- The need to develop a better understanding of sources of fine-grained sediments, to explain the relatively high suspended sediment concentrations observed in the Solent.
- Studies are required to examine the role of turbulence (notably salinity/freshwater - driven turbulence), in controlling estuarine transport.
- Models need to continue to improve their representation of non-linearities and include physical, chemical and biological interactions.
- There is a need to develop an understanding of what is meant by the "structure and function" of a site in relation to conservation, measurement and monitoring; in particular, the examination of potential changes to the structure and function within an ever-changing environment.
- Work is required to understand the role of sub-components of a site and inter-relationships between sites e.g. due to bird movements. (Note: this need may well be encompassed within a proper understanding of structure and function).
- Develop a better understanding of system response times.

Overall, there is a need to avoid attempting to isolate the "natural" system; humans are an integral part of it and that is what has to be managed. The potential exists to use information, technology and partnerships to promote a multi-disciplinary initiative, founded on a "systems approach," to bring together the many strands of good science that are presently on-going.

MANAGEMENT/TRAINING/PROMOTION

- Train those using the outputs from models, in processes and modelling techniques.
- Create opportunities to make use of new research e.g. in the way tenders are evaluated, or the budget allocated to a particular initiative.
- Improve the presentation of scientific information.
- Maximise the opportunity to publish work carried out and data held commercially, and presently available only through the 'grey literature'.
- Review outputs from the first round of the SMPs.
- Increase requirement for post-project/development evaluation and, at the same time, ensure that lessons (positive and negative) are disseminated e.g. controls imposed which have not had the desired effect.
- Identify opportunities within the Solent for habitat improvement and creation.
- Improve access to information; this may require lobbying Government to change the present requirement for quangos/agencies to derive a proportion of their income from the exploitation of data banks, established through public funds.
- Raise public awareness; this is a very broad requirement, which needs to be targeted at a range of different interest groups, with knowledge ranging from the well-informed to the complete novice.
- The Solent Forum should examine whether there is scope to fund long-term studentships and whether it could provide the front to consortia bids, seeking EU funding for studies of direct relevance to the Solent.

Some key messages need to be communicated as widely as possible e.g. the importance of mud within the overall Solent system.

CONCLUSIONS

For mangers and conservationists – there is a good science base available, which is being disseminated; to obtain maximum benefit from this, there needs to be more attention given to staff training and to providing the working environment that enables and facilitates the initiation of new ideas.

For the scientists – there is a great deal of good scientific research already in progress, but few attempts have been made to synthesise the findings; a framework is needed to promote a systems view, as a mechanism for integrating these multi-faceted components.

Conclusions and Close

Maldwin Drummond, OBE, JP, DL, Hon.DSc.

Chairman, Solent Forum

You will remember our Conference aims:

- to raise awareness and understanding of scientific information;
- to bring together scientists, academics, planners and managers to discuss the use of information; and
- to develop a co-ordinated research agenda.

I believe the Conference achieved these three goals and that it has been an event that may well be judged to have been fundamental, a base line study to further detailed scientific and management initiatives in the future.

The presentations have been carefully prepared by acknowledged authorities, currently practising in their respective fields. The work is, therefore, up to date. The publication that will result from the Conference needs to reflect this excellence and it will.

Contributors were asked to recommend actions and detail research gaps; they have undertaken this robustly, as you have seen.

The organisation of the Conference has been of a high standard. I would like to add my congratulations and thanks, as Chairman of the Solent Forum, to Michael Collins and his team and to Kate Ansell and Zoe Hughes. Alan Inder, our *eminence grise*, has provided polished leadership and advice. They have all been tireless and good humoured. There has been excellent technical back-up, too, with "leading edge" facilities and comforts well-associated with this building, the Southampton Oceanography Centre (SOC). We are all grateful to Professor John Shepherd, the Director. I personally liked the innovation of *edible policy workshops* - working lunches - a useful innovation. I would like to thank those responsible for providing the material for the Poster Sessions, which enabled close examination and questioning on a large number of subjects. This concept, together with the Policy Workshops and the tea and coffee breaks, provided opportunities for networking and *together* provided a model, in my view, for future conferences.

This leads me to forward ideas for the future, beyond the publication of the Proceedings and the energetic follow up. A way forward could be to take subjects addressed at this Conference as a series of science and management workshops. This approach would enable us to monitor progress against the 1998 base line and push ahead in a series of specific focuses. If you have ideas or any feedback on how this Conference went, for the future, please let Kate (Ansell) know.

Thanks to all who chaired, spoke, helped or attended, but before you go, let me share with you David Tomalin's assessment, written on the ferry this morning.

Mistress Solent, She captivates us all
We have stripped her down beyond the point of dignity.
We have examined and debated her internal workings
(The Chirp sub-bottom profiler of Southampton has become her barium meal).
Since her future has been questioned we have
Viewed her as a patient - discussing her health.
Yet while symptoms and diagnoses are earnestly discussed,
There are some questions we still demure to ask.

Just how old is the patient?
Is she fully developed?
And, in future, will she still tolerate the
The cosy intimacy we have nurtured over the past 5,000 years.

David Tomalin, Isle of Wight Council

Again, my thanks to the SOC, to the organisers, the speakers and to all of you for making this Conference such a success.

INDEX

Location Index

Alum Bay, 23, 74
Ashlett Creek, 181

Barton-on-Sea, 90, 167
Beaulieu River, 15, 21, 33, 137, 151-154, 173, 234, 236, 264, 275, 334
Bembridge, 11, 13, 24, 30, 74, 167, 172, 234, 239, 248, 250, 253-255, 307, 308
    Harbour, 234
    Ledges, 11, 74, 248, 254
Bouldnor, 16, 73, 75
Bracklesham Bay, 34, 248, 249, 253, 315
Brading Harbour, 236
Brambles Bank, 32, 104, 152
Browndown, 238
Bursledon, 272, 273

Cadland Creek, 305
Calshot, 34, 56, 57, 58, 152, 180, 235, 254, 263
    Castle, 200, 203
    Spit, 21, 28, 45-48, 77, 89, 201, 256, 276
Chichester Harbour, 21, 23, 28, 30, 56, 74, 75, 108, 135, 137, 146, 171, 173, 218, 226, 233, 234, 247, 248, 254, 255, 263-267, 312, 315, 316, 319, 324, 335, 358, 360, 361, 362
    Dell Quay, 170
    East Head, 321, 323
    Emsworth Channel, 170
    Thorney Island, 34
Christchurch Harbour, 15-17, 26, 28, 32, 34, 56, 90, 97-99, 103, 105, 272, 273, 309
Colwell Bay, 73
Cowes, 12, 33, 164, 167, 271-273, 312
Culver Cliff, 236, 237, 248, 249

Dock Head, 48, 64, 152
Dolphin Bank, 32, 303

East Cowes, 172, 312
Eastern Solent, 13, 14, 73, 105
Eastney, 167, 296, 297, 311
Eling, 176, 177, 352
Ethel Point, 308

Farlington Marshes, 236, 266, 267
Fawley, 3, 14, 74, 127, 151, 154-157, 180, 271, 303, 305, 334, 349
Fort Albert, 275
Freshwater Bay, 16, 237, 248, 249, 296, 297

Gosport, 14, 349
Gurnard, 165

Hamble Estuary, 11, 64, 67, 68, 115, 137, 144, 154, 155, 158, 170, 255, 283, 284, 285, 287, 293, 296, 311-313
Hamstead, 13
Hanover Point, 308
Hayling Island, 21, 34, 235, 238, 271
Headon Warren, 73
Hill Head, 173, 275
Horse Ledge (Shanklin), 308
Hurst
    Castle, 200, 201, 238
    Channel, 10, 73
    Narrows, 21, 26, 28, 30, 33, 73, 105
    Spit, 10, 11, 12, 14, 28, 34, 60, 73, 77, 104, 105, 248, 261, 263, 275
Hurst-Pennington, 15
Hythe, 64, 159, 178, 181, 183, 235, 334, 351, 352

Itchen River, 64, 67, 68, 74, 137-140, 143, 149, 150, 151, 154, 155, 158, 168, 170, 204, 205, 206, 277, 349, 352

Keyhaven, 73, 173, 234, 358, 360, 362

Langstone Harbour, 21, 28, 34, 75, 108, 127, 135, 137, 146, 155, 170, 171, 173, 209, 218, 226, 233, 234, 236, 247, 248, 263, 264, 265, 267, 276, 296, 323, 358, 359, 362
   East Winner Bank, 104
Lee-on-the Solent, 34, 115
Lepe Middle Bank, 169
Lymington, 13, 15, 16, 21, 33, 171, 173, 234, 236, 272, 273, 275, 358, 360, 362
   Flats, 21
   River, 15, 16, 173, 234, 236

Marchwood, 156, 157, 176, 235, 256, 271, 277, 278, 352
Medina Estuary, *see* Medina River, 61, 170
Medina River, 12, 61, 74, 168, 169, 170, 173, 234, 275, 307
Medmerry Bank, 32, 107, 108
Mixon Hole, 248, 251, 316

Newtown Creek, 15, 16, 30, 33, 74, 75, 77, 169, 173, 234, 248, 275, 334, 358
Northney, 312
Norton, 172

Pennington, 13, 14, 171, 173
Pilsey Sands, 321
Poole Harbour, 26, 28, 56, 97, 98, 99, 156, 159, 254, 334, 357, 358, 361, 362
Portland Bill, 34, 268
Portsmouth Harbour, 3, 21, 28, 30, 33, 34, 69, 80, 101, 103, 108, 115, 123, 135, 137, 146, 155, 159, 170, 171, 172, 173, 209, 213, 218, 221, 233, 234, 247, 248, 255, 256, 257, 263, 264, 265, 267, 271, 272, 273, 275, 311, 312, 323
Princessa Shoal, 108, 250

Ryde
   Middle Bank, 32, 61, 173
   Sands, 104

Seagrove, 165
Seaview, 275
Selsey Bill, 107, 247, 248, 249, 250, 251, 268, 315, 316
Shanklin, 308
Shingles Bank, 11, 32, 105
Slowhill, 137, 352
Southampton Docks, 251
Southampton Water, 45, 64, 137, 141, 175, 176, 199, 201, 256, 271,
Southsea, 165, 187, 193, 255, 275
Sowley Ground, 16, 73, 169, 173
St. Catherine's Point, 307
St. Helen's Fort, 238, 248, 253, 254
Stansore Point, 74, 275
Stanswood Bay, 169, 173, 200, 248, 276
Stone Point, 4, 200
Swanage Bay, 283

Test River, 11, 67, 68, 137, 138, 139, 140, 143, 149, 150, 155, 158, 177, 204, 238, 256, 277, 349, 350
The Nab, 56, 250, 349, 350, 351
The Needles, 10, 23, 74, 232, 236, 237, 248, 249
Totland Bay, 165, 167, 173

Ventnor, 167, 307

Western Solent, 11, 13, 14, 16, 73, 74, 75, 77, 88, 104, 105, 358
Whitecliff Bay, 253
Woodmill (Itchen River), 168, 170, 205
Woodvale, *see also* Gurnard, 172
Wootton Creek, 75, 77, 78, 275
Wootton-Quarr, 74, 75

Yar River, 15, 16, 21, 33, 248, 296, 358, 359
Yarmouth, 13, 16, 73, 74, 77, 78, 169, 171

Subject Index

Aggregates, *see also* Dredging
21, 73, 111, 339, 340, 341, 343, 344, 345, 346
Cumulative Effects Assessments (CEA), 339
Extraction, 21, 238, 334, 339, 340, 341, 342, 346
Legislation, 339
Licences, 339, 340, 341, 345, 346, 349, 350, 351, 368
Algae, *see also* Phytoplankton
52, 138, 145, 175, 248, 250, 251, 253,
Brown, 249, 250
Growth, 146, 206, 249, 250
Macro, 135, 146, 218, 283, 284
Red, 250
Alien Species, 226, 232, 254, 293
Ammonia, 138, 143, 146, 217
Anenomes, 239, 250, 251, 253
Anthropogenic
Adjustment, 67
Discharge, 136
Factors, 69, 181, 182, 185, 226, 232
Inputs, 151
Processes, 141
Sources, 178
Antifouling, *see also* Tributyltin, 151, 158, 254, 307
Paints 151, 158, 254, 307
Aquaculture, 171, 211, 271, 334, 343, 358
Archaeological, 344
Survey, *see* Survey
Evidence, 78
Preservation, 90
Resources, 80
Sites, 21, 71, 72, 344
Structures, 75, 77
Artificial substrates, 251
Ascidians, 250, 251

Bacteria, 163-165, 167, 169, 170, 172, 189, 190, 205
Attachment, 214
Sulphate-Reducing, 213, 214
Barnacles, 209, 251, 307
Bathing Waters, *see also* Water Quality, 163, 165, 166, 167, 168, 187, 191, 221, 226
Beach Erosion, *see* Erosion
Beach Replenishment, 104, 105
Bedforms, 31, 32, 35, 108, 109, 113
Asymmetry, 35, 113
Fields, 31, 32
Benthic communities, 78, 316, 343, 344, 345
Biodiversity, 217, 225, 227, 228, 229, 232, 235, 237, 238, 240, 241, 242, 329, 331, 361
Bioindicators, 163, 309
Biotopes, 247, 256, 332
Bird(s), 261-269, 320-324, 353, 360
Breeding, 236, 262, 266
Counts, 269, 323
Feeding, 323
Harbour ecosystem, 319, 323
Important Bird Area (IBA), 263, 265
Migrating, 319
Population, 231, 234, 241, 265, 267, 319
Roost sites, 321, 323, 324
Roosting, 261, 303, 319, 321, 323
Underhill Index, 319
Wintering, 233, 262, 266
Species
Bar-Tailed Godwit, 321, 322
Brent Geese, 135, 262, 264, 267, 321
Dunlin, 323
Redshank, 320, 321
Ringed Plover, 321, 323
Sandwich Tern, 265
Bivalves, *see also* Clams, 173, 225, 253, 305, 334
Burrowing (*Mya Arenaria*), 158
British Marine Aggregates Producers Association (BMPA), 111
British Trust for Ornithology, 266, 319, 324
Bryozoa, 250, 251, 253
Buried River Valleys, 12, 26, 28, 30, 32, 98

Chalk Basement, 23
Chalk Ridge, 10
Chlorophyll, 140, 141, 142, 144, 145, 219, 287, 290, 291
Clam (*Scobicularia Plana*), *see also* Fisheries, 158
Clay Minerals, 32
Cliffs
   Chalk, 237
   Clay, 236, 237, 316
   Gravel, 115
   Sea, 236
Climate
   Boreal, 247
   Change, 67, 101, 102, 106, 125, 227, 333
   Lusitanean, 247
   Variability, 293
Coastal
   Defence, 5, 72, 78, 86, 88, 91, 92, 103, 104, 105, 106, 121, 125, 126, 128, 226, 234, 236, 240, 241, 355, 356, 357, 360
   Development, 17, 85, 226
   Impact Studies, 343
   Planning, *see also* Management 72, 333, 355
   Processes, 10, 17, 55, 72, 73, 79, 80, 86, 88, 121, 127, 232, 237, 238, 240, 241, 268, 333, 342, 343, 344, 358
   Protection, 30, 72, 117, 235, 237, 333
   Squeeze, 101, 127, 241, 267, 355
   Coastal Plain, 233, 358
Cockle, common (*Cerastoderma edule*), 156, 157, 305
Common Fisheries Policy, 228, 272
Conservation, 3, 5
   Archaeological, 71, 72, 79, 333,
   Birds, 261, 262, 265, 266, 268, 269, 329, 330
   Fauna, 295, 297
   Habitat, 21, 125, 128, 225-229, 231-243, 247, 248, 253, 256, 315, 318, 319, 355-357, 349, 360, 362
   Land, 87, 89, 90, 92, 341-345, 368
   Management, 91, 232, 351

Construction Industry Research and Information Association (CIRIA), 111
Containerisation, 347
Contaminants, 45, 137, 343
Convention for the Protection of the Marine Environment, 228
Coral, 225
Cord Grass, *see* Marsh Grass
Currents, *see also* Tidal
   Near-bed flow, 46, 51, 113
   Residual, 49, 52, 107, 199, 200, 202, 204
   Gyre, *see also* Eddies 61, 64, 200, 202
   Eulerian, 199
   Lagrangian, 199, 200, 201, 202, 204
   Transport, 50, 61
   Wakes, 107
   Surface, 34, 47
   Tidal, 28, 33, 34, 45, 46, 49, 50, 52, 64, 107, 112, 169, 201, 232, 247
   Wave-induced, 33, 60, 62, 112

Deltas (tidal), 30, 32, 36, 98, 105
Department of the Environment, Transport and the Regions (DETR), 111, 136, 154, 165, 339, 340, 341, 342, 343, 351
Diatoms, 16, 52, 81, 144, 293
Dinoflagellates, 144, 291, 293
Dog-Whelk (*Nucella lapillus*), 156, 254, 307, 308
   Imposex, 156, 254, 307
   Juvenile Dispersal (Rafting), 309
Double High Water, *see* Tides
Dredgers, 350, 351, 352, 353
Dredging, *see also* Aggregates, 5, 33, 45, 105, 128, 129, 159, 220, 238, 278, 323, 340, 345, 346, 360
   Channel, 47, 67, 347, 349
   Dump Sites, 219, 344, 350
   Licences, 339, 349, 351, 368
   Maintenance, 30, 33, 64, 127, 339, 349, 352
   Material, 104, 113, 117, 334, 341, 342, 349, 351
   Navigational, 77, 78, 121, 128
   Siltation, 64, 351-353

EC Directive
 Bathing Water, 163, 167, 187
 Environmental Impact Assessment
  (EIA), 152, 340, 341, 342
 Habitats, 227, 228, 239, 261, 265, 268,
  269, 296, 342, 368
 Nitrates, 135
 Shellfish Hygiene, 169
 Shellfish Waters, 153, 169, 171, 219, 368
 Water Framework, 228, 368
Echinoderms, 225
Ecology, 257, 267, 269, 305, 311, 314
Ecosystem, 135, 137, 220, 225, 231, 324
Eddies, *see also* Currents *and* Tidal,
 33, 49, 50, 107, 108, 131
Effluent,
 Industrial, 154, 303-305
 Sewage, 45, 52, 56, 135-138, 141, 143, 146,
  150, 155, 163-169, 171, 172, 176, 177,
  187, 188, 190-193, 217, 219, 220, 226
 Emergency Planning, 91
English Nature, 72, 87, 111, 225, 243, 247,
 257, 267, 315, 332, 343, 351, 352, 357, 360
Environmental
 Change, 73, 74, 78, 79, 80, 229, 333, 355
 Impacts, 71
 Quality, 149, 152, 153, 158, 159, 305
 Risk, 209
 Risk Assessment, 339, 342, 346, 353,
  360, 362
Epidemiological Investigations, 188, 190
Erosion, *see also* Saltmarsh
 5, 24, 30, 78, 180, 241, 352, 358, 362
 Beach, 101, 105
 Cliff, 10, 73, 90, 102, 103, 105, 121, 127,
  236, 237
 Coastal, 13, 34, 74, 115, 117, 321, 333, 344
 Seabed, 28
Estuary, *see under* River *in* Location Index
Estuarine
 Circulation, 49-51, 204
 Flushing, 48, 50, 57, 137, 156
 Morphology, 21, 24, 26, 33, 36, 66

Estuarine (*cont.*)
 Hydrodynamics, 46, 57, 61, 64, 66, 67, 69,
  101, 136, 141, 199, 202-204
 System, 21, 184, 199, 200, 202, 204, 235,
  332
 Transport, 368
European Union Assessment Project
 (PINTO), 211
Eutrophication, 135, 136, 146, 217, 218, 226,
 234, 324

Fisheries
 Activities, 159
 Clam, 158, 276, 278, 334, 352
 Commercial, 342, 343, 344, 345
 Landings, 219, 272, 274, 275, 344
 Oyster, 169, 272, 274, 275, 276, 278
 Policy, 228
 Salmon, 128, 349, 350
 Shellfish, 172
 Stocks, 226, 341
 Hygiene Regulations, 171
 Smolt, 350, 353
Flood Tide, *see* Tides
FLUXMANCHE, 49, 111
Fossils, 90
Fouling, *see also* Antifouling
 209, 211, 213, 221, 254
Freshwater, 12, 23, 24, 30, 48, 49, 50, 69,
 137, 140, 146, 176, 184, 204, 229, 235, 239,
 241, 368
 Discharge, 206
 Habitats, *see also* Marshes, 229
 Inflow, 52, 233

Geographical Information System (GIS),
 79, 80, 87, 91, 256, 257, 329
Geology, 16, 21, 25, 35, 130, 232, 237
 Barton Group, 24, 25
 Bembridge Formation, 24
 London Clay, 23, 25
 Lower Greensand, 23, 25, 349, 351, 352
 Lower Headon Beds, 24
 Solent Group, 24, 25

Geology (*cont.*)
  Upper Chalk, 23, 30
  Upper Greensand, 23, 25
  Upper Headon, 24
Geomorphology, 9, 16, 21, 28, 34, 35, 80, 232, 237, 350
Global Ocean Observing System (GOOS), 146
Global Warming, 4, 101
Gravel
  Banks, 32, 30
  Dunes, 32
  Fluvial, 28
  Terraces, 14, 27, 28, 98, 238
Groynes (Rock), 117
Gulls, *see also* Birds, 262, 265

Habitats, *see also* Marshes, Heathland, Intertidal flats, Intertidal Wetlands, Maritime Grasslands, Mudflats *and* Saltmarsh,
  80, 90, 126, 225, 228, 234, 239, 251, 256, 265, 266, 287, 297, 315, 316, 318, 321, 324, 330-332, 335, 342, 345, 367-369
  Cliff, 236, 237
  Coastal, 74, 227, 231-233, 235, 240-243, 261, 263, 266, 268
  Estuarine, 233
  Intertidal, 77, 232, 241, 267
  Loss, 78, 241, 268
  Re-creation, 242, 360
  Sand, 237, 238, 240, 248, 251, 253
  Sea wall, 262, 263, 267, 269, 295, 361
  Sublittoral, 250
  Subtidal, 231, 241
  Types, 249
 Hampshire Basin, 23, 26
 Heathland, 235, 236, 240
 Holocene, 12, 17, 26, 28, 71, 74, 77, 80, 97, 105, 333
 Hydrocarbons, 154, 157, 159, 175, 176, 178, 179, 180, 181
  Anthropogenic Sources, 178
  Polynuclear Aromatic (PAH), 178

Hydrodynamics, *see also* Currents, Estuarine Circulation, Modelling, Tides *and* Water Circulation,
  33, 36, 48, 55, 57, 69, 88, 128, 136, 233, 344
Hydroids, 250
Hypernutrification, *see* Nutrients

Ictis (Prehistoric British Island), 9, 12
Intertidal
  Flats, 21, *see also* Mudflats
  Wetlands, 355, 357, 358, 360, 361, 362, 363
Invertebrate
  Biodiversity, 235
  Cliff fauna, 237
  Communities, 236, 237, 238
Irradiance, *see* Underwater light

Joint Nature Conservation Committee, 243, 247, 266, 296, 315, 319

Kelp (*Undaria pinnatifida*), 254, 255, 311, 312, 313, 314

Lagoons, 176, 239, 240, 241, 254
Laminaria, *see* Kelp
Land Reclamation, 28, 67, 104, 266, 355, 360
Landslide, 103
Linear Furrows, 33
Lithology, 23, 25, 29, 32, 107, 108
Local Nature Reserves (LNR), 21, 231
Looping Snail *(Truncatella subcylindrica)*, 255, 295, 296, 297

Managed Retreat, 91, 102, 121, 123, 125, 235, 241, 242, 330, 362
Management
  Coastal, 48, 72, 79, 92, 309, 355, 356
  Disturbance, 320
  Risk, 360
  Shoreline, 85, 86, 89, 91, 92, 126, 237
  Unit, 86, 88, 121

Management Plans 229
  Biodiversity Action Plan (BAP), 296
  Shoreline (SMP), 72, 85-92, 103, 105, 121, 268, 318, 333, 357, 358, 369
  Shoreline Strategies, 86
Mariculture Ponds, 232, 240
Marine Conservation Society, 315
Marine Mammals, 254, 256, 344
Marine Nature Reserves (MNR), 228, 341
Marine Protected Areas, 228
Marshes, see also Saltmarsh
  13, 73, 142, 154, 247, 303
  Freshwater, 235
  Grazing, 233, 234, 235, 236, 239, 240, 241, 361
  Marsh Grass (Cord)
    Die-Back, 77, 127, 235
    Spartina alterniflora, 235
    Spartina anglica, 234, 235, 303
Maritime Grassland, 236, 237
Microbiology, 163, 192
Microflagellates, 291
Ministry of Agriculture, Fisheries and Food (MAFF), 73, 80, 86, 88, 111, 124, 179, 272, 273, 276, 343, 349, 351
Mitigation, 345
Models
  Beach Plan, 117
  Continental Shelf, 57
  Hydraulic, 67
  Hydrodynamic, 60, 61, 107, 111, 199
  Baroclinic, 199
  Mathematical, 115
  Numerical, 56
  Process, 66
  Regime, 66
  Tidal, 57
  Wave, 60, 111
Modelling, 52, 55, 66, 69, 74, 113, 115, 116, 117, 127, 268, 330, 331, 349, 353, 369
  Boundary Conditions, 56, 57, 66, 67
  Frictional Effect, 57, 60, 64, 107
  Hydrodynamic, 56, 111
  Morphological, 66, 67
  Numerical, 35, 36, 49, 52, 55, 56, 57, 60, 69

Modelling (*cont.*)
  Nutrient uptake, 145
  Physical, 55, 56
  Tidal, 46, 55
  Wave, 60, 111
Molluscs, 30, 159, 175, 225, 250, 253, 295, 307
Mud, 30, 32, 60, 77, 159, 248, 249, 253, 254, 295, 304, 315, 333, 355, 369
  Contaminated, 128
  Fluid, 353
Mudflats, see also Intertidal Flats
  104, 105, 247, 261, 262, 303, 352, 362

National Nature Reserves (NNR), 21, 231
Nitrification, 146, 217
Nitrite, 138
Nitrogen, 34, 115, 135, 153, 205
Nutrients, 135, 136, 137, 138, 140, 205, 213, 218, 219, 220, 357
  Availability, 136
  Concentrations, 135, 136, 138, 140, 205
  Denitrification, 137, 142, 218
  Discharge, 136
  Dissolved, 137
  Enrichment, see also Eutrophication, 135
  Hypernutrification, 135, 136
  Inputs, 135
  New, 141

Oil, 235
  Exploration, 21
  Pollution, 176
  Refining, 303
  Spills, 45, 303
Oil Pollution Research Unit (OPRU), 303
Organic Carbon, 151, 175, 178, 185
  Plant Litter, 175, 176, 184, 185
  Total (TOC), 176
  Volatile (VOC), 175, 182, 183, 184
  Organic Matter, 177, 304
Ornithology, see also Birds
  261, 262, 344
Oxygen, dissolved, 136, 144